D1219064

Topology

AN INTRODUCTION WITH APPLICATION TO TOPOLOGICAL GROUPS

INTERNATIONAL SERIES IN PURE AND APPLIED MATHEMATICS

WILLIAM TED MARTIN AND E. H. SPANIER, CONSULTING EDITORS

AHLFORS · Complex Analysis
BELLMAN · Stability Theory of Differential Equations
BUCK · Advanced Calculus
BUSACKER AND SAATY · Finite Graphs and Networks
CHENEY · Introduction to Approximation Theory
CODDINGTON AND LEVINSON · Theory of Ordinary Differential Equations
DETTMAN · Mathematical Methods in Physics and Engineering
EPSTEIN · Partial Differential Equations
GOLOMB AND SHANKS · Elements of Ordinary Differential Equations
GRAVES · The Theory of Functions of Real Variables
GREENSPAN · Introduction to Partial Differential Equations
GRIFFIN · Elementary Theory of Numbers
HAMMING · Numerical Methods for Scientists and Engineers
HILDEBRAND · Introduction to Numerical Analysis
HOUSEHOLDER · Principles of Numerical Analysis
LASS · Elements of Pure and Applied Mathematics
LASS · Vector and Tensor Analysis
LEPAGE · Complex Variables and the Laplace Transform for Engineers
MC CARTY · Topology: An Introduction with Application to Topological Groups
NEHARI · Conformal Mapping
NEWELL · Vector Analysis
RALSTON · A First Course in Numerical Analysis
ROSSER · Logic for Mathematicians
RUDIN · Principles of Mathematical Analysis
SAATY AND BRAM · Nonlinear Mathematics
SIMMONS · Introduction to Topology and Modern Analysis
SNEDDON · Elements of Partial Differential Equations
SNEDDON · Fourier Transforms
STOLL · Linear Algebra and Matrix Theory
STRUBLE · Nonlinear Differential Equations
WEINSTOCK · Calculus of Variations
WEISS · Algebraic Number Theory
ZEMANIAN · Distribution Theory and Transform Analysis

Topology

AN INTRODUCTION
WITH APPLICATION TO TOPOLOGICAL GROUPS

GEORGE MC CARTY
Department of Mathematics
University of California, Irvine

MC GRAW-HILL BOOK COMPANY
New York
San Francisco
St. Louis
Toronto
London
Sydney

TOPOLOGY: AN INTRODUCTION WITH APPLICATION TO TOPOLOGICAL GROUPS

Library of Congress catalog card number: 67-12628

1234567890 MP 72106987

44815

Preface

THIS TEXT introduces the student to that part of geometry which is generally labeled "topology." It will give him that familiarity with elementary point set topology, including an easy acquaintance with the line and the plane, which has become prerequisite to most graduate programs in mathematics. Nevertheless, it is not a collection of such topics; rather, it early employs the language of point set topology to define and discuss topological groups. These geometric objects in turn motivate a further discussion of set-theoretic topology and of its applications in function spaces. An introduction to homotopy and the fundamental group then brings the student's new theoretical knowledge to bear on very concrete problems: the calculation of the fundamental group of the circle and a proof of the fundamental theorem of algebra. Finally, the abstract development is brought to a satisfying fruition with the classification of topological groups by equivalence under local isomorphism.

There is general agreement that every serious student of mathematics should take part in some sustained, deep geometric development. This text is quite close, for instance, to the recommendations of CUPM for a one-year course on set-theoretic topology.* Such a sustained development might be considered complete, for some students, at the end of Chapter IX, with the study of local isomorphism classes omitted. However, much significance will be lost to the student who fails to reach the fundamental

* These recommendations are outlined on pages 61–64 of the pamphlet *Preliminary Recommendations for Pregraduate Preparation of Research Mathematicians*, published by the Committee on the Undergraduate Program in Mathematics of the Mathematical Association of America, May, 1963.

theorem of algebra (that is, to finish Chapter IX). Thus an instructor may need considerable flexibility of material to adjust the global schedule of his course. Many extra topics are provided, any of which may be omitted without disturbing the continuity of reasoning; these are gathered at the very end of each chapter as PROBLEMS. Much of this material may be discussed in class during a one-year course, or it might be omitted entirely for a shorter presentation. The easy questions which every student should consider are labeled EXERCISES and grouped in front of the PROBLEMS. Some, but not all, of the EXERCISES are cited at the ends of those sections where they may first be conveniently assigned.

Linear algebra is drawn upon for many examples in later chapters; although this material could be deleted, an introduction to vectors and matrices will normally be prerequisite to the course. With the thought that some excellent young students might wish to enroll in the prerequisite course concurrently, I have made no mention of linear algebra prior to Chapter VII. However, an instructor can easily add linear examples and exercises to the chapters on groups and on topological groups. An understanding of mathematical induction and of the completeness of the reals is assumed, with the first usage of each concept signaled by footnote. A knowledge of modern algebra is not presumed; of course, a class trained in elementary group theory could regard Chapter II as a review (although its emphasis on infinite groups may be news). At the other extreme, this text may be used as a short course on topological groups for a class already familiar with point set topology; parts of Chapters V and VII through X will be pertinent.

A feature of the text is its emphasis on the quotient-function–equivalence concept; a uniform treatment in the contexts of sets, groups, spaces, and then topological groups stresses aspects common to all these settings. Although functors and categories are nowhere defined, at each appearance of a new functor its algebraic properties are derived and emphasized. Additionally, several problems explore categorical characterizations of various concepts.

There are a few didactic novelties here. A "Cayley theorem" is offered for topological groups. The triangle law for a product metric is proved in a way new to me. The fact that locally isomorphic groups have globally isomorphic based-path groups is made the basis of the classification theorem. And a canonical factorization of functions (and morphisms) results in the composite of a quotient function, a 1-1 correspondence, and an inclusion function.

As did many topologists, I cut my teeth on Pontrjagin's *Topological Groups*. Among my numerous other debts are those to Richard Arens,

Saunders Mac Lane, and C. B. Tompkins; each of these men has, by his unique example, taught me to think.

Students at Harvey Mudd College and at the University of California at Irvine who studied early versions have, by their kind suggestions, improved this book in many ways. Edwin Spanier has helped me with several comments on mathematical exposition. The entire manuscript was read by Robert F. Brown; without his many wise criticisms and his corrections of errors of mathematics, exposition, style, and English, this book would not be complete.

George McCarty

Table of Contents

Chapter VII FUNCTION SPACES 165

Chapter VIII THE FUNDAMENTAL GROUP 186

Chapter IX THE FUNDAMENTAL GROUP OF THE CIRCLE 215

Topology

AN INTRODUCTION WITH APPLICATION TO TOPOLOGICAL GROUPS

Introduction

This book is about topology. You may have read popular articles about "rubber-sheet geometry"; at this point it is difficult to say more precisely just what is studied in this branch of mathematics. We shall examine many mathematical objects, both familiar and strange: the real, complex, and quaternionic number systems, universal covering spaces and fundamental groups, spheres and projective spaces, deck transformations, and the compact-open topology (to name a few you might already be curious about). At the end, you will be able to form your own partial answer to the question "what is topology?"

But between here and the end lies hard work for you, as well as fun. Topology is a part of mathematics; this theory is applied in many fields, from quantum mechanics to sociology, and we shall point out some of these applications. But we must first build up the vocabulary of a theoretical structure. We shall construct our theory abstractly, with axioms, and our major effort will be the exploration of the theoretical consequences of those axioms. This contrasts with the calculus, where usually one strives mainly to acquire a competence in solving specific problems; a theorem is often regarded as a recipe for applying the formulas to special cases. Here you will strive to understand the theorems so thoroughly that you yourself can invent proofs for new theorems. In short, really study the proofs!

EXERCISES AND PROBLEMS

You will get exercise, sometimes vigorous exercise, as you read this book. It requires work to fill in the detailed reasoning from one sentence to the next; when the jump is particularly large we shall sometimes signal this with such euphemisms as "it is easy to see that . . ." and "it is obvious (or clear) that . . ."; other times the word "why?" may be interjected in parentheses to remind you that a few details are missing. Fill in these details mentally as you read; you will find that you become better and better at this with practice.

However, you want to learn to use mathematics, not just to read it. And you cannot learn that by osmosis, by watching someone else do it, any more than you could learn to play chess or football by close observation. There are EXERCISES at the end of each chapter; *do them all.* They are not repetitive drills; you can expect some of the joy of discovery and creation with each solution you construct. Furthermore, you will learn topology, chapter by chapter, through your exercises; we shall count on your efforts by presuming, as each new topic is presented, that you understand the results of those exercises that have appeared earlier. With experience, you will know which you can do in your head and which are difficult for you; write out detailed solutions for the harder ones.

In each chapter, following the exercises, there is a set of PROBLEMS. You will need neither the results of the problems nor the exercise of working them to continue your reading of the book. Instead, they contain interesting applications and further theory, a sort of payoff for your work in the chapter.

INTERNAL REFERENCES

Throughout the book we have tried to minimize the number of formal references to previous material, preferring instead to refer to a theorem by name or a brief statement of the result. However, we have found it necessary to number some of the statements in Chaps. I, II, and IV. A reference in Chap. II, for instance, to Theorem 3 means Theorem 3 of that chapter, while a reference there to Theorem I.3 means Theorem 3 of Chap. I. Similarly, a reference in Chap. II to Exercise A means Exercise A of Chap. II, while reference there to Exercise I.B means the second exercise of Chap. I. Problems are labeled with doubled letters to distinguish them from exercises; a reference to Prob. CC is to the third problem at the end of the chapter where the reference appears.

Should you not recall the content of a theorem, such as the Quotient Theorem for Groups, when it is referred to, you will find it listed in the index.

DEFINITIONS

We shall define new words in two ways: A defined word in boldface (a **thingamabob** is an orange whatsit) is part of our minimal vocabulary; it will be used later and *must be memorized.* A word appearing in quotation marks in its definition (a set of three skeletons is called a "full closet") will not be required later in this book. It is provided for your reference use; it may appear in a parallel text you use, or may be preferred by your instructor. Quotation marks are also used to set off "suggestive" statements which are not part of a formal argument; this usage will be self-explanatory.

SET-THEORETIC NOTATION

The concept of a **set** and operations involving sets, along with logical arguments expressed in ordinary English, form the language of this book. A set (synonyms: **class, family, collection**) is, intuitively, a bunch, aggregate, flock, etc.† All statements about sets are to be made within some "large" set called the **universe,** which contains every set in view. The particular choice of universe may change from topic to topic, but either it will always be clearly understood or it will be made explicit just which universe is being used in the discussion at hand. For example, if S is defined to be the set of all positive real numbers in a context where the universe is understood to be the set **R** of all real numbers, then the complement of S is the set of real numbers which are either negative or zero. But if, in a different context, the real numbers are themselves considered to be a subset of the universe **C** of all complex numbers (that is, **R** is the x-axis), then the complement of S in **C** is the set $\{x + iy\colon \text{either } y \neq 0 \text{ or } x \leq 0\}$, that is, the whole plane except for S.

The "set-braces" notation of the sentence above will be used frequently.

† A pod (of whales) or an exaltation (of larks)!

That is, let $P(x)$ be a statement involving the variable x such that for each particular value a of the variable x either $P(a)$ is true or $P(a)$ is false. Then $\{x: P(x)\}$ denotes the set of all those elements a of the universe for which $P(a)$ is true. The braces are also used to contain implicit or explicit lists of the elements of a set. Self-explanatory examples of this in the universe **R** of real numbers are $R = \{1,3,10\}$, $S = \{1, 2, 3, \ldots, 27\}$, and $T = \{1, \frac{1}{2}, \frac{1}{4}, \frac{1}{8}, \ldots\}$.

We now offer a condensed description of the elementary set-theoretic operations and their nomenclature. This is provided as a rapid review and reference list. If you are not quite familar with some of the ideas involved, consult the references listed at the end of Chap. I (on page 24) for more leisurely introductions.

The membership of an object x in a set S is denoted by $x \in S$. While the meaning of this symbol is fixed, you may read it (and many other mathematical symbols) in various ways in English sentences, depending on context; examples are ". . . x, which is a member of S, . . . ," ". . . x is a member of S . . . ," ". . . (let) x be a member of S . . . ," etc. If $x \in S$, then x is a **member, element,** or **point** of S; because of this nomenclature, sets are sometimes referred to as "point sets." If every member of a set S is also a member of

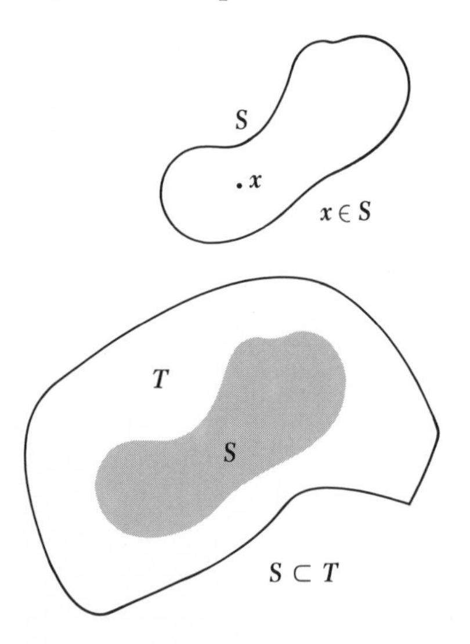

the set T, then S is said to be **contained** (or **included**) in T, written $S \subset T$, or T **contains** (or **includes**) S, written $T \supset S$; S is then a **subset** (or **subfamily**) of T and T is a "superset" of S. Two sets S and T are defined to be **equal,** written $S = T$, if both $S \subset T$ and $T \subset S$. This *definition* of equality of sets seems to subvert the usual convention that "$A = B$" means that "A" and "B" are two names for the same object. If you cleave to that meaning of equality, this definition of equality of sets may be understood, in our intuitive set theory, to say exactly what a set is: the totality of its members and nothing else.

Each of the above symbols may be negated by the addition of a slanted stroke: $\notin$, $\not\subset$, $\not\supset$, $\neq$. The definitions of the resulting symbols are clear; for instance, $S \neq T$ means that either $S \not\subset T$ or $T \not\subset S$. (In this last sentence and throughout mathematics, the word "or" is used in the nonexclusive sense; the words "or both" are understood.) If $S \neq T$ and $S \subset T$, then S is a **proper subset of** T. The **complement of** S **in** T, $T - S$ (read "T minus S"), is the set of elements of T which are not elements of S,

$$T - S = \{x: x \in T \text{ and } x \notin S\}.$$

The **complement** S' of a set S is the set of nonelements of S; if U denotes the universe, then $S' = U - S$.

The **intersection** of two sets S and T, $S \cap T$ (read "S intersect T"), is the set of all elements common to both S and T,

$$S \cap T = \{x: x \in S \text{ and } x \in T\}.$$

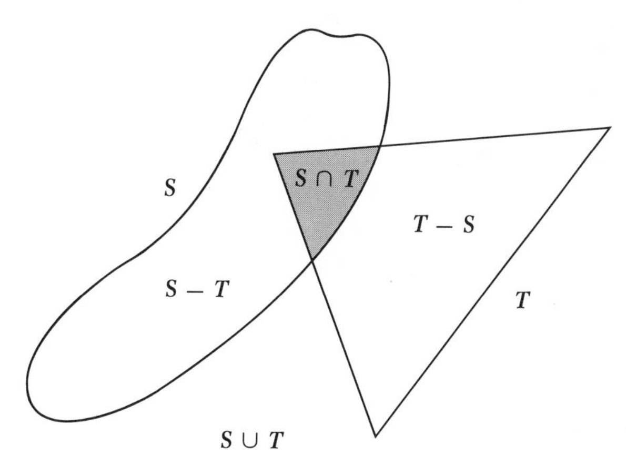

The **union** of S and T, $S \cup T$ (read "S union T"), is the set of all elements which belong to either S or T,

$$S \cup T = \{x: x \in S \text{ or } x \in T\}.$$

For any objects a and b, let (a,b) denote the **ordered pair,** with a first and b second. Two ordered pairs (a,b) and (a',b') are equal if and only if both $a = a'$ and $b = b'$. The **direct** (or "cartesian") **product** of S and T is the set $S \times T$ of all ordered pairs whose whose first element is a member of S and whose second element is a member of T,

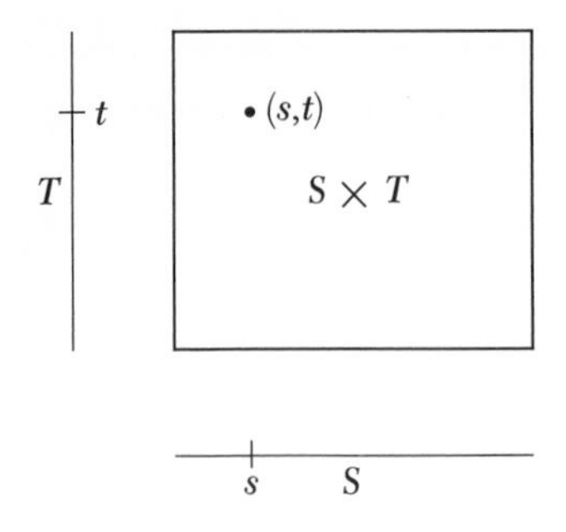

$$S \times T = \{(s,t): s \in S \text{ and } t \in T\}.$$

The sets S and T are called the **factors** of $S \times T$.

LOGIC

The word **implies** is used in this book in its mathematical sense: it is *not* intended to have its English-language meaning. Rather, if we say "P implies Q," where P is a mathematical statement which might be either true or false, and Q is another such statement, we mean "if P is true, then Q is true." The statement "P implies Q" is true unless both P is true and Q is false; that is, "P implies Q" is true if either Q is true or P is false. This precise definition results in some true statements being outrageous when read in the common language; "$1 = 0$ implies $1 = 1$" is an instance, as is "$1 = 0$ implies $1 = 3$." Notice that the statements "P implies Q," "if Q is false, then P is false," and "P is true only if Q is true" are logically equivalent.

SPECIAL SYMBOLS

We shall frequently use the symbol **iff** to stand for the words "if and only if"; thus "P iff Q" means both P implies Q and Q implies P.

The shorthand symbol ■ has become more fashionable than Q.E.D. as a signal to the reader that a proof is concluded. We shall use it for that; however, its appearance immediately following the statement of a theorem will indicate either that the proof preceded the statement or that the proof is omitted. Whenever a proof is omitted, it is easy; construct it in your head at once.

A few particular sets will be discussed enough to make special symbols helpful for their recognition. Some of these are

$\varnothing$ the **empty** ("void," "null") set, $\varnothing = \{x: x \neq x\}$,
Z the set of all integers, $\mathbf{Z} = \{\ldots, -1, 0, 1, 2, 3, \ldots\}$,
R the set of all real numbers,
C the set of all complex numbers,
I the closed interval of real numbers between 0 and 1,
$\mathbf{I} = [0,1] = \{x: x \in \mathbf{R} \text{ and } 0 \leq x \leq 1\}$.

A complete index of special symbols is given immediately in front of the index, along with a Greek alphabet, for your reference.

Sets and Functions

CHAPTER I

The concept of a function, together with its related ideas of quotient and inclusion, is perhaps the most important and pervasive idea in modern mathematics. However, the intuitive definition of a function as a rule or correspondence is not sufficiently precise for us to use in exploring the concept. In this chapter, after broadening the notions of union and intersection to consider infinite families of sets, we shall define a function to be a special kind of relation. With this concise set-theoretic definition we shall be able to examine the structure of functions in general, not just the behavior of a particular one.

Our new knowledge will thenceforth be put to daily use as we study groups and topologies.

UNIONS AND INTERSECTIONS

It is easy to show that both the union and the intersection operations are both commutative and associative: for all sets R, S, and T in a given universe U,

1

$$\begin{aligned} R \cap S &= S \cap R, \\ R \cup S &= S \cup R, \\ (R \cap S) \cap T &= R \cap (S \cap T), \\ (R \cup S) \cup T &= R \cup (S \cup T). \end{aligned}$$

If $\mathcal{S}$ is a class whose members are sets in a given universe U, $\bigcap\mathcal{S}$ denotes the **intersection** of all the members of $\mathcal{S}$,

$$\bigcap\mathcal{S} = \{x: x \in S \text{ for every } S \in \mathcal{S}\}.$$

Similarly, $\bigcup\mathcal{S}$, the **union** of $\mathcal{S}$, is the set of all members of members of $\mathcal{S}$,

$$\bigcup\mathcal{S} = \{x: x \in S \text{ for some } S \in \mathcal{S}\}.$$

If $\mathcal{S}$ is a finite collection of sets, with at least two members in $\mathcal{S}$, then the import of statements 1 is that these notions of intersection and union agree with the results of applying the earlier defined binary operations of intersection and union *in any order whatsoever* to the members of $\mathcal{S}$. If $\mathcal{S}$ has exactly one member S, then $\bigcap\mathcal{S} = \bigcup\mathcal{S} = S$. Further, if $\mathcal{S} = \varnothing$, then examination of the definitions shows that $\bigcap\mathcal{S} = U$, the universe, and $\bigcup\mathcal{S} = \varnothing$. A natural mnemonic for these extreme cases is that $\bigcap\mathcal{S}$ "grows larger" as $\mathcal{S}$ "grows smaller," and $\bigcup\mathcal{S}$ grows smaller as $\mathcal{S}$ grows smaller. No other convention is possible, but the case $\mathcal{S} = \varnothing$ will often be treated redundantly by itself in definitions and proofs, as a reminder of the null case.

To form an example, let $S_a = \{x \in \mathbf{R}: x > a\}$ for each real number a. (The set-braces notation for S_a has been slightly abbreviated here; properly, it should be $\{x: x \in \mathbf{R} \text{ and } x > a\}$.) Define the family $\mathcal{S}$ to be $\{S_a: a \in \mathbf{R}\}$. (This is another abbreviation; properly, $\mathcal{S} = \{S: \text{there exists } a \text{ with } a \in \mathbf{R} \text{ and } S = S_a\}$.) If $\mathcal{F}$ is a *finite* subfamily of $\mathcal{S}$, then $\bigcup\mathcal{F}$ is a proper subset of $\mathbf{R}$, and $\bigcap\mathcal{F} \neq \varnothing$ (why?). Nevertheless, $\bigcup\mathcal{S} = \mathbf{R}$ and $\bigcap\mathcal{S} = \varnothing$.

2 THEOREM *If $\mathcal{S}$ is a class of sets in a universe U and T is a fixed set in U; then*

- **i** $T \cap \bigcup\mathcal{S} = \bigcup\{T \cap S: S \in \mathcal{S}\}$,
- **ii** $T \cup \bigcap\mathcal{S} = \bigcap\{T \cup S: S \in \mathcal{S}\}$, *(distributive laws)*
- **iii** $T - \bigcup\mathcal{S} = \bigcap\{T - S: S \in \mathcal{S}\}$,
- **iv** $T - \bigcap\mathcal{S} = \bigcup\{T - S: S \in \mathcal{S}\}$. *(De Morgan formulas)*

Only formulas i and iii are proved here; ii and iv are left as exercises for the student.

Proof of i Let $x \in T \cap \bigcup\mathcal{S}$; since $x \in \bigcup\mathcal{S}$, there is some $S \in \mathcal{S}$ with $x \in S$. But $x \in T$ also, so $x \in T \cap S \subset \bigcup\{T \cap S: S \in \mathcal{S}\}$. Thus

$$T \cap \bigcup\mathcal{S} \subset \bigcup\{T \cap S: S \in \mathcal{S}\}.$$

Now assume $x \in \bigcup\{T \cap S: S \in \mathcal{S}\}$; there is then an element $S \in \mathcal{S}$ with $x \in T \cap S$. Hence $x \in T$ and $x \in S \subset \bigcup\mathcal{S}$, so $x \in T \cap \bigcup\mathcal{S}$. Therefore $\bigcup\{T \cap S: S \in \mathcal{S}\} \subset T \cap \bigcup\mathcal{S}$. ∎

Proof of iii Let $x \in T - \bigcup \mathcal{S}$; thus $x \in T$ and $x \notin \bigcup \mathcal{S}$. For each $S \in \mathcal{S}$, then, $x \notin S$, and so $x \in T - S$. Therefore $x \in \bigcap \{T - S: S \in \mathcal{S}\}$, and $T - \bigcup \mathcal{S} \subset \bigcap \{T - S: S \in \mathcal{S}\}$.

On the other hand, if $x \in \bigcap \{T - S: S \in \mathcal{S}\}$, then for every $S \in \mathcal{S}$, $x \in T - S$, and thus $x \in T$ and (for every S) $x \notin S$. Clearly, $x \notin \bigcup \mathcal{S}$, so $x \in T - \bigcup \mathcal{S}$. ∎

Exercises A, B, C, and D

RELATIONS

If A and B are sets, a subset $R \subset A \times B$ will be called a **relation** (or "binary relation") **in** A **to** B.† For instance, the entire direct product $A \times B$ is a relation in A to B, as is its empty subset. If S is the set of all human females and T is the set of all human beings, then $\{(s,t): (s,t) \in S \times T, s \text{ is a sister of } t\}$ is a relation in S to T.

If R is a relation in A to A, then R will be called a **relation in** A. A relation R in A is **reflexive** if the "diagonal" set $\{(a,a): a \in A\} \subset R$. It is

The shaded regions represent relations

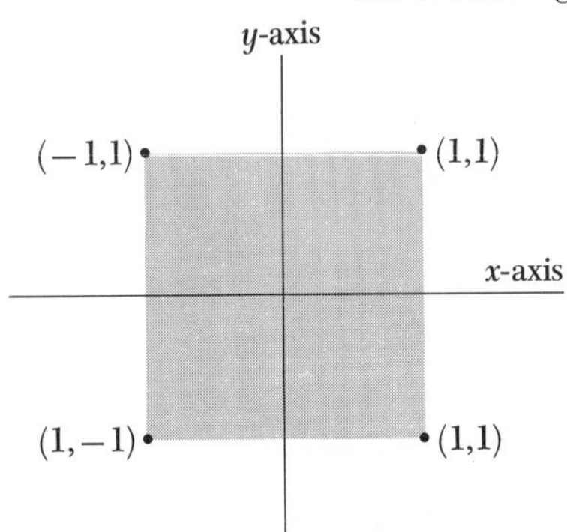

Symmetric and transitive, but not reflexive

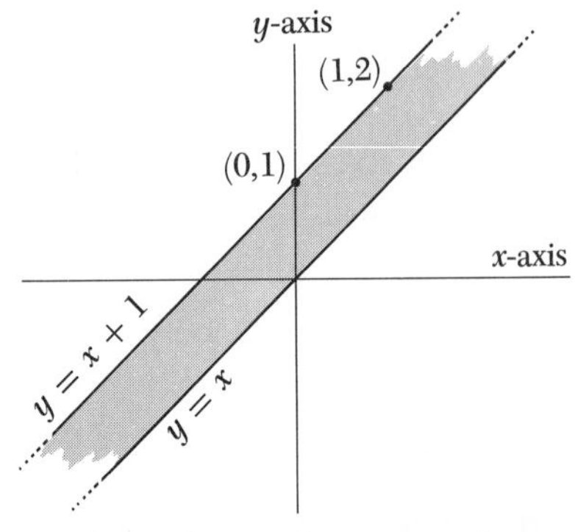

Reflexive, but neither symmetric nor transitive

symmetric if $\{(b,a): (a,b) \in R\} = R$, and it is **transitive** if $\{(a,c):$ for some $b \in A$ both (a,b) and $(b,c) \in R\} \subset R$. You should verify the following statements.

† Notice that, strictly speaking, the expression $R \subset A \times B$ should have parentheses inserted thus: $R \subset (A \times B)$. However, "$(R \subset A) \times B$" makes no sense at all, since $(R \subset A)$ is not a set and therefore has no direct product with B. Henceforth we shall write such an expression, which makes sense in only one way, without parentheses or further comment.

3 THEOREM *Let R be a relation in the set A; R is reflexive iff $(a,a) \in R$ for every $a \in A$, R is symmetric if $(a,b) \in R$ whenever $(b,a) \in R$, and R is transitive if whenever (a,b) and (b,c) are members of R, then $(a,c) \in R$.* ∎

If, for example, the set A is the set of all human beings (living or not), then $\{(a,b): a$ and b are brothers$\}$ is symmetric; it is neither reflexive nor transitive (assuming that no man is his own brother). The relation $\{(a,b): a$ is not a descendent of $b\}$ is reflexive, but not symmetric and not transitive. Can you think of a transitive, nonreflexive, nonsymmetric relation between real numbers?

A relation R in a set A which is reflexive, symmetric, and transitive is called an **equivalence relation** (or just an **equivalence**) **on** A. The relation $R = A \times A$ is always one such, as is the **diagonal relation** $\Delta \subset A \times A$, $\Delta = \{(a,a): a \in A\}$ (check these examples in your mind). If $A = \mathbf{Z}$, the integers, and $n \in \mathbf{Z}$, another equivalence on A is $\{(x,y): n$ divides $x - y\}$; this relation is called **congruence modulo** n. If R is an equivalence relation on A and $a \in A$, let $[a]$ denote the set of elements of A which are equivalent to a; $[a] = \{b: (a,b) \in R\}$ is called the **equivalence class** of a.

4 THEOREM *Let R be an equivalence relation on A and $a, b \in A$. Then either $[a] \cap [b] = \varnothing$ or $[a] = [b]$.*

Proof Assume that $[a] \cap [b] \neq \varnothing$ and choose an element $c \in [a] \cap [b]$. Then (a,c) and (b,c) are members of R, so (a,c) and (c,b) are both in R; hence $(a,b) \in R$. However, if $d \in [b]$, then $(b,d) \in R$; since $(a,b) \in R$, so is (a,d), and $d \in [a]$. This shows $[b] \subset [a]$. By the symmetry of the assumption, $[a] \subset [b]$, and $[a] = [b]$. ∎ The last sentence of this proof means that if we interchange the symbols a and b in the hypothesis, the same statement results. That is, we assumed R was an equivalence relation, that a and b were in A, and then that $[a] \cap [b] \neq \varnothing$. We concluded that $[b] \subset [a]$. If we interchange the symbols a and b everywhere in the assumption, the proof, and the conclusion, then the assumption remains the same; yet the proof from that assumption now concludes that $[a] \subset [b]$.

Therefore the equivalence classes constitute a division of A into nonoverlapping subsets. To be more specific, let a family $\mathcal{S}$ of sets be termed **mutually disjoint** if for every pair S and T of members of $\mathcal{S}$, $S \neq T$ implies $S \cap T = \varnothing$. In particular, two sets are termed **disjoint** if they have an empty intersection. A **partition** of a set U is a mutually disjoint family $\mathcal{S}$ of subsets of U such that $\bigcup \mathcal{S} = U$.

5 THEOREM *If R is an equivalence relation on A, then the family $\mathcal{S}$*

of equivalence classes of elements of A, $\mathcal{S} = \{[a]: a \in A\}$, *is a partition of A. Conversely, if* $\mathcal{T}$ *is a partition of a set A, then there exists a unique equivalence relation R on A for which* $\mathcal{T}$ *is exactly the family of R-equivalence classes.*

Proof Clearly, $\bigcup\mathcal{S} = A$, and $\mathcal{S}$ is mutually disjoint, by Theorem 4, so $\mathcal{S}$ is a partition. On the other hand, if $\mathcal{T}$ partitions A, define the relation R on A by $R = \{(a,b)$: both $a \in T$ and $b \in T$ for some $T \in \mathcal{T}\}$. The verification that R is an equivalence relation and that the family of R-equivalence classes is indeed $\mathcal{T}$ is relegated to the exercises. ■

It will often be convenient when we are discussing a particular equivalence R on A to say that elements a and b of A are **equivalent,** or to write aRb, to indicate that $(a,b) \in R$. As an example, let R be the relation of congruence modulo 3: $(x,y) \in R$, or xRy, iff 3 divides $x - y$, so that there exists an integer k with $x - y = 3k$. Then $x = y + 3k$; conversely, whenever a multiple $3k$ of 3 is added to y the sum is equivalent to y. The equivalence class of y is exactly $[y] = \{y + 3k: k \in \mathbf{Z}\}$; it is infinite. The class $[17]$, for instance, is the set $\{\ldots, -1, 2, 5, \ldots, 14, 17, 20, 23, \ldots, 62, \ldots\}$; this is exactly the set of integers which leave a remainder of 2 when divided by 3. It is easy to see that the partition $\mathcal{S}$ defined by R is $\mathcal{S} = \{[0],[1],[2]\}$; it has exactly three members.

Another example is given by the measurement of angles (in radians). Such statements as "the angle 0 is the same as the angle 2π" mean that there is defined an equivalence relation on the real numbers $\mathbf{R}$ (the set of "measurements" of angles) with xRy iff x and y are two measurements of the same angle. Thus the equivalence classes are like $[0] = \{2k\pi: k \in \mathbf{Z}\}$ and $[\pi/2] =$

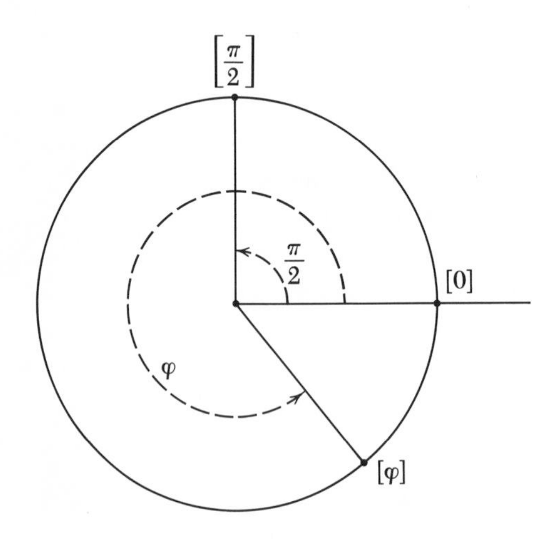

$\{[(4k+1)/2]\pi: k \in \mathbf{Z}\}$. Each class is infinite (it has one member for each integer k). Furthermore, the number of equivalence classes is itself infinite; each class corresponds to exactly one angle. The set $\mathcal{S}$ of equivalence classes can be corresponded one to one with the points on a circle (see Exercise L.)

Exercises E and F

FUNCTIONS

The familiar function x^2, the squaring function, which has real arguments (x is real) and real values (x^2 is real) has as its graph the set of points (x,y) in the plane for which $y = x^2$, that is, the set $\{(x,x^2): x \in \mathbf{R}\}$. There is

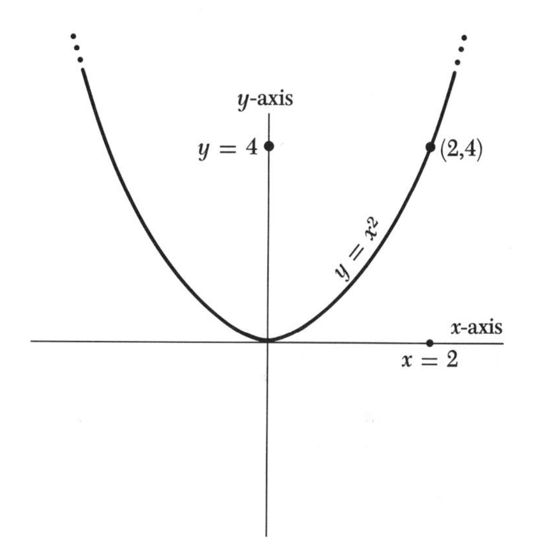

exactly one point on the graph lying on each vertical line; there is exactly one pair (x,y) in the graph for each real x. We now draw a formal definition of function which says essentially that a function *is* its own graph.

A function is a special kind of relation: a relation R in A to B is a **function** (or **transformation**) if each member of A occurs as the first element of exactly one member of R. Thus if $a \in A$, then there exists an element $b \in B$ such that $(a,b) \in R$; and furthermore, if $(a,b) \in R$ and $(a,b') \in R$, then $b = b'$. Then b is denoted $R(a)$ and called the **value of** R **at** a or the **image of** a **under** R. The set A is the **domain** of R, B is the **range** of R, and $\{R(a): a \in A\}$, the set of second elements of members of R, is the **image of** R. The domain of R is denoted $Dom\,(R) = A$, the range $B = Rng\,(R)$,

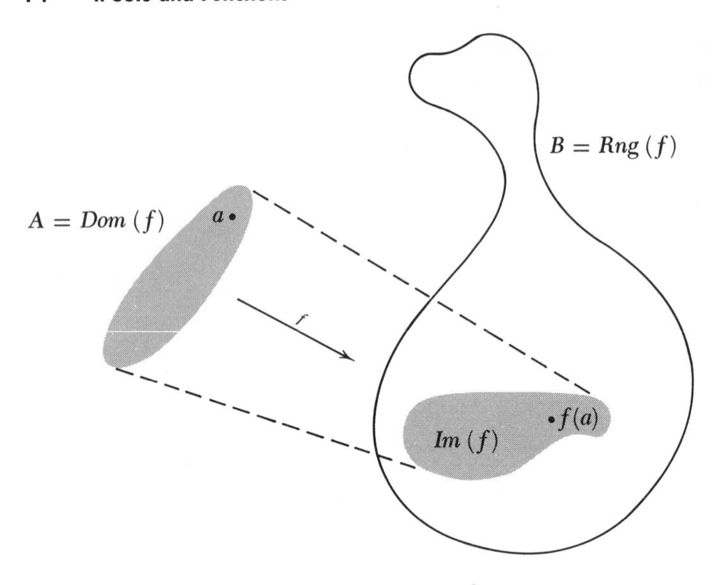

and the image $Im\,(R)$. This situation is sometimes denoted $R: A \to B$, and sometimes $A \xrightarrow{R} B$; R is called a function **on** A **into** B or **from** A **to** B. Applying this vocabulary to the squaring function $f(x) = x^2$, we see that $f = \{(x,x^2): x \in \mathbf{R}\} \subset \mathbf{R} \times \mathbf{R}$, $f(3) = 9$ is the value of f at 3, or the image of 3, $\mathbf{R}$ is the domain set and also the range set of f, and the set of nonnegative reals $\{x \in \mathbf{R}: x \geq 0\}$ is the image of f. With a transparent addition to our notation for functions, we write

$$f: \mathbf{R} \to \mathbf{R}: f(x) = x^2.$$

The trivial relations $A \times B$ and $\varnothing$ in A to B are not, in general, functions. For each set A, the diagonal relation $\Delta \subset A \times A$ is always a function from A to A; it is the **identity function** on A, and it is often denoted 1_A: $1_A(a) = a$ for each $a \in A$. Among the relations between human beings (living or not), $\{(\alpha,\beta): \beta$ is the mother of $\alpha\}$ is a function, but $\{(\gamma,\delta): \delta$ is the nephew of $\gamma\}$ is not. If A and B are any sets and $B \neq \varnothing$, then there is, for each member $b \in B$, a **constant function** $f_b: A \to B$ whose value is that fixed b for all $a \in A$, $f_b = \{(a,b): a \in A\} \subset A \times B$. Thus a function is constant iff its image is a set having exactly one element (such a set is called a **singleton**). Note that for each range set B there exists one function having domain $\varnothing$, namely, the "empty function" $\varnothing \subset \varnothing \times B = \varnothing$. However, if $A \neq \varnothing$, then there is no function at all with domain A and range $\varnothing$, since the set of image points could not be empty.

The **power set** $\mathcal{P}(A)$ of a set A is the set of all subsets of A, $\mathcal{P}(A) = \{S: S \subset A\}$. A function $f: A \to B$ defines (or **induces**) in a natural way a new function, usually denoted by the same symbol f, from $\mathcal{P}(A)$ to $\mathcal{P}(B)$; if $S \subset A$, then $f(S) = \{f(a): a \in S\} \in \mathcal{P}(B)$. Hence $f(S)$ is just the set of image points of elements of S; we may write $Im(f) = f(A)$, for example. The same function $f: A \to B$ also induces a function, usually denoted f^{-1} (read "f inverse"), from $\mathcal{P}(B)$ to $\mathcal{P}(A)$; if $T \subset B$, then $f^{-1}(T) = \{a \in A: f(a) \in T\}$. Thus $f^{-1}(T)$, the **inverse image** of T, is the subset of A consisting of all those points which f carries into T. If, for example, f is our squaring function and T is the interval $[4,9]$, then $f^{-1}(T) = [-3,-2] \cup [2,3]$; if $T' = [-3,-2]$, then $f^{-1}(T') = \varnothing$. If $f: A \to B$ and b is a point of B, we shall write $f^{-1}(b)$ for $f^{-1}(\{b\})$. Similarly, if $f^{-1}(b)$ is a singleton set $\{a\}$, we shall say $f^{-1}(b) = a$. (This common elision of notation is related to the function which sends a set S into its power set $\mathcal{P}(S)$ by assigning to each $s \in S$ the subset $\{s\} \subset S$.)

Exercise M

QUOTIENT FUNCTIONS

Since all functions are relations, we may ask whether a given function $f: A \to A$ is also an equivalence relation; certainly 1_A is (can you find some others?). However, there is another, deeper correspondence between functions and equivalences. If R is a relation on a set A, let $\mathcal{S}$ be the partition of A associated with R; $\mathcal{S}$ is the set of R-equivalence classes of A (see Theorem 5). The assignment of the class $[a]$ to each element a in A defines a function $q: A \to \mathcal{S}: q(a) = [a]$.

The partition $\mathcal{S}$ is called the **quotient** (or "factor") **set of A modulo R**, and q is the **quotient function.** The set $\mathcal{S}$ is often denoted A/R, and then $q: A \to A/R$.

If, for instance, $R = \Delta$ is the diagonal relation on A, then q assigns to each $a \in A$ the singleton $\{a\} = [a] \subset A$; it is a 1-1 correspondence, called the **trivial quotient of** A. We may also write $q(a) = a$, to emphasize this triviality. A nontrivial example of a quotient set was given in the above discussion of equivalence relations; if R is the relation of equivalence modulo 3 between integers, then the quotient set is $\mathbf{Z}/R = \{[0],[1],[2]\}$ and the quotient function is $q: \mathbf{Z} \to \mathbf{Z}/R: q(z) = [z] = \{z + 3k: k \in \mathbf{Z}\}$. Another quotient example is also given there: a quotient set member is an equivalence class of all those real numbers which are measurements of a fixed angle; the quotient set is called "the reals modulo 2π."

COMPOSITION OF FUNCTIONS

If $f: A \to B$ and $g: B \to C$, then we may form a new function, the **composite** $g \circ f: A \to C$ of f and g, by setting $(g \circ f)(a) = g[f(a)]$ for each $a \in A$. This situation may be diagramed as follows:

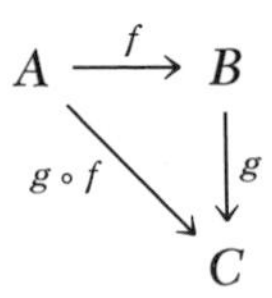

If the domain of g is not the same set as the range of f, then $g \circ f$ is not defined. When $g \circ f$ is defined, its domain is that of f and its range is that of g.

For example, if $f: A \to B$ is some function, and $1_A: A \to A$ is the identity function on the domain of f, then $f \circ 1_A = f$. Similarly, $1_B \circ f = f$.

6 THEOREM *The composition of functions is an associative binary operation; that is, if* $A \xrightarrow{f} B \xrightarrow{g} C \xrightarrow{h} D$, *then* $(h \circ g) \circ f = h \circ (g \circ f)$.

Proof By inspection, the two composite functions $(h \circ g) \circ f$ and $h \circ (g \circ f)$ have the same domain and the same range. Therefore they are equal iff they have the same value at each point in the common domain. But for each $a \in A$, $[(h \circ g) \circ f](a) = (h \circ g)[f(a)] = h(g[f(a)])$, and $[h \circ (g \circ f)](a) = h[(g \circ f)(a)] = h(g[f(a)])$. ■ Note that here and elsewhere the use of lightface parentheses, ( ), and brackets, [ ], may aid your eye in sorting out their pairings. In these cases, the *mathematical* meaning of each enclosure is the same. Confusion should not result from the simultaneous use of set

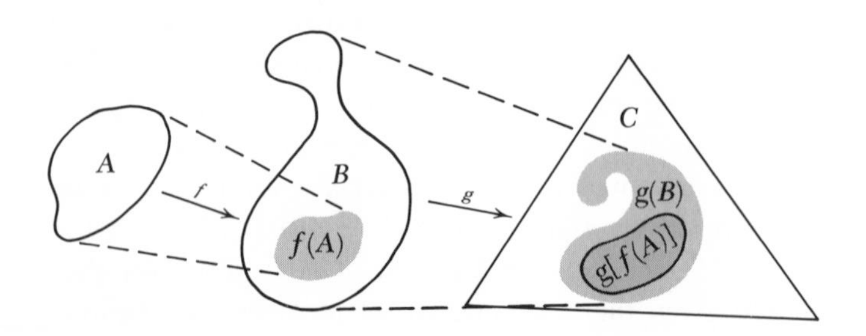

braces, for example, with their special meaning, or the use of boldface brackets, as in **[**a**]**, to denote the equivalence class of a.

A function $f: A \to B$ is **one to one** (or 1-1) if for all pairs a, $a' \in A$, $f(a) = f(a')$ implies $a = a'$. That is, if $a \neq a'$, then $f(a) \neq f(a')$. We have many examples at hand of 1-1 functions: each inclusion $S \subset T$ of a subset in a superset is one, and a special case of this is the identity function 1_T on a set T. We shall sometimes emphasize that a function $f: S \to T$ is an inclusion function, $f(x) = x$ for each $x \in S$, by writing $f: S \subset T$ or $S \overset{f}{\subset} T$. A nontrivial example is the cubing function on the real line: if $x < y$, then $x^3 < y^3$; hence $x \neq y$ implies $x^3 \neq y^3$. Such functions may be characterized as follows.

7 THEOREM *A function $f: A \to B$, $A \neq \varnothing$, is 1-1 iff there exists a function $g: B \to A$ such that $g \circ f = 1_A$.*

Proof Assume there exists g such that $g \circ f = 1_A$, and let a and a' be elements of A. If $f(a) = f(a')$, then $a = g \circ f(a) = g \circ f(a') = a'$, so f is 1-1.

Conversely, assume that f is 1-1. A function $g: B \to A$ may be defined as follows: Since $A \neq \varnothing$, choose $a_0 \in A$, and for all $b \in B - Im\,(f)$ let $g(b) = a_0$. But for each $b \in Im\,(f)$ there exists $a_b \in A$ such that $f(a_b) = b$, and a_b is unique, since f is 1-1. Define $g(b)$ to be a_b. Now, given $a \in A$, it is clear that $g \circ f(a) = g[f(a)]$ is the unique element of A, namely, a, which f carries to $f(a)$; thus $g \circ f = 1_A$. ∎

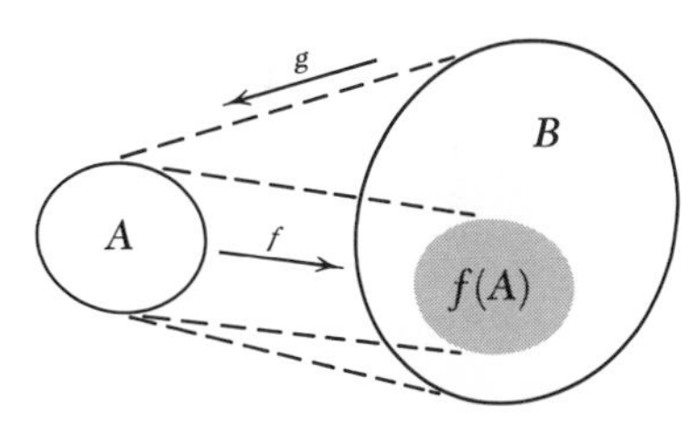

The function g above is called a **left inverse** for f; so the theorem may be restated as "a function with nonempty domain is 1-1 iff it has a left inverse." Do you see why the hypothesis $A \neq \varnothing$ was made?

It is worth noting that the equation $g \circ f = 1_A$ is independent of the values of g for elements of $B - Im\,(f)$. For example, if f is the square-root function, with domain the nonnegative reals and range the reals, then the squaring function g, with domain R and *range the nonnegative real numbers*, is a left inverse for f. But consider the function g' defined by

$$g'(x) = \begin{cases} x^2 & \text{if } x \geq 0 \\ 17 & \text{if } x < 0. \end{cases}$$

It is clear that $g' \circ f = 1_A$, yet $g' \neq g$. In fact, you can easily see that there is only one left-inverse function for a given 1-1 function f iff $Im(f) = Rng(f)$.

The "mirror image" of the above property of functions is the property of having a right inverse. Formally, if $f: A \to B$, then g is a **right-inverse function** for f if $f \circ g = 1_B$. Then g must be a function from B to A, since

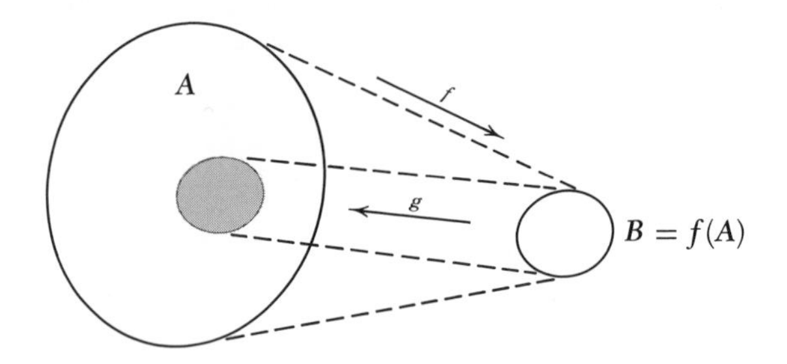

$Dom(f \circ g) = Dom(1_B) = B$, and $f \circ g$ is not defined unless $Rng(g) = Dom(f) = A$. The characterization corresponding to Theorem 7 is as follows.

8 THEOREM *A function $f: A \to B$ has a right inverse iff $Im(f) = B$. (Such a function f is called* **onto,** *or "surjective.")*

Proof Assume f is onto; that is, $f^{-1}(b) \neq \varnothing$ for all $b \in B$. It seems intuitively reasonable that an element a_b may be chosen from the nonempty set $f^{-1}(b)$ for every b; this defines the desired function g: $g(b) = a_b$. The other part of the proof is left to the reader. ∎

Our set theory is intuitive; had we begun with axioms for a set theory, the guarantee of the existence of a right-inverse function g for every onto function f would have been called "the axiom of choice." For discussions of a system of axioms for set theory, see the references at the end of this chapter (for example, Halmos).

We can diagram the situation of Theorems 7 and 8 for 1-1 and onto functions, respectively, as follows:

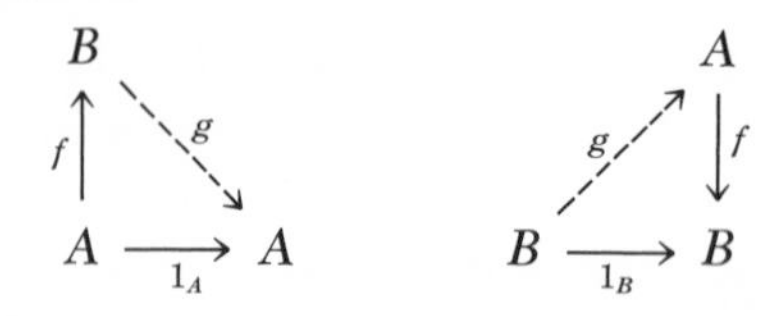

(The import of this and subsequent such diagrams is that the various possible composite routes obtained by following arrows forward from one set to another yield identical functions; the claimed existence of a function is

indicated by a dotted arrow. Thus the left figure above says "given $f: A \to B$, there exists $g: B \to A$ such that $g \circ f = 1_A$," which is true when f is 1-1. Such handy diagrams are suggestively called **commutative.**)

An example of an onto function is given above. Let f be the squaring function, thought of as having the nonnegative reals as its range.

One right-inverse function for f is the real-valued function g of a nonnegative real variable defined by

$$g(x) = \sqrt{x}.$$

Another possible right inverse for f is defined by

$$g'(x) = -\sqrt{x}.$$

Yet a third right inverse for f is given by

$$g''(x) = \begin{cases} \sqrt{x} & \text{if } x \text{ is an integer} \\ -\sqrt{x} & \text{if } x \text{ is not an integer.} \end{cases}$$

Note that each of these right-inverse functions, g, g', and g'', is 1-1. This is no accident.

It is clear that if f has a right inverse g, then g has a left inverse, f. Therefore, a right inverse g (for an onto function f) is always 1-1, and a left-inverse function is always onto. We now examine the happy event that a function is both 1-1 and onto.

9 THEOREM *A function $f: A \to B$ is both* 1-1 *and onto iff there exists a function g which is both a right inverse and a left inverse for f.*

Proof It is obvious that the existence of g shows that f is 1-1 (by Theorem 7) and onto (by Theorem 8).

Assume now that f is 1-1 and onto; if $A = \varnothing$, then $B = \varnothing$, and the empty function $\varnothing = \varnothing \times \varnothing$ is a **two-sided inverse** for f (that is, both left and right). If $A \neq \varnothing$, then there is a g_1 such that $g_1 \circ f = 1_A$ and a g_2 such that $f \circ g_2 = 1_B$ (by Theorems 7 and 8). But then

$$g_1 = g_1 \circ 1_B = g_1 \circ (f \circ g_2) = (g_1 \circ f) \circ g_2 = 1_A \circ g_2 = g_2,$$

so $g = g_1 = g_2$ is the required two-sided inverse for f. ∎

The above proof shows more than was claimed. Each right inverse is the same function as every left inverse, so there is a unique inverse on either side; this justifies the notation f^{-1}, read "f inverse," for g. But g has an inverse, f, so $(f^{-1})^{-1} = f$, and f^{-1} is always 1-1 and onto. Such a function is sometimes referred to as a **one-to-one correspondence** between its domain and its range. It may also be called a **set isomorphism** of its domain with its range, or a "bijection."

Examples of these 1-1 correspondences abound: For each set S the identity function 1_S is one, and $(1_S)^{-1} = 1_S$. The inverse of the cubing function f: $\mathbf{R} \to \mathbf{R}$ is the cube-root function g, $g(x) = \sqrt[3]{x} = x^{1/3}$, each real number has a real cube root, and $(\sqrt[3]{x})^3 = x = \sqrt[3]{x^3}$. The negation function function h with real (or complex) domain and range, $h(x) = -x$, is its own inverse, trivially, and hence is a set isomorphism. To count a set S of 23 ducks is to make a set isomorphism between S and the set of the first 23 positive integers.

Certainly a set isomorphism does not "preserve" algebraic properties. This assertion means, for instance, that the cubing function does not preserve addition; $(x + y)^3$ need not be $x^3 + y^3$; $(1 + 1)^3 \neq 1^3 + 1^3$. Nevertheless, if we ignore such structures as addition and distance, the pairing off of members of one set with those of another can be thought of as a "renaming" of elements of the first set using the second set as a source of names. This is the spirit of counting: there is a 1st duck, a 2nd duck, . . . , a 23d duck. Similarly, the negation function on the real line might be thought of as an interchange of left and right; the number named "3 units to the right of zero" becomes renamed "3 units to the left of zero."

To emphasize this aspect of a set isomorphism f: $A \to B$, we may denote it f: $A \cong B$ or $A \overset{f}{\cong} B$.

Exercises G, H, J, Q, and R

FACTORING FUNCTIONS

Each function f: $A \to B$ may be expressed as the composite of (or **factored** into) an onto function g followed (in its action) by a 1-1 function i. Take the onto function to be g: $A \to Im\,(f)$: $g(x) = f(x)$, and let i: $Im\,(f) \to B$: $i(x) = x$ be the inclusion $Im\,(f) \subset B$. The diagram is

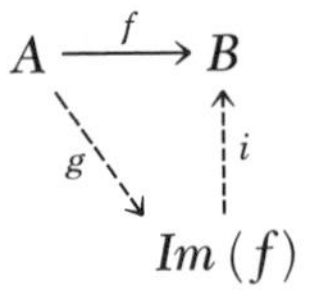

$$\begin{array}{ccc} A & \xrightarrow{f} & B \\ & {\scriptstyle g}\searrow & \uparrow{\scriptstyle i} \\ & & Im\,(f) \end{array}$$

There is another such factorization of f into a composite of a 1-1 function with an onto function which is less obvious than this but just as natural. To describe it, we need the following facts.

10 THEOREM *A function $f: A \to B$ defines an equivalence relation R on its domain A by xRy iff $f(x) = f(y)$. Further, f induces a function h from the quotient set $\mathcal{S}$ defined by R into B, $h: \mathcal{S} \to B$: $h([x]) = f(x)$.*

Proof That R is an equivalence is trivial (check it in your head). There remains only the question of whether h is well defined, that is, whether the recipe $h([x]) = f(x)$ indeed defines a function at all. It must be shown that if another *name* is chosen for the class $[x]$, say $y \in [x]$, so that $[x] = [y]$, then $h([x])$ and $h([y])$ are not defined in a conflicting fashion. A glance at the definition verifies that if $y \in [x]$, then $f(y) = f(x) = f([x]) = f([y])$. ∎ (Incidentally, it is worth noting that $[x] = f^{-1}[f(x)]$.)

Now it is clear that $f = h \circ q$, where q is the quotient function from A to $\mathcal{S}$. And q is onto, as is every quotient function. But h is 1-1, for if $[x] \neq [y]$, then $h([x]) = f(x) \neq f(y) = h([y])$. Hence the diagram

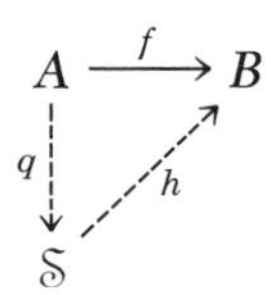

exhibits a factorization of the function f through a special sort of onto function (a quotient) followed by a 1-1 function. This is a "symmetric" result to the factorization of f through an onto function followed by a special sort of 1-1 function (an inclusion) which we saw earlier.

You may have observed above that $\mathcal{S}$ and $Im\ (f)$ were very much alike. More specifically, the function $r: \mathcal{S} \to Im\ (f)$: $r([x]) = f(x)$ is a set isomorphism, or "renaming"; it is simple to check that r is 1-1 and onto, and its inverse sends $f(x)$ to $[x]$, of course. The function r bridges the two factorizations discussed above; in that notation, $g = r \circ q$ and $h = i \circ r$, so $f = i \circ r \circ q$.

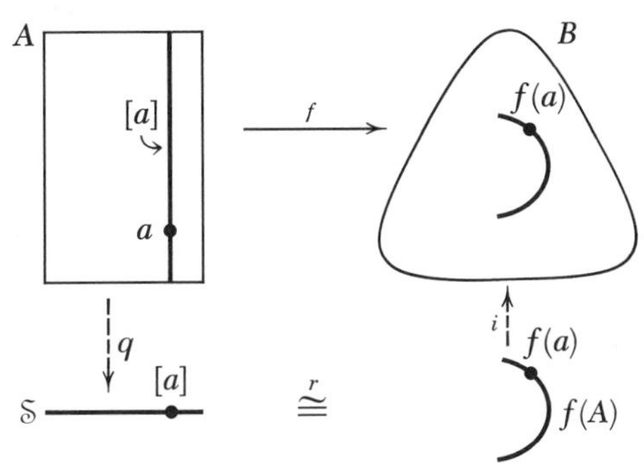

As an example, consider the squaring function f from **R** to **R**. First the quotient function q puts each real number x in its class $\{x, -x\}$, which has two members unless $x = 0$. Geometrically, q might be described as a folding of the real line at zero. Then the set isomorphism r renames each class $\{x, -x\} = [x]$ with the label x^2; it might be said to stretch the folded line and then pin it down along a real half-line. Finally, i is the inclusion of the nonnegative reals in the reals, the half-line in the whole line.

Returning to the general case, we collect the facts in a teaching statement which systematizes several related results.

11 THE QUOTIENT THEOREM FOR SETS *Each function $f: A \to B$ has a unique factorization $f = i \circ r \circ q$, or*

$$\begin{array}{ccc} A & \xrightarrow{f} & B \\ {\scriptstyle q}\downarrow & & \uparrow{\scriptstyle i} \\ \mathcal{S} & \overset{r}{\cong} & Im\,(f) \end{array}$$

where q is a quotient, r is a set isomorphism, and i is an inclusion. The function $r \circ q: A \to Im\,(f)$ is onto, and $i \circ r: \mathcal{S} \to B$ is 1-1. ■

We omit the proof of the uniqueness of the factorization; it is easy. The other assertions have already been established. The fact that $r \circ q$ is onto can be generalized: every composite of onto functions is onto (why?). Similarly, the composite of 1-1 functions is always a 1-1 function.

Exercises K and L

RESTRICTIONS AND EXTENSIONS

There is one glaring flaw in our definition of composite functions. If $f: A \to B$ and $g: C \to D$, it is natural to be satisfied with a composite function defined only on $f^{-1}(B \cap C)$, yet we have offered no definition of $g \circ f$ if $B \neq C$. This may be remedied by forming a new function $F: f^{-1}(B \cap C) \to C$, which has the same values as f at each point in its domain. Then $g \circ F: f^{-1}(B \cap C) \to D$ is the hoped-for composite. Formally, if $f: A \to B$ with $X \subset A$ and $f(X) \subset Y \subset B$, then the **restriction** of f to $X \times Y$, $f|_{X \times Y}: X \to Y$, is $f \cap X \times Y$. Usually the range will be clearly enough understood; we shall then speak simply of the restriction of f to X, $f|_X$. (Many writers understand the range of $f|_X$ to be B.) On the other hand, if $g: X \to Y$ and $f: A \to B$ are functions, $X \subset A$, $Y \subset B$, and $f|_{X \times Y} = g$, then f is said to **extend** (or be

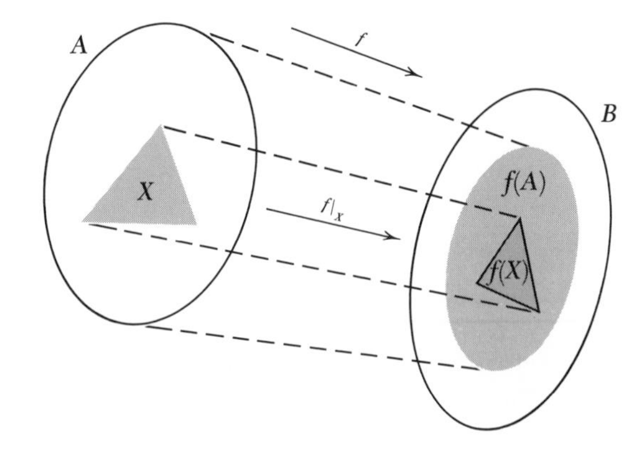

an **extension** of) g. You can easily manufacture examples of these notions.

Let $f_1: A_1 \to B_1$ and $f_2: A_2 \to B_2$ be functions, and let $C \subset A_1 \cap A_2$. If $f_1(x) = f_2(x)$ for each $x \in C$, then f_1 and f_2 **agree** on C. If f_1 and f_2 agree on $A_1 \cap A_2$, they are said to be **combinable.** In that case, one may form the **combined** (or "union") **function** $f: A_1 \cup A_2 \to B_1 \cup B_2$, which has values

$$f(x) = \begin{cases} f_1(x) & \text{if } x \in A_1 \\ f_2(x) & \text{if } x \in A_2. \end{cases}$$

That f_1 and f_2 are combinable it is necessary and sufficient that f be well defined.

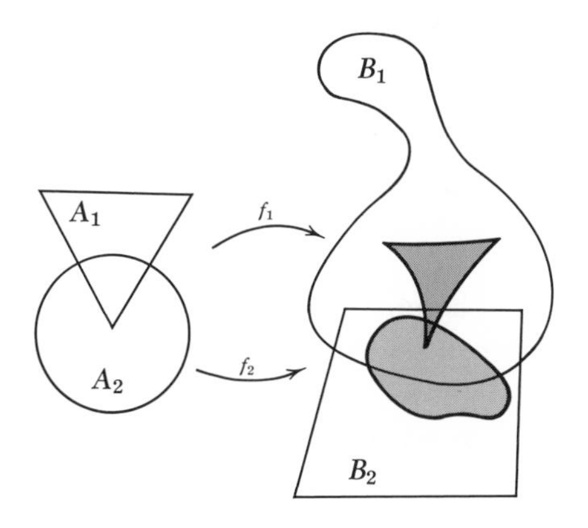

For an example of real-valued combinable functions, let $f_1(x) = x$ for each $x \geq 0$ and $f_2(x) = -x$ for each $x \leq 0$; the combined function f is the absolute-value function $f(x) = |x|$ for all real x.

The definition of combinable and the combined function for a finite family of functions is immediate; each pair of members of the family must be combinable, and the combined function is defined on the union of the member's domains. This definition works as well for an arbitrary family of functions, but we shall not require that.

REFERENCES AND FURTHER TOPICS

There are many introductory texts for elementary set theory. Two of these, available in paperback, are

R. R. Christian, *Introduction to Logic and Sets*, 2d ed. (New York: Blaisdell, 1965).

R. R. Stoll, *Sets, Logic and Axiomatic Theories* (San Francisco: Freeman, 1961).

You will also find such material in texts introducing other subjects. Two of the many such are

S. Feferman, *The Number Systems*, chap. 2 (Reading, Mass.: Addison-Wesley, 1964).

R. C. James, *University Mathematics*, chap. 2 (Belmont, Calif.: Wadsworth, 1963).

For a treatment very close in spirit to ours, see

S. -T. Hu, *Elements of General Topology*, chap. 1 (San Francisco: Holden-Day, 1964).

J. L. Kelley, *General Topology*, pp. 1–13 (Princeton: Van Nostrand, 1955).

The presentation of this chapter is work-a-day. We shall use the language of set theory in every definition and argument ahead, but many fascinating questions have been ignored here (remember, for example, the remarks we made concerning the equality of sets and the axiom of choice). Certainly this fundamental language of mathematics can be made more rigorous by introducing a system of axioms; there is a long history of the development of such systems. And of course, it is reasonable to ask what different set theories arise when the axioms are changed, or a certain axiom is not assumed at all. More sophisticated approaches to the theory of sets may be found in

A. A. Fraenkel, *Abstract Set Theory* (Amsterdam: North-Holland, 1953).
P. R. Halmos, *Naive Set Theory* (Princeton: Van Nostrand, 1960).
F. Hausdorff, *Set Theory*, 2d ed. (New York: Chelsea, 1962).

Each of these books is quite readable. A very condensed, axiomatic presentation may be found in the appendix of Kelley, cited above.

EXERCISES

A For each positive integer $n = 1, 2, \ldots,$ let the subset T_n of $\mathbf{R}$ be defined by

$$T_n = \left\{ y\colon |y - 1| < n \text{ and } |y + 1| > \frac{1}{n} \right\}.$$

If $\mathfrak{T} = \{T_n\colon n = 1, 2, \ldots\}$, find $\bigcup \mathfrak{T}$ and $\bigcap \mathfrak{T}$.

B Show that for all sets A, B, and C in a given universe U,

- **i** $\varnothing' = U$; $U' = \varnothing$; $A \subset A$; $A \cap A = A \cup A = A$.
- **ii** $\varnothing \subset A$; $A \subset U$; $\varnothing \cap A = \varnothing$, $\varnothing \cup A = A$, $U \cap A = A$; $U \cup A = U$.
- **iii** $A \subset B$ and $B \subset C$ implies $A \subset C$.
- **iv** $A \subset B$ iff $A \cap B = A$ iff $A \cup B = B$.
- **v** $A \cap A' = \varnothing$; $A \cup A' = U$.
- **vi** $A \subset B$ iff $A \cap B' = \varnothing$ iff $B \cup A' = U$.
- **vii** $A \subset C$ and $B \subset C$ implies $(A \cup B) \subset C$.
- **viii** $A \subset B$ and $A \subset C$ implies $A \subset (B \cap C)$.
- **ix** $A = (A - B) \cup (A \cap B)$.
- **x** $A \subset B$ iff $B' \subset A'$; $A - B = A \cap B'$.

C Prove Theorem 2.ii.

D Prove Theorem 2.iv.

E Finish the proof of Theorem 5.

F Find the flaw in the following argument, which purports to show that if a relation is both symmetric and transitive, then it is reflexive:

Since R is symmetric, if aRb, then bRa. Since R is transitive, aRb and bRa together imply aRa. Therefore R is reflexive.

G Give an example of a function with finite domain and range sets which has two different right inverses, and also a function with two left inverses. Can you find one function which will serve for both examples?

H Show that if $f \circ g$ is 1-1, then so is g, and if $f \circ g$ is onto, then so is f. What can be said if $f \circ g$ is both 1-1 and onto?

J Prove that if both f and g have two-sided inverses, then so has $f \circ g$ (if it is defined, of course), and $(f \circ g)^{-1} = g^{-1} \circ f^{-1}$.

K Supply a detailed proof of the uniqueness assertion of Theorem 11.

L Let the function $f\colon \mathbf{R} \to \mathbf{C}$ be defined by $f(x) = \cos x + i \sin x$ for each real x. If $a + ib$ is a complex number, then its absolute value is $|a + ib| = (a^2 + b^2)^{1/2}$; it is clear that $|f(x)| = 1$ for each real x. Hence the image of f lies on the circle of unit radius which is centered at the origin in the plane. The function f may be visualized as assigning to each real number the point on that circle arrived at by going counterclockwise around the circle a distance x, starting at $1 = 1 + i0$; $x < 0$ requires going the distance $|x|$ in the clockwise direction. This makes it clear that the image of f equals the unit circle; can you prove it rigorously?

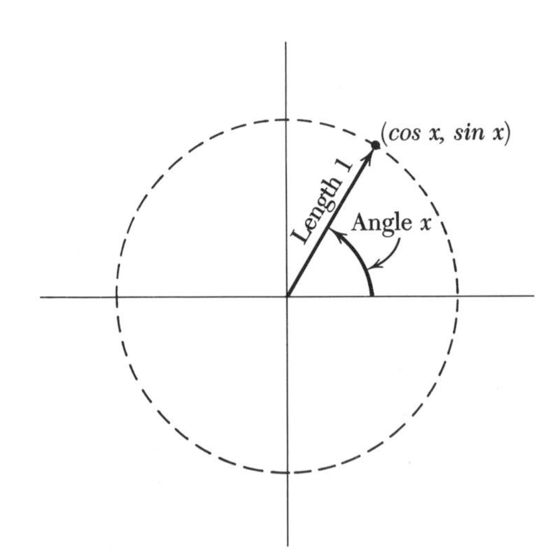

Discuss the function f in the light of the Quotient Theorem (11); that is, find the equivalence relation that f defines on $\mathbf{R}$, and define the quotient q, the set isomorphism r, and the inclusion i for which $f = i \circ r \circ q$.

M Draw a picture or graph of the relation which the squaring function defines on its domain **R**. The relation is, of course, a subset of $\mathbf{R} \times \mathbf{R}$; depict that subset.

N If A and B are finite sets with M and N members, respectively, how many members has $A \times B$? What if $M = 0$? What if one factor set is infinite?

P If A and B are sets, give a set isomorphism from $A \times B$ to $B \times A$. An expression of this result is "formation of direct products is a commutative (binary) operation, up to set isomorphism." Show also that this operation is associative, up to isomorphism; in other words, that $(A \times B) \times C$ is set isomorphic to $A \times (B \times C)$.

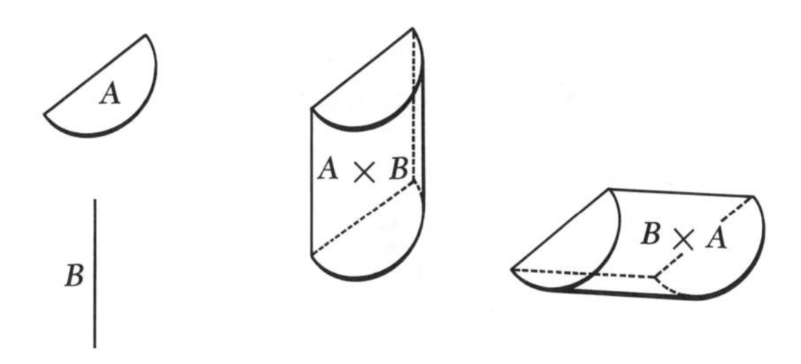

Q The usual identification of complex numbers with points in the plane regards $x + iy$ as *being* the point having coordinates (x,y) in the plane. This may be rephrased as an *identification* of **C** with $\mathbf{R}^2$; we shall use these two systems of notation interchangeably, sometimes referring to the origin (0,0) of the plane as 0, to (1,0) as 1, etc. With this understanding, the function $Re\colon \mathbf{C} \to \mathbf{R}$, which assigns to $x + iy$ its real part x, goes from a direct product onto one of the factors of that product. Exhibit two distinct right-inverse functions for Re. Also give two distinct left inverses for the inclusion of **R** in **C** as the real axis.

R The function $p_1\colon A \times B \to A\colon p_1(a,b) = a$ is called the **projection** of the direct product on its first factor; similarly, $p_2\colon A \times B \to B\colon p_2(a,b) = b$ is the projection on the second factor. Show that projections are always onto functions, and give a right-inverse function for p_1 and also for p_2 (assume here that $A \times B \neq \varnothing$). When can two distinct right inverses for p_1 be found?

PROBLEMS

AA Relations Generalize the notion of composition of functions to include all relations: For given relations R in A to B and S in B to C, define $S \circ R = \{(a,c): a \in A,\ c \in C$; there exists $b \in B$ such that $(a,b) \in R$ and $(b,c) \in S\}$, a relation in A to C. The "inverse" (or "converse") of the relation R is $R^{-1} = \{(b,a): (a,b) \in R\}$, and if $D \subset A$, then $R(D) = \{b$: there exists $d \in D$ such that $(d,b) \in R\}$. Show that $(R^{-1})^{-1} = R$, $(R \circ S)^{-1} = S^{-1} \circ R^{-1}$, and $(R \circ S) \circ T = R \circ (S \circ T)$ for all relations R, S, and T for which the composites are defined. If f is a function, f^{-1} is a relation; what is $f^{-1} \circ f$? Is $R \circ S(D) = R[S(D)]$? Do relations preserve unions or intersections? That is, is $R(D \cup D') = R(D) \cup R(D')$, or is $R(D \cap D') = R(D) \cap R(D')$?

Prove that a relation R is an equivalence iff $(\Delta \cup R^{-1} \cup R \circ R) \subset R$.

BB Categorical Matters We defined an identity function by specifying that its domain and range sets were equal and that its value at each point x was x, for every x in its domain. An identity function 1_S has the property that if $f: S \to T$, then $f \circ 1_S = f$, and also if $g: U \to S$, then $1_S \circ g = g$. However, an identity function can be "characterized" (or defined) without mention of its domain, range, or values. Suppose F is a function with the property that for each function f, if $f \circ F$ is defined, then $f \circ F = f$, and if $F \circ f$ is defined, then $F \circ f = f$. We assert that F must be an identity, for if $D = Dom\,(F)$, then $F \circ 1_D$ is defined, so $F \circ 1_D = 1_D$. However, 1_D is an identity, so $F \circ 1_D = F$.

Although the property of a function's being onto was defined in terms of the elements of its domain and range sets, we may, using Theorem 8, characterize the "ontoness" of a function by looking only at compositions of functions, and not at values. A function G is onto iff there exists a function g such that $G \circ g$ is defined and $G \circ g$ is an identity function. To establish another such description of onto functions, a function f is onto iff for every pair (u,v) of functions $u \circ f = v \circ f$ implies $u = v$ (so f is right-cancellable). Similarly, show that f is 1-1 iff $f \circ u = f \circ v$ implies $u = v$ (or f is left-cancellable).

In the same vein, assume that A_1, A_2, and B are sets and that $p_1: B \to A_1$ and $p_2: B \to A_2$ are functions with the property that, given any set C and functions $f_1: C \to A_1$ and $f_2: C \to A_2$, there exists a unique function $f: C \to B$ such that $f_1 = p_1 \circ f$ and $f_2 = p_2 \circ f$. Show that there is a set isomorphism of B with $A_1 \times A_2$. Thus a direct-product set can be characterized, up to set isomorphism, without referring to its elements. The diagram below illustrates the situation. The unique map f is usually denoted $f_1 \times f_2$ and called the "direct product of f_1 and f_2." If $B = A_1 \times A_2$, then $(f_1 \times f_2)(c) = [f_1(c),f_2(c)]$. The functions p_1 and p_2 are called the "projections" of the direct product; if $B = A_1 \times A_2$, then $p_i(a_1,a_2) = a_i$; $i = 1$ and 2:

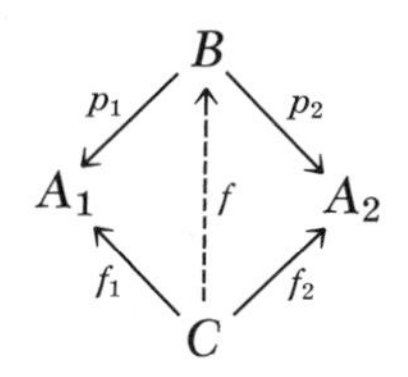

CC Factoring through Quotients Suppose that R is an equivalence relation on the domain of a function $f: A \to B$, with $q: A \to A/R$ the quotient function defined by R. When is it possible to factor f through q? That is, what requirements need be placed on q and f to ensure the existence of a function $g: A/R \to B$ with $f = g \circ q$?

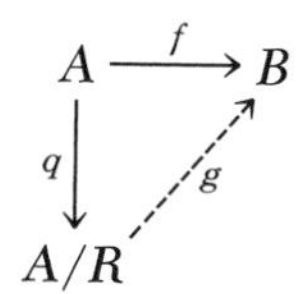

Use your result above to establish necessary and sufficient conditions that if, for $i = 1$ and 2, $q_i: A_i \to A_i/R_i$ is a quotient and $f: A_1 \to A_2$, there exists a function g which makes the following diagram commutative:

$$\begin{array}{ccc} A_1 & \xrightarrow{f} & A_2 \\ {\scriptstyle q_1}\downarrow & & \downarrow{\scriptstyle q_2} \\ A_1/R_1 & \xrightarrow[g]{} & A_2/R_2 \end{array}$$

Groups

CHAPTER II

Time after time, as you have studied each new set of mathematical objects, ways have arisen naturally to put together two set members to get a third one. Two numbers can be added to get a third number, whether the number system in question is the integers or the rational, real, or complex numbers. Multiplication is another operation pairing two numbers to get a third number. Similarly, a study of sets begins with the "binary operations" of intersection and union. The operation of composition yields a new function for *some* pairs of functions, though not every pair of functions can be composed.

These operations are certainly objects of interest in their own right, and they may or may not have certain "properties" which can be defined for binary operations; for instance, each of the above operations has the property of associativity. Commutativity is another property of operations. While most of the above examples are commutative, we shall see that the composition of functions is not a commutative operation. We now begin the study of a new property of operations, the group property. The definition of the group property is more complex than that of the above properties; in fact, as part of the definition we shall require that a group operation be an associative operation. But the underlying idea is not a bit deeper; an operation either does or does not have the group property, and that can be decided by comparing the operation with the definition of the group property.

THE GROUP PROPERTY

The set **R** of real numbers has the algebraic "operation" of addition defined on it; if x and y are reals, then $x + y$ is a uniquely determined real number. A restatement of this fact is that "there is a function called addition on $\mathbf{R} \times \mathbf{R}$ with range **R**; the value of this function at (x,y) is denoted $x + y$" (see Chap. I for definitions of function, value, etc.).

Similarly, if $S = \{a,b,c\}$ is a set of three elements, and T is the set of all functions whose domain and range are S, then the composition of members of T defines a function $\kappa: T \times T \to T$, with $\kappa(f,g) = f \circ g$. That is, the members of the set T are functions from S to S (how many members has T?), and so members of $T \times T$ are ordered pairs (f,g) of functions in T. The image of each pair (f,g) under κ is the composite $f \circ g$, which is also a member of T.

In each of these examples the operation is associative; for instance, for every triple f,g,h of members of T, $f \circ (g \circ h) = (f \circ g) \circ h$ (see Theorem I.6). An equivalent statement is that $\kappa[\kappa(f,g),h] = \kappa[f,\kappa(g,h)]$ for all triples f,g,h. Also, in each example there is a unique element which acts as an "identity" for the operation; the real number 0 has the property that for all real x, $0 + x = x + 0 = x$. The function 1_S has a similar property with respect to composition; for all $f \in T$, $1_S \circ f = f \circ 1_S = f$, or equivalently, $\kappa(f,1_S) = \kappa(1_S,f) = f$.

It is also true that for each real x there exists another real number y such that $x + y = y + x = 0$; $y = -x$ will do nicely, and of course, no real other than $-x$ has this property (for a given x). However, if f is the constant function from S to S which has the value a for each point in its domain, then there exists no function $g \in T$ for which either $f \circ g = 1_S$ or $g \circ f = 1_S$. In fact, Theorem I.9 teaches that an element $f \in T$ has an inverse f^{-1} if f is both 1-1 and onto, and then f^{-1} is also a member of T. Thus the set $V = \{f \in T: f \text{ is 1-1 and onto}\}$ contains, for each of its elements f, an element g such that $f \circ g = g \circ f = 1_S$, and g is unique. Furthermore, $1_S \in V$, and if f and g are members of V, so is $f \circ g$ (see Exercise I.J). Again, the composition operation is associative on V, since the members of V are functions. Elements of V are called permutations of S; if A is any set, then a **permutation** of A is a 1-1 onto function from A to A.

The set of real numbers together with the operation of addition, or the set V of permutations of S with the composition operation, forms an example of a group ("group" is *not* a synonym of "set").

A **group** (G,m) is a set G together with a function $m: G \times G \to G$ such that if the value $m(g,h)$ of m at a pair (g,h) of elements of G is denoted simply by juxtaposition, $m(g,h) = gh$, then

i $f(gh) = (fg)h$ for all f, g, and h in G,
ii There exists an element $e \in G$ such that for all $g \in G$, $eg = ge = g$;
iii For each $g \in G$ there exists an element $h \in G$ such that $gh = hg = e$, where e is the element of G guaranteed by property ii.

An example of a set G with a function m such that (G,m) is not a group is given by (T,κ) above; it does not have property iii. Another example is the set $\mathbf{R}$ with the operation of real multiplication.

The element e of property ii of a group is unique, for if the elements e and e' of G both satisfy property ii, then ee' is equal to e and also equal to e', by property ii. ∎ This unique element e of each group is called the **identity.** Furthermore, for each $g \in G$ the element h guaranteed by property iii is unique, since, if $hg = gh = e$ and also $h'g = gh' = e$, then

$$h = he = h(gh') = (hg)h' = eh' = h'.$$

This unique element h is denoted g^{-1}, read "g **inverse.**" It is now apparent that $(g^{-1})^{-1} = g$. These facts can be reinforced as follows.

1 THEOREM *If g and h are elements of a group* (G,m), *then*

i $gh = g$ *implies* $h = e$,
ii $hg = g$ *implies* $h = e$,
iii $gh = e$ *implies* $h = g^{-1}$,
iv $hg = e$ *implies* $h = g^{-1}$.

Proof of i and iii Let $gh = g$; then

$$e = g^{-1}g = g^{-1}(gh) = (g^{-1}g)h = eh = h.$$

Similarly, if $gh = e$, then

$$h = eh = (g^{-1}g)h = g^{-1}(gh) = g^{-1}e = g^{-1}. \quad ∎$$

The proofs of parts ii and iv are quite similar. Let us use the theorem to show that *for all members* g_1 *and* g_2 *of a group* (G,m),

$$(g_1g_2)^{-1} = g_2^{-1}g_1^{-1};$$

note that

$$(g_1g_2)(g_2^{-1}g_1^{-1}) = [g_1(g_2g_2^{-1})]g_1^{-1} = (g_1e)g_1^{-1} = g_1g_1^{-1} = e.$$

But if we apply Theorem 1.iii, with $g = g_1g_2$ and $h = g_2^{-1}g_1^{-1}$, the result follows. ∎

If (G,m) is a group, the function m is often called the **group multiplication,** or **product,** in analogy to the group of nonzero reals under real multiplication (verify that this is a group). The value $m(g,h)$ is usually called the **product** of g and h. The last result above may now be paraphrased as "the inverse of a product is the product of the inverses, in the reverse order." When only one function m is being considered for a given set G, we shall often speak of "the group G" rather than "the group (G,m)." By this time it is scarcely apparent that the group property belongs to the operation at all. A group now seems to be the set G, called the **underlying set** of the group, which is the domain of the operation m. But this is inevitable; a glance at Theorem 1 shows that the group property of m is quite special, and many statements can be made about such an operation. These statements are phrased in terms of elements of G, however, and the function m has disappeared altogether from our notation.

If, for every pair g,h of members of a given group G, $gh = hg$, then G is an **abelian,** or **commutative,** group. In this case the notation $g + h$ rather than gh is often used for the product of g and h; we then speak of the **sum** of g and h. This convention has no new meaning; it merely emphasizes that G is abelian. An example is the abelian group $(\mathbf{R},+)$ of real numbers under addition. The group (V,κ) of isomorphisms of the set $S = \{a,b,c\}$ which was discussed at the beginning of the chapter is not abelian; the function which interchanges a and b and leaves c fixed does not commute (under composition) with the function interchanging b and c and leaving a fixed (check this yourself). Groups which are definitely not abelian are termed **nonabelian,** or "noncommutative."

Exercises A, B, and D

SUBGROUPS

A **subgroup** H of a group G is a subset H of the *set* G with the property that the multiplication of G, considered only on H, makes H a group. More specifically, if (G,m) is a group, $H \subset G$, $m(H \times H) \subset H$ and $m|_{H\times H}$ is a group multiplication for H, then $(H,m|_{H\times H})$ is a subgroup of (G,m). Every group G has the two trivial subgroups G and $\{e\}$. A nontrivial example is the subset of positive real numbers in the group of all nonzero real numbers under real multiplication. The set of integers divisible by 4 is a subgroup of the additive group of all integers $(\mathbf{Z},+)$. Is it clear that the identity of a subgroup H must be the identity of the group G?

If a subset H of a group G is **closed under multiplication** [that is, if $m(H \times H) \subset H$], then $m|_{H\times H}$ is necessarily an associative operation, since m is associative on all of G. In this case H is a subgroup iff $e \in H$ and for every $h \in H$, $h^{-1} \in H$ (that is, H is **closed under inversion**). Of course, if h and h^{-1} are in H and H is closed under multiplication, then $hh^{-1} = e$ is always in H. A new abbreviation will be handy for describing this state of affairs: if A and B are subsets of G and (G,m) is a group, AB means the set $m(A \times B) \subset G$; thus $AB = \{ab\colon a \in A \text{ and } b \in B\}$. If A is a singleton set (that is, if A has exactly one member), write aB for $\{a\}B$. Further, let $A^{-1} = \{a^{-1}\colon a \in A\}$. Some examples of this in the multiplicative group of nonzero real numbers are

$$\{-1,2\}\{1,2,3\} = \{-1,-2,-3,2,4,6\},$$
$$\{3\}\{-1,2\} = 3\{-1,2\} = \{-3,6\}, \quad \text{and} \quad \{-1,2\}^{-1} = \{-1,\tfrac{1}{2}\}.$$

In this notation a subset $H \neq \varnothing$ of G is a subgroup of G iff both $HH \subset H$ (closure under multiplication) and $H^{-1} \subset H$ (closure under inversion).

2 THEOREM *A subset H of a group G is a subgroup of G iff $H \neq \varnothing$ and $HH^{-1} \subset H$.*

Proof It is obvious that if H is a subgroup, then H contains e and $HH^{-1} \subset H$. Conversely, if $H \subset G$ is nonempty and $HH^{-1} \subset H$, then there is an element $h \in H$ and $hh^{-1} = e \in H$. Therefore H is a subgroup if it is closed under multiplication and inversion. But $H^{-1} = eH^{-1} \subset HH^{-1} \subset H$. Similarly, if h_1 and h_2 are members of H, then $h_2{}^{-1} \in H$, so $h_2 = (h_2{}^{-1})^{-1} \in H^{-1}$ and $h_1h_2 \in HH^{-1} \subset H$; hence $HH \subset H$ and H is a subgroup. ■

A **left coset** of a subgroup H of G is a set gH for some $g \in G$. It is easy to see that if g_1 and g_2 are members of G, with $g_2 \in g_1H$, then $g_2H \subset g_1H$, since g_2 is of the form $g_2 = g_1h_1$ for some $h_1 \in H$; thus each element $g_2h_2 \in g_2H$ is of the form $g_2h_2 = g_1h_1h_2$, and $h_1h_2 \in H$, so $g_2h_2 \in g_1H$. (Notice that the use of parentheses to indicate the association of products has been abandoned; see Exercise C.) On the other hand, if $g_2 \in g_1H$, then $g_1 \in g_2H$, since $g_2 = g_1h$ implies $g_1 = g_2h^{-1} \in g_2H$. Putting these two results together, we see that $g_2 \in g_1H$ implies that the two left cosets g_1H and g_2H are equal. This surprise leads immediately to a greater one: if g_1 and g_2 are members of G and $g_1H \cap g_2H \neq \varnothing$, then $g_1H = g_2H$! This follows from the existence of an element g_3 in $g_1H \cap g_2H$; $g_3 \in g_1H$ implies $g_3H = g_1H$, and $g_3 \in g_2H$ implies $g_3H = g_2H$, so $g_1H = g_2H$. Another way to state this result is that *the family* $\mathcal{S} = \{gH: g \in G\}$ *of left cosets of a subgroup* H *of* G *is a family of mutually disjoint subsets of* G. ∎

Since every $g \in G$ is in gH, the family $\mathcal{S} = \{gH: g \in G\}$ is a partition of G; $\mathcal{S}$ is called the **left-coset family of** G **modulo** H and is denoted $\mathcal{S} = G/H$. Thus the left-coset family of a subgroup of G is a quotient set of G. (Here the word "family" means no more than "set.") The equivalence relation associated with this partition is that g_1 is related to g_2 iff $g_1H = g_2H$. But we have seen that $g_1H = g_2H$ iff $g_2 \in g_1H$, and this happens iff there exists $h \in H$ with $g_2 = g_1h$, or $g_1^{-1}g_2 = h$. Hence g_1 is related to g_2 iff $g_1^{-1}g_2 \in H$.

If the group G is abelian, then each left coset gH is also the set Hg; however, these sets may differ when G is nonabelian. It would have been possible to partition G into the "right-coset family" $\{Hg: g \in G\}$ of the subgroup H in G. In general, this partition of G is different; it gives rise to an equivalent theory, however. Henceforth we shall consider only left cosets; the word "left" will be suppressed, and we shall speak simply of "the cosets" and "the coset family" of H in G.

An example of a coset family is now in order. Let G be the set of nonzero complex numbers with multiplication as the operation. Each element x of G is an ordered pair (x_1,x_2) of real numbers, and the product xy of two members of G is $(x_1,x_2)(y_1,y_2) = (x_1y_1 - x_2y_2, x_1y_2 + x_2y_1)$. Clearly, $(1,0)$ acts as an identity for this multiplication; it lies on the real axis and will usually be denoted 1. The associativity can be checked by brute force; use the product recipe to show that $(xy)z$ and $x(yz)$ are equal. The **conjugate** $\bar{x}$ of a complex number $x = (x_1,x_2)$ is the complex number $(x_1,-x_2)$, and the product $x\bar{x} = x_1^2 + x_2^2$ is always real; we write $x\bar{x} = |x|^2$. We denote the origin $(0,0)$ by 0, and in general regard the real numbers as the subset of the complex numbers lying along the horizontal axis of the plane. If $x \neq 0$, then $|x|^2 > 0$, so $1/|x|^2$ is a real number, and $x\bar{x}(1/|x|^2) = 1$, or $x[\bar{x}(1/|x|^2)] = 1$. Hence $\bar{x}(1/|x|^2)$ is an inverse for x, that is, $x^{-1} = [x_1/(x_1^2 + x_2^2), x_2/(x_1^2+x_2^2)]$. This completes the verification that the set G of nonzero complex numbers is a group under complex multiplication.

The real number $|x| = (x_1^2 + x_2^2)^{1/2}$ is called the **absolute value** of x (see Exercise I.L); it is just the distance from the point (x_1,x_2) to the origin, and $|x| > 0$ iff $x \neq 0$. It is easy to check that $|xy|^2 = |x|^2|y|^2$ for all complex numbers x and y; since absolute values are nonnegative, this implies that $|xy| = |x|\,|y|$. Thus, if $|x| = 1$ and $|y| = 1$, then $|xy| = 1$. Further, if $|x| = 1$, then $|\bar{x}| = 1$ and $x\bar{x} = |x|^2 = 1$, so $|x^{-1}| = |\bar{x}| = 1$ (note here that if $|x| = 1$, then $x^{-1} = \bar{x}$). This shows that the subset $H = \{x \in G: |x| = 1\}$ of G is a subgroup; it is closed under multiplication and inversion. If $x \in G$ and $y \in H$, then $|xy| = |x|\,|y| = |x|$, so each

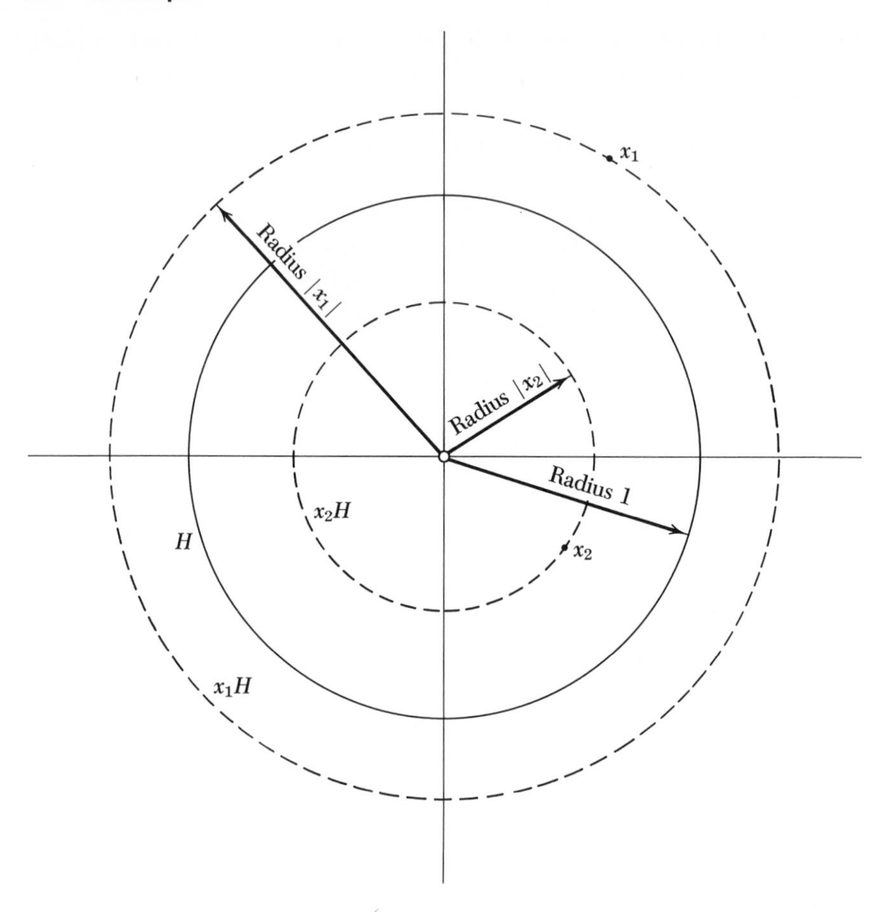

element xy of the coset xH has absolute value the same as that of x. Conversely, suppose $|x_1| = |x_2|$; then $|x_1||x_1^{-1}| = 1 = |x_2||x_2^{-1}|$, so $|x_1^{-1}| = 1/|x_1| = 1/|x_2|$ and $|x_1^{-1}x_2| = (1/|x_2|)|x_2| = 1$; $x_1^{-1}x_2$ is therefore a member of H. This says that $x_1H = \{x_2 \in G\colon |x_2| = |x_1|\}$. Since the absolute value of an element x of $\mathbf{C}$ is just the distance from x to the origin 0, we may picture H as the circle of radius 1 (the **unit circle**) in the plane; each coset xH of H in G is just the circle of radius $|x|$. The coset family is the family of all circles in the plane (of positive radius) centered at the origin. There is exactly one such circle passing through each point of the plane. There is an obvious 1-1 correspondence between the coset family and the set of positive real radii.

Exercises E, G, and K

MORPHISMS

The absolute-value function $f: G \to \mathbf{R} - \{0\}: f(x) = |x|$ described above has the interesting property that $f(xy) = f(x)f(y)$. That is, first multiplying x by y in G and then applying f to the product xy gives the same result $f(xy)$ as finding the values $f(x)$ and $f(y)$ first and then multiplying in the range group of nonzero reals to get $f(x)f(y)$. This is reminiscent of the logarithm function, whose domain D is the group of positive real numbers under real multiplication and whose range is the additive group of all real numbers. It is a fact that $log\ (xy) = log\ (x) + log\ (y)$. In other words, we get the same result by first multiplying and then finding the logarithm as by finding logarithms first and then adding. (The usefulness of all this is due to another feature of the *log* function; it is a 1-1 correspondence, and so has an inverse. The equation above may therefore be rewritten as $xy = log^{-1}[log\ (x) + log\ (y)]$, and the mechanics of real addition are easier than those of positive real multiplication.) This special property of these two functions—the absolute value and the logarithm—of "relating" the group operations of their domains and their ranges deserves a name of its own plus further study.

A function $f: G \to H$ from the underlying set of one group to that of another is called a **morphism** (or "homomorphism") of the group G into the group H if f preserves products, that is, if for all g_1 and $g_2 \in G$, $f(g_1g_2) = f(g_1)f(g_2)$. Note that the group operation on the left of the equality is in the domain and that on the right in the range of f. If m is the product of G and n is the product of H, then the requirement on f can be expressed by saying that the following diagram commutes:

$$\begin{array}{ccc} G \times G & \xrightarrow{f \times f} & H \times H \\ {\scriptstyle m}\downarrow & & \downarrow{\scriptstyle n} \\ G & \xrightarrow{\quad f \quad} & H \end{array}$$

Here $(f \times f)(g_1,g_2) = [f(g_1),f(g_2)]$,† and the requirement that the diagram commute is that the function $n \circ (f \times f) = f \circ m$. You may wonder why a morphism is not required to preserve identities and inverses as well as

† This definition of the product $f \times g$ of two functions f and g differs from that given in Prob. I.BB. What is the relationship of the two?

products. It turns out, however, that it is not necessary; if $f: G \to H$ is a morphism and e is the identity of G, then

$$f(e) = f(ee) = f(e)f(e);$$

by Theorem 1.i, $f(e)$ must be the identity of H; *morphisms preserve identities.* ■ Similarly, if $g \in G$, then (if we denote the identity of H by the same symbol e as the identity for G)

$$e = f(e) = f(gg^{-1}) = f(g)f(g^{-1});$$

by Theorem 1.iii, then, $f(g^{-1})$ must be the inverse $f(g)^{-1}$ of $f(g)$; *morphisms preserve inverses.* ■ There is another way of expressing this last result: for each group G let an inversion function ι_G be defined by $\iota_G: G \to G: \iota_G(g) = g^{-1}$. Then for each morphism $f: G \to H$ we have $\iota_H \circ f = f \circ \iota_G$, or

$$\begin{array}{ccc} G & \xrightarrow{f} & H \\ \downarrow \iota_G & & \downarrow \iota_H \\ G & \xrightarrow{f} & H \end{array}$$

We have mentioned the absolute value and logarithm functions as examples of morphisms. Another example is given in Exercise I.L: $f: \mathbf{R} \to \mathbf{C} - \{0\}: f(x) = (\cos x, \sin x)$ is a function from the additive group

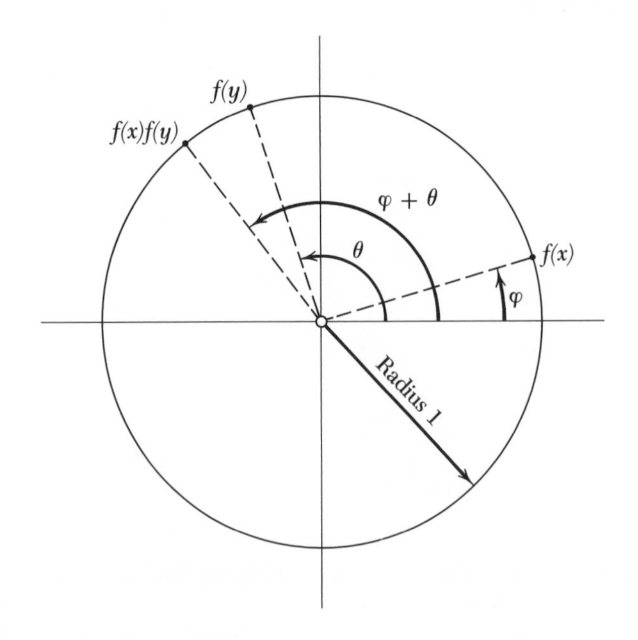

of reals to the multiplicative group of nonzero complex numbers. It is an easy exercise (see Exercise F) to show that for all real numbers x and y, $f(x + y) = f(x)f(y)$, where the addition $x + y$ takes place in the domain group and the multiplication $f(x)f(y)$ is carried out between complex numbers. You have often used the familiar fact that if the values of f, which are the points on the unit circle in the plane, are thought of as angles, then the product $f(x)f(y)$ is, geometrically, just the "sum" of the angles $f(x)$ and $f(y)$. Thus our fact can be restated as "the sum of the measurements of two angles is one measurement of the sum of those angles." For instance, $(3\pi/2) + \pi = 5\pi/2$, which is the same angle as $\pi/2$. In terms of the function f this becomes $f(3\pi/2)f(\pi) = f(5\pi/2) = f(\pi/2)$.

If H is a subgroup of G, then the inclusion i: $H \subset G$ is a morphism; in particular, the identity function 1_G: $G \to G$ on a group is always a morphism. If G and H are groups, there always exists a unique constant morphism c: $G \to H$: $c(g) = e$ whose only value is the identity of H.

If f: $G \to H$ is a morphism, then $f^{-1}(e)$ is a subgroup of G, the subgroup of elements of G which f carries to the identity. This is a simple consequence of Theorem 2; $e \in f^{-1}(e) \neq \varnothing$, and if $f(g_1) = f(g_2) = e$, then

$$f(g_1g_2^{-1}) = f(g_1)f(g_2^{-1}) = f(g_1)f(g_2)^{-1} = ee^{-1} = e,$$

so $g_1g_2^{-1} \in f^{-1}(e)$. The subgroup $f^{-1}(e) \subset G$ is called the **kernel** of f, *Ker* (f). The kernel of a morphism is a special sort of subgroup. If *Ker* (f) is denoted K, then for all $g \in G$, $g^{-1}Kg \subset K$, since if $f(k) = e$, then $f(g^{-1}kg) = f(g^{-1})f(k)f(g) = f(g^{-1})f(g) = f(g^{-1}g) = f(e) = e$. A subgroup with this property is termed **normal;** a *subset* H of a group G is normal if each **conjugate** $g^{-1}Hg$ of H by an element $g \in G$ is a subset of H. Equivalently, for each $g \in G$, $g^{-1}Hg = H$ (why?). Every subgroup of an abelian group is normal, but it is not difficult to find nonnormal subgroups of the group (V,κ) of permutations of $\{a,b,c\}$. If f_c interchanges a and b and fixes c, while r sends a to b, b to c, and c to a, then a quick check shows that $r^{-1}fr$ is not equal to f_c or to the identity function on $\{a,b,c\}$. Hence the subgroup consisting of f_c and the identity function is not normal in the group V of all permutations of $\{a,b,c\}$.

The image *Im* (f) of a morphism f is also a subgroup (of its range). $f(e) = e \in Im\ (f) \neq \varnothing$, and if $f(g_1)$ and $f(g_2)$ are in *Im* (f), then $f(g_1)f(g_2)^{-1} = f(g_1)f(g_2^{-1}) = f(g_1g_2^{-1}) \in Im\ (f)$. By Theorem 2, $Im(f)$ is a subgroup. Can you give an example showing that the image of a morphism need not be a normal subgroup of the range?

Exercises F, J, L, and M

A LITTLE NUMBER THEORY

Suppose H is a nontrivial subgroup of the additive group of integers. Since $H \neq \{0\}$, let $x \in H$ and $x \neq 0$, so either x or $-x$ is positive. Then the nonempty set of positive elements of H has a least member $n \in H$.† We claim that H is just the set of all multiples of n. It is clear that $\{nk: k \in \mathbf{Z}\} \subset H$, since $n \in H$ and H is a subgroup. But if $m \in H$, then there is a least element nk_0 of $\{nk: k \in \mathbf{Z}\}$ which is larger than m. The multiples of n are scattered along the real line n units apart, so the distance $nk_0 - m$ is no greater than $n - 1$. But $nk_0 - m$ is a nonnegative member of H no bigger than $n - 1$; since n is the least positive member of H, $nk_0 - m = 0$, and each element m of H is a multiple of n. We have just shown that *every nontrivial subgroup of the integers is the set of all multiples of some fixed integer.* ■ The trivial subgroups are $\{0k: k \in \mathbf{Z}\} = \{0\}$ and $\{1k: k \in \mathbf{Z}\} = \mathbf{Z}$. If for each positive integer n we denote such a subgroup by $n\mathbf{Z} = \{nk: k \in \mathbf{Z}\}$, then the coset of $n\mathbf{Z}$ which contains $a \in \mathbf{Z}$ is $n\mathbf{Z} + a = \{nk + a: k \in \mathbf{Z}\}$, the set of integers which leave the same remainder as does a when divided by n. There are exactly n different possible remainders, $0, 1, 2, \ldots, n - 1$, so there are n different cosets $n\mathbf{Z}, n\mathbf{Z} + 1, n\mathbf{Z} + 2, \ldots, n\mathbf{Z} + (n - 1)$. The associated equivalence relation is that two integers are related (that is, in the same coset of $n\mathbf{Z}$) iff their difference is in $n\mathbf{Z}$. This happens iff n divides their difference. The equivalence classes (or cosets) are $[0], [1], [2], \ldots, [n-1]$.

Now, let $nk_1 + a$ and $nk_2 + b$ be two integers (which are in the cosets $[a]$ and $[b]$ respectively); the integer $(nk_1 + a) + (nk_2 + b) = n(k_1 + k_2) + (a + b)$ is clearly in the coset $[a + b]$. This suggests that we define an addition of cosets, $[a] + [b] = [a + b]$; the argument above shows this addition to be well defined. That is, the coset of the sum is the same, regardless of which elements $nk_1 + a$ and $nk_2 + b$ are chosen from the cosets $[a]$ and $[b]$ to find the coset $[a + b] = [(nk_1 + a) + (nk_2 + b)]$. With this addition, the family of cosets is a group, denoted $\mathbf{Z}_n$. If $n > 0$, then $\mathbf{Z}_n$ has n elements, $[0]$ is its identity, and the inverse $-[a]$ of an element $[a] \in \mathbf{Z}_n$ is $[-a] = [n - a]$. This group $\mathbf{Z}_n$ is called the **cyclic group of order** n, and $\mathbf{Z}$ is called the **infinite cyclic group.** (In general, the "order" of a finite group, a group with a finite number of elements, is the number of elements in the group. A group G is called "cyclic" if there is a single element g of G such that the set of all products $g^0 = e$, $g^1 = g$, $g^2 = gg, \ldots, g^n, \ldots$ and

† This can be proved by use of mathematical induction.

$g^{-1}, g^{-2} = g^{-1}g^{-1}, \ldots, g^{-n}, \ldots$ includes every element of G. In the case at hand, $\mathbf{Z}_n = \{[0], [1], [1] + [1], [1] + [1] + [1], \ldots\}$, and every element of $\mathbf{Z}$ is clearly a sum of 1's or of -1's.)

For each integer $n \geq 0$, the quotient function $q_n: \mathbf{Z} \to \mathbf{Z}_n: q_n(a) = [a]$ is a morphism. In fact, the requirement that q_n be a morphism is just that $[a + b] = [a] + [b]$, and this is the definition of addition in $\mathbf{Z}_n$, which is the coset family $\mathbf{Z}/n\mathbf{Z}$.

For instance, the cyclic group of order 2 is $\mathbf{Z}_2$, the arithmetic of parities. Let $\mathcal{E}$ stand for the set of even integers $[0]$; $\mathcal{O} = [1]$ is the set of odd numbers. The sum of two even numbers is even, or $\mathcal{E} + \mathcal{E} = \mathcal{E}$; the sum of an odd number and an even number is odd, or $\mathcal{O} + \mathcal{E} = \mathcal{O}$; the sum of two odd numbers is even, or $\mathcal{O} + \mathcal{O} = \mathcal{E}$. The quotient function $q: \mathbf{Z} \to \mathbf{Z}_2$ assigns to each integer n its "parity," $\mathcal{E}$ or $\mathcal{O}$, according to whether n is even or odd.

THE ADDITION TABLE FOR $\mathbf{Z}_3$

$a \backslash b$	[0]	[1]	[2]
[0]	[0]	[1]	[2]
[1]	[1]	[2]	[0]
[2]	[2]	[0]	[1]

The entry in the *a-th* row and *b-th* column is $a + b$.

Exercise Y

QUOTIENT GROUPS

Perhaps the trick above can be generalized. If H is a subgroup of G, can we imitate the definition of addition in $\mathbf{Z}_n$ to bestow a group product on the coset family G/H? The product of two cosets would then be defined by $(g_1H)(g_2H) = (g_1g_2)H$. But there is an immediate objection to this: if it were to work, so that G/H were a group, the quotient function $q: G \to G/H: q(g) = gH$ would be a morphism. Then the kernel of q, which would be H (why?), would be a *normal* subgroup of G. Since we know that there are nonnormal subgroups, our proposed "definition" of the multiplication of cosets must be faulty. On reflection, if H is not normal in

G, then there must be two cosets g_1H and g_2H for which the product is not well defined [or, there exist h_1 and $h_2 \in H$ such that $g_1h_1g_2h_2 \notin (g_1g_2)H$]. And there are. If H is not normal, then there is a member g of G with $g^{-1}Hg \not\subset H$, so there is a member h of H with $g^{-1}hg \notin H$; then $(g^{-1}h)(ge) \notin (g^{-1}g)H = H$. To put it another way, $(g^{-1}h)H = g^{-1}H$, but our "definition" says $(g^{-1}hH)(gH) = (g^{-1}hg)H$ and $(g^{-1}H)(gH) = (g^{-1}g)H = H$, yet $(g^{-1}hg)H \neq H$.

However, if the subgroup H is normal in G, then the proposed definition does work, for then the product of the cosets g_1H and g_2H, regarded as *subsets* of G, is

$$(g_1H)(g_2H) = \{g_1h_1g_2h_2\colon h_1 \text{ and } h_2 \in H\}.$$

Since H is normal, each conjugate $g_2{}^{-1}Hg_2$ is contained in H, so for each element $h_1 \in H$ there is $h_3 \in H$ with $g_2{}^{-1}h_1g_2 = h_3$. If $g_1h_1g_2h_2$ is an element of $(g_1H)(g_2H)$, then

$$g_1h_1g_2h_2 = g_1g_2g_2{}^{-1}h_1g_2h_2 = g_1g_2h_3h_2 \in (g_1g_2)H.$$

This proves that $g_1Hg_2H \subset (g_1g_2)H$. Of course, if $g_1g_2h \in (g_1g_2)H$, then $g_1eg_2h \in g_1Hg_2H$, so $g_1Hg_2H = g_1g_2H$.

3 THEOREM *If H is a normal subgroup of a group G, then the product $(g_1H)(g_2H) = g_1g_2H$ defines a group multiplication on the left-coset family G/H of G modulo H. The quotient function $q\colon G \to G/H$ is a morphism onto G/H, and Ker $(q) = H$.*

Proof That the product on G/H is well defined is shown above, and products of cosets are again cosets. Clearly $eH = H$ is an identity, and the product is associative, since G is a group. But $(gH)(g^{-1}H) = gg^{-1}H = H$, so $(gH)^{-1} = g^{-1}H$ is an inverse for $gH \in G/H$, which is therefore a group.

The function q assigns to $g \in G$ its equivalence class $gH = q(g) \in G/H$, and $q(g_1g_2) = g_1g_2H$ is the product $(g_1H)(g_2H)$ in G/H, so q preserves multiplication. By definition, q is onto, and *Ker* $(q) = H$, since $g \in H$ iff $gH = H$ (G/H is a partition of G). ∎

One consequence of this theorem is that *a subgroup H of G is normal in G iff it is the kernel of some morphism with domain G.* ∎

The group G/H is called the **quotient** (or **factor**) **group of G modulo H**, and q is the **quotient** (or "natural") **morphism.**

Every morphism is a function, and as such, it may be 1-1 or onto, or neither, or both. Unfortunately, other terms are also commonly used for these properties of morphisms of groups. We shall say a morphism f is **epic** (or onto, or an **epimorphism**) if f is an onto function, that is if *Im* $(f) =$

Rng (f). A morphism is **monic** (or 1-1, or a **monomorphism**) if its kernel is e; this is equivalent to its being a 1-1 function, since the inverse image of each element of its image is a coset of the kernel. An **isomorphism** is a morphism which is both epic and monic. Every quotient morphism is epic, and every inclusion of a subgroup in a group is a monomorphism. Observe that if $f: G \to H$ is an isomorphism, then the inverse function f^{-1} is also an isomorphism. This will be clear if only f^{-1} preserves products, but if $x, y \in H$, then $f^{-1}(x)f^{-1}(y)$ is the unique element of G which f carries to

$$f[f^{-1}(x)f^{-1}(y)] = f[f^{-1}(x)]f[f^{-1}(y)] = xy = f[f^{-1}(xy)].$$

Exercises H and P

FACTORING MORPHISMS

If f and g are morphisms of groups, $G \xrightarrow{f} H \xrightarrow{g} J$, then the composite function $g \circ f: G \to J$ is also a morphism, since

$$g \circ f(xy) = g[f(xy)] = g[f(x)f(y)] = g[f(x)]g[f(y)] = [g \circ f(x)][g \circ f(y)].$$

Many facts about functions translate naturally to true statements about morphisms. For example, since composites of 1-1 functions are again 1-1 functions, *composites of monomorphisms are monic.* In the same way, *composites of epimorphisms are epic,* and thus *composites of isomorphisms are isomorphisms.* ∎

4 THE QUOTIENT THEOREM FOR GROUPS *Each morphism $f: G \to H$ has a unique factorization $f = i \circ r \circ q$, or*

$$\begin{array}{ccc} G & \xrightarrow{f} & H \\ {\scriptstyle q}\downarrow & & \uparrow{\scriptstyle i} \\ G/Ker\,(f) & \overset{r}{\cong} & Im\,(f) \end{array}$$

where q is a quotient morphism, r is a group isomorphism, and i is the inclusion of a subgroup.

Proof Since f is a function, Theorem I.11, the Quotient Theorem for Sets, teaches that there is a unique factorization $f = i \circ r \circ q$ into the composite of a quotient *function,* a *set isomorphism,* and a *set inclusion.* Hence the claimed factorization of f into *morphisms* will exist and be unique iff each of these functions is a morphism whenever f is a morphism.

The inclusion i of $Im\,(f)$ is the inclusion of a subgroup into H when f is a morphism.

The quotient q was defined in Theorem I.11 by f; the associated equivalence classes were, for each $g_1 \in G$, $[g_1] = \{g_2 : f(g_2) = f(g_1)\}$. But if $f(g_1) = f(g_2)$, then

$$f(g_1 g_2^{-1}) = f(g_1) f(g_2^{-1}) = f(g_1) f(g_2)^{-1} = f(g_1) f(g_1)^{-1} = e,$$

so $g_1 g_2^{-1} \in Ker\,(f)$ and g_1 and g_2 lie in the same coset of $Ker\,(f)$ in G. Conversely, if g_1 and g_2 are in the same coset, then there is an element k of the subgroup $Ker\,(f)$ for which $g_1 = g_2 k$ (why?), so

$$f(g_1) = f(g_2 k) = f(g_2) f(k) = f(g_2) e = f(g_2).$$

This shows that the equivalence classes of the quotient q defined by the *function* f are exactly the cosets of the subgroup $Ker\,(f)$; hence q is a quotient morphism.

Now, the renaming function r is a set isomorphism. By definition, if $[g] \in G/Ker\,(f)$ is the class (or coset) of an element g of G, then $r([g]) = f(g)$. But is r a morphism? This is trivial:

$$r([g_1][g_2]) = r[g_1 g_2] = f(g_1 g_2)$$

and

$$r[g_1] r[g_2] = f(g_1) f(g_2);$$

since f preserves products, so does r. Hence r is a group isomorphism. ■

This important statement fits together three facts: the image of a morphism is a subgroup of its range, the inverse images under a morphism of the points in the image set are the cosets of the kernel, and the quotient group of the domain modulo the kernel is **isomorphic** in a natural way with the image subgroup.

If H and K are normal subgroups of G and $H \subset K$, then the quotient morphism $q: G \to G/H$ carries K to a normal subgroup $q(K)$ of G/H. That $q(K)$ is a subgroup is clear; but if $gH \in G/H$, then the conjugate

$$g^{-1}H[q(K)]gH = q(g^{-1}) q(K) q(g) = q(g^{-1}Kg) = q(K),$$

so $q(K)$ is normal in G/H. Conversely, if L is a normal subgroup of G/H, then $q^{-1}(L)$ is a normal subgroup of G and it contains H. Thus q sets up a 1-1 correspondence between the set of (normal) subgroups of G which contain H and the set of (normal) subgroups of G/H. (Here we mean that this sentence can be read with or without the two instances of the word "normal.") This fact yields the following consequence of the Quotient Theorem.

COROLLARY *Let $f: G \to G'$ be a morphism with kernel K, and let H be normal in G. Then there exists a factoring of f through G/H iff $K \supset H$; that is, there exists a morphism $\varphi: G/H \to G'$ such that $f = \varphi \circ q$ iff $K \supset H$:*

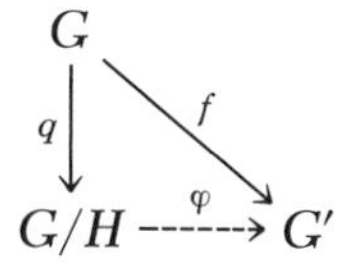

$$\begin{array}{ccc} G & & \\ {\scriptstyle q}\downarrow & \searrow^{f} & \\ G/H & \overset{\varphi}{\dashrightarrow} & G' \end{array}$$

If $K \supset H$, then the image of f is isomorphic to the quotient of G/H modulo $q(K)$; that is,

$$\frac{G}{K} \cong \frac{G/H}{K/H}.$$

Proof If φ exists, then $K = f^{-1}(e) = q^{-1}[\varphi^{-1}(e)]$; hence $\varphi^{-1}(e)$ is a normal subgroup of G/H and it corresponds (via q^{-1}) to K, which must then be a normal subgroup of G containing H.

Conversely, if $K \supset H$ we define φ by $\varphi(gH) = f(g) \in G'$. This works: φ is well defined, since $h \in gH$ implies $h^{-1}g \in H \subset K$, so $f(h^{-1}g) = f(h)^{-1}f(g) = e$ and $f(h) = f(g)$.

In this case, where $K \supset H$, the Quotient Theorem says that $Im\ (f)$ is isomorphic to G/K. Since $Im\ (f) = Im\ (\varphi \circ q) = Im\ (\varphi)$ and the kernel of φ is $K/H = \{kh: k \in K\}$, we have that $Im\ (\varphi)$ is isomorphic to G/K and also to the quotient of the domain G/H of φ modulo the kernel K/H of φ. ∎ This corollary, in the case $K \supset H$, is sometimes called the "first isomorphism theorem."

As an example, let $f: \mathbf{Z} \to \mathbf{Z}_n$ be the quotient of the integers modulo n, and let $q: \mathbf{Z} \to \mathbf{Z}_m$ be the quotient of $\mathbf{Z}$ modulo m. Then f factors through $\mathbf{Z}_m$ iff $n\mathbf{Z} \supset m\mathbf{Z}$, and this occurs just when $m \in n\mathbf{Z}$, or m is a multiple of n. The final statement of the corollary says that, for instance, if $m = 6$ and $n = 3$, $\mathbf{Z}_3$ is isomorphic to the quotient of $\mathbf{Z}_6$ modulo the two-element subgroup $\{[0],[3]\}$ of $\mathbf{Z}_6$ which is the image of $3\mathbf{Z}$ in $\mathbf{Z}_6$.

Exercises S and Z

DIRECT PRODUCTS

To add complex numbers we add real parts to real parts, imaginary to imaginary: $(x_1,x_2) + (y_1,y_2) = (x_1 + y_1, x_2 + y_2)$. The identity of $\mathbf{C}$ is the pair $(0,0)$ of coordinate identities (for addition). Addition in $\mathbf{C}$ is commutative and associative simply because addition in **R** has these properties; for instance,

$$(x_1,x_2) + (y_1,y_2) = (x_1 + y_1,\ x_2 + y_2)$$
$$= (y_1 + x_1,\ y_2 + x_2) = (y_1,y_2) + (x_1,x_2).$$

Since $\mathbf{C} = \mathbf{R} \times \mathbf{R}$, we might try to imitate the definition of addition in the plane to get a group operation for the direct product of an arbitrary pair of groups; this attempt turns out to be completely successful!

If G and H are groups, then the set $G \times H$ has a natural group structure with the product $(g_1,h_1)(g_2,h_2) = (g_1g_2,h_1h_2)$. The identity is (e,e), the paired identities of G and H (check in your head that this works). The inverse of (g,h) is (g^{-1},h^{-1}), where the inverses of the coordinates g and h are found in the groups G and H. For associativity in $G \times H$, observe that

$$(g_1g_2,h_1h_2)(g_3,h_3) = [(g_1g_2)g_3,(h_1h_2)h_3]$$

and

$$(g_1,h_1)(g_2g_3,h_2h_3) = [g_1(g_2g_3),h_1(h_2h_3)];$$

the right-hand sides are equal, since G and H are groups.

The group $G \times H$ is called the **direct product** of the groups G and H. The projections p_G: $G \times H \to G$: $p_G(g,h) = g$ and p_H: $G \times H \to H$: $p_H(g,h) = h$ are epimorphisms. There is a natural monomorphism i_G: $G \to G \times H$ which assigns to each $g \in G$ the pair (g,e); similarly, i_H: $H \to G \times H$: $i_H(h) = (e,h)$ is a monomorphism. These monomorphisms are right inverses for the projection functions; for example, $p_G \circ i_G = 1_G$. The image of i_G is exactly the kernel of p_H; thus the image $i_G(G)$ is a normal subgroup of $G \times H$, as is $i_H(H)$.

If f: $J \to G$ and g: $J \to H$ are morphisms on some group J, then $f \times g$: $J \to G \times H$: $(f \times g)(j) = [f(j),g(j)]$ is also a morphism, called the **direct product** of f and g. Clearly, $p_G \circ (f \times g) = f$ and $p_H \circ (f \times g) = g$.

The projections of the plane assign to a complex number its real or imaginary part, and the monomorphism corresponding to the first factor **R** of $\mathbf{R} \times \mathbf{R} = \mathbf{C}$ is the familiar embedding of the real line in the plane as the horizontal axis.

You should notice here that the direct-product group structure of the plane $\mathbf{R}^2$ is just that of the additive complex numbers **C**. However, the multiplicative structure of **C** is not a direct product of that of **R**, and the group $\mathbf{C} - \{0\}$ is not the same as $\mathbf{R} - \{0\} \times \mathbf{R} - \{0\}$ (be sure you understand this). This difference is customarily reflected in the notation; **C** is *not* the same algebraic object as $\mathbf{R}^2$. With this made explicit, we shall continue to identify the two where convenient, using the symbol **C** sometimes to emphasize the multiplicative structure.

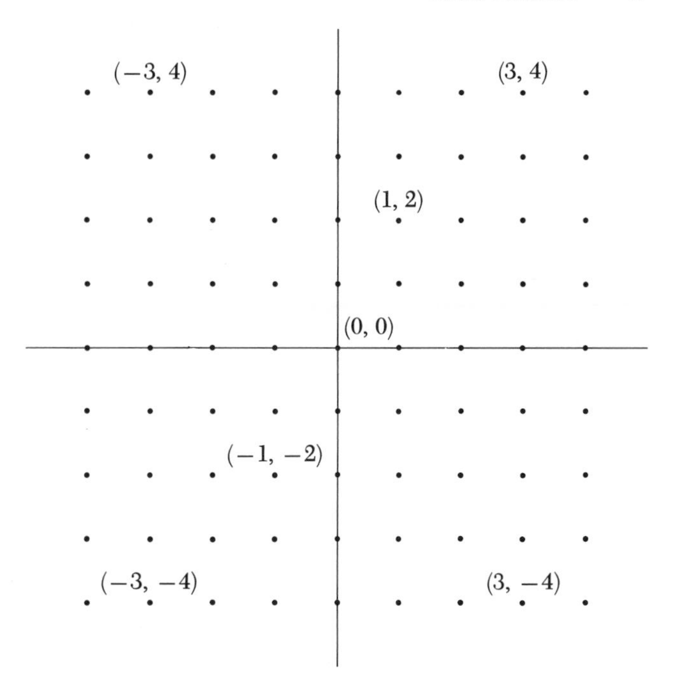

The direct product of two copies of the familiar group of integers (under addition) is $\mathbf{Z} \times \mathbf{Z}$, the group of **gaussian integers.** It may be thought of as a subgroup of **C**, the additive subgroup of those complex numbers with both coordinates integral (is this really a subgroup?) Its picture is a lattice of points in the plane.

Another direct-product example is the torus. If S^1 denotes the group of angles (the unit circle in $\mathbf{C} - \{0\}$), then $S^1 \times S^1$ is called the **torus group.** Its members are ordered pairs of angles (φ,θ), where each angle is

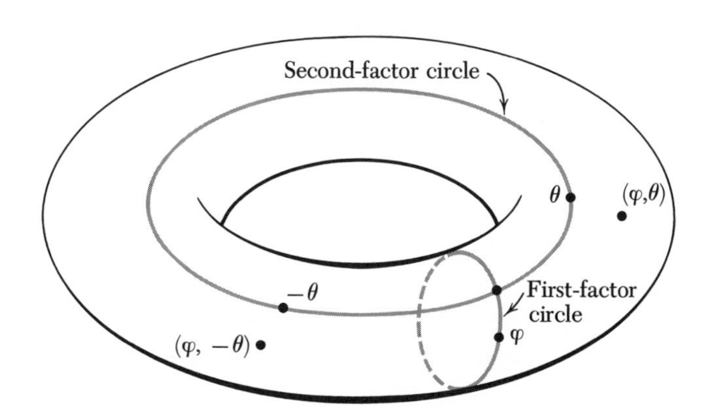

a point on the circle. The word "torus" means a geometric shape like an innertube. We picture our torus group as a circle of circles; through each point of a circle along the rim (of the absent wheel) goes a circle through the spokes (of the wheel).

Exercise N

REFERENCES AND FURTHER TOPICS

The use of group theory as a language is widespread in modern mathematics, perhaps second only to set theory. The study of the theory for its own sake is, of course, a large and important part of the work of today's mathematicians. The exposition of this chapter is only skeletal; it contains the minimum needed for our work in topology. For instance, no glimpse is provided of the rich theory of finite groups. A reference for further topics, as well as an elegant and extremely readable introduction, is

A. G. Kurosh, *The Theory of Groups* (New York: Chelsea, 1956).

The second edition of this book is in two volumes; vol. 1, part 1, contains the elementary material. You will enjoy Kurosh's technique of using as few symbols as possible; instead he achieves a verbal exposition like that of a good lecturer with a small blackboard.

Another good reference on a high level is

M. Hall, Jr., *The Theory of Groups* (New York: Macmillan, 1959).

More elementary presentations abound, some with "group theory" in their titles and some embedded in discussions of "modern algebra" or "abstract algebra." A few are listed below.

Birkhoff and Mac Lane, *A Survey of Modern Algebra*, 3d ed., chap. VI (New York: Macmillan, 1965).

I. N. Herstein, *Topics in Algebra,* chap. 2 (New York: Blaisdell, 1964).

S. -T. Hu, *Elements of Modern Algebra,* chaps. II, III, and IV (San Francisco: Holden-Day, 1965).

Mostow, Sampson, and Meyer, *Fundamental Structures of Algebra,* chaps. 1 and 10 (New York: McGraw-Hill, 1963).

L. Pontrjagin, *Topological Groups,* chap. 1 (Princeton, N. J.: Princeton University Press, 1939).

In the introduction we advertised some applications of our theory to problems outside mathematics. The algebra of groups has many such applications; one which can be easily digested at this stage is crystallography. Crystals can be studied via their groups of "symmetries" (the word is used loosely here); differing crystal forms have different groups associated with them. A good introduction to all this is offered by

M. Hamermesh, *Group Theory and Its Application to Physical Problems,* chaps. 1 and 2 (Reading, Mass.: Addison-Wesley, 1962).

Another and very different presentation of crystallographic groups is offered in

Hilbert and Cohn-Vossen, *Geometry and the Imagination,* chap. II (New York: Chelsea, 1952).

This book presents the material of a series of lectures given by David Hilbert, one of this century's broadest and deepest mathematicians. Its object, as the title suggests, is to present geometric topics in their visual intuitive aspects, avoiding formal abstract structure.

EXERCISES

A The group V of permutations of the set $S = \{a,b,c\}$ has exactly six members ($3! = 6$); let us give names to these functions by describing their values in a list. In the column labeled f appears the name of each permutation, and in the same row with its name appear the values of each member of V.

THE MEMBERS OF V

f	$f(a)$	$f(b)$	$f(c)$
1	a	b	c
f_a	a	c	b
f_b	c	b	a
f_c	b	a	c
r	b	c	a
r^2	c	a	b

Do you remember the addition and multiplication tables of elementary school? The same scheme may be used to display the values of the multiplication function on a group: V has six members, so $V \times V$ has

THE MULTIPLICATION TABLE FOR V

Left factors \ Right factors	1	f_a	f_b	f_c	r	r^2
1		f_a				
f_a	f_a	1		r^2		
f_b		r^2				
f_c		r				
r		f_c			r^2	1
r^2		f_b				

thirty-six members (see Exercise I.N), one corresponding to each little box in the multiplication table above. If f and g are members of V, then the composite $f \circ g$ [the value of the multiplication function $m(f,g)$] is given at the intersection box of the row of f with the column of g. A few entries have already been made in the table; check these for correctness. Then complete the table, filling in all the products of V. (*Caution:* The composite function $f \circ g$ is computed by *first* applying g, *then* applying f.) The existence of a left inverse for each $f \in V$ is equivalent to the appearance of 1 in each column of the table. The fact that left inverses are also right inverses can be seen from the table, since the scattering of 1's in it is symmetric with respect to the diagonal line which goes downward from left to right. Unlike the multiplication table for the integers 1 through 12, each entry in the table is among the right factors listed across the top of this table; that is, products of pairs of members of V are again members of V. How does the table show the existence of right inverses, the nonabelian nature of V, and the fact that 1 is a left (or right) identity? A description of associativity in terms of the geometry of the table seems obscure to me; can you see a simple one?

B Do the following sets and operations form groups? If not, why not?

i The set of complex numbers $\{0 + yi: y \in \mathbf{R}\}$ with complex addition.
ii The set of complex numbers $\{0 + yi: y \in \mathbf{R}, \text{ and } y \neq 0\}$ with complex multiplication.
iii The set of all functions $f: \mathbf{C} \to \mathbf{C}$ which preserve distances, that is, such that $|f(a) - f(b)| = |a - b|$ for all pairs a and b of complex numbers, under composition.

iv The set of functions $\{f: \mathbf{R} \to \mathbf{R}$ such that f is of the form $f(x) = rx + s$, with r and s real numbers and $r \neq 0\}$ (such a function is called **affine,** or sometimes, "linear"), with the operation of composition.

v The set $\{1,2,3,4\}$ with the multiplication given by the following table. The element in the box in the i-th row and in the j-th column is the product ij.

2	3	4	1
3	4	1	2
4	1	2	3
1	2	3	4

vi The same set as in v above, with the multiplication given in the following table.

1	2	3	4
2	4	3	1
3	2	4	3
4	3	1	2

C Assume that a binary operation on a set S (that is, a multiplication function from $S \times S$ to S) satisfies the equation $(ab)c = a(bc)$ for all triples (a,b,c) of elements of S. Show that whenever $a_1, a_2, \ldots, a_n$ are all elements of S, the product $a_1a_2 \ldots a_n$ is the same, regardless of how parentheses are sprinkled in to indicate the order in which pairs of objects are to be multiplied. (*Hint:* Use mathematical induction on n.)

D Prove parts ii and iv of Theorem 1 in detail.

E Which of the following sets are subgroups of the indicated groups? If not, why not?

i The set of nonnegative integers in the additive group of all integers $(\mathbf{Z},+)$,

ii The set of even integers in $(\mathbf{Z},+)$,

iii The set of odd integers in $(\mathbf{Z},+)$,

iv The set of permutations which interchange just two elements (or none) in the group of permutations of $\{a,b,c\}$ (under composition), v The positive real axis $\{(x,0): x > 0\}$ in the group of nonzero complex numbers under multiplication.

F Show that the function $f: \mathbf{R} \to \mathbf{C} - \{0\}: f(x) = (\cos x, \sin x)$ is a morphism from the group of additive reals to the group of nonzero complex numbers.

G Verify for the multiplication of *subsets* of a group that

$$(AB)C = A(BC) \qquad (AB)^{-1} = B^{-1}A^{-1} \qquad A \subset B \text{ implies } CA \subset CB.$$

H Let $\mathcal{G}$ be a set whose elements are groups, and let R be the relation on $\mathcal{G}$ such that $(G,H) \in R$ iff there exists an isomorphism $f: G \to H$. Show that R is an equivalence relation. Elements of the same equivalence class are said to be isomorphic; they are alike as far as group theory goes (that is, except for the names of the elements and the name for the product).

J Show that if $\mathcal{S}$ is a family of subgroups of a group G, then $\bigcap \mathcal{S}$ is a subgroup of G. If each member of $\mathcal{S}$ is normal, so is $\bigcap \mathcal{S}$. Give an example of a family $\mathcal{S}$ of subgroups of a group G such that $\bigcup \mathcal{S}$ is not a subgroup.

K Show that if A is a subset of a group G, the set of all products of elements chosen (with repetition allowed) from $A \cup A^{-1}$, and taken in all the various permutations if G is nonabelian, is a subgroup of G. It is called the **subgroup generated by** A. Since intersections of subgroups are always subgroups, $\bigcap\{S: S \text{ is a subgroup of } G \text{ and } S \supset A\}$ is a subgroup; it is the **smallest** subgroup containing A. Show that this is exactly the subgroup that A generates.

L Which functions from the integers $\mathbf{Z}$ into $\mathbf{Z}$ itself are morphisms of the additive group of integers? Similarly, what are the possible morphisms of the additive group of rationals into itself?

M Describe the family of cosets of the subgroup $\{(x,0): x > 0\}$ of positive reals in the multiplicative group of nonzero complex numbers (it may help to plot a few points of a coset on graph paper). Then construct a morphism of the group $\mathbf{C} - \{0\}$ into itself whose kernel is exactly the positive real axis.

N Let G be the set $\{(a,b) \in \mathbf{R}^2: a \neq 0\}$ of ordered pairs of real numbers with the first one nonzero, so that $G = (\mathbf{R} - \{0\}) \times \mathbf{R}$, a direct product of sets. Define a product between members of G by $(a_1,b_1)(a_2,b_2) = (a_1a_2, a_1b_2 + b_1)$. Show that this is a group product for G, and that G is not abelian. (This group is isomorphic to that of Exercise B.iv.)

There are projection functions $p_1: G \to \mathbf{R} - \{0\}: p_1(a,b) = a$ and $p_2: G \to \mathbf{R}: p_2(a,b) = b$ of the direct-product *set* G; is either p_1 or p_2 a morphism? Show that both $p_1^{-1}(1)$ and $p_2^{-1}(0)$ are subgroups. Are they normal? What are the corresponding families of cosets like? Show that each $g \in G$ can be expressed uniquely as a product $g = hk$, where $h \in p_1^{-1}(1)$ and $k \in p_2^{-1}(0)$.

P It was remarked in the text that the logarithm function is an isomorphism between the multiplicative positive reals and the additive reals. Does there exist an isomorphism from the additive reals to the group of *all* nonzero reals?

Q Let us denote the unit circle in the plane by S^1; it is also called the **1-sphere.** A rotation R_θ of S^1 by an angle θ is a "rigid motion" of the circle which carries it clockwise, each point moving a distance θ. Thus R_θ is a function from S^1 to S^1. Show that the set of all rotations of S^1 forms a group under composition, and give an isomorphism of this group with the circle subgroup $\{x: |x| = 1\}$ of the multiplicative nonzero complex numbers.

R Prove that if H is a subgroup of G and $g \in G$, the conjugate gHg^{-1} of a subgroup is always a subgroup. Then find that subgroup if, in the notation of Exercise A, $g = r$ and $H = \{1,f_a\}$ in the group G of permutations of the set $\{a,b,c\}$.

S Describe the functions i, r, and q which are guaranteed by the Quotient Theorem for Groups to give $i \circ r \circ q = f$ when $f: \mathbf{R} \to \mathbf{C} - \{0\}$ is the morphism which takes x to $f(x) = (\cos x, \sin x)$.

T State why the intersection and union operations on the power set $\mathcal{P}$ of all subsets of a set S do not have the group property. Then show that the operation Δ on $\mathcal{P}$, $X \Delta Y = X \cup Y - X \cap Y$, does make of $\mathcal{P}$ a group. This group is abelian, trivially. Prove that if each element of a given group is its own inverse, then that group must be abelian.

U Let C be the set of all continuous real-valued functions of a real variable. For f, $g \in C$ define $(f + g)(x) = f(x) + g(x)$; $f + g$ is again a member of C, called the **pointwise sum** of f and g. Prove that with this addition F is a group. Then show that the function $\sigma: C \to \mathbf{R}: \sigma(f) = \int_0^1 f(x)\, dx$ is a morphism to the additive reals, and also that the function $\tau: C \to C: \tau(f) = \int_0^x f(x)\, dx$ is a morphism. Can you phrase similar statements involving differentiation?

W Let G be a nonempty set of functions, each member of G having domain and range equal to a fixed set S, such that if $g_1, g_2 \in G$, then $g_1 \circ g_2 \in G$ and also that $g_1 \in G$ implies that there is a left inverse g_2 in G for g_1; $g_2 \circ g_1 = 1_S$. Show that composition is a group operation for G.

Y Prove that every subgroup of a cyclic group must itself be cyclic. Also show that each cyclic group is isomorphic to a quotient group of $\mathbf{Z}$, that is, isomorphic to $\mathbf{Z}_n$ for some integer $n \geq 0$.

Z Let $f: A_1 \to A_2$ be a morphism of groups with kernel K, and let $q_1: A_1 \to A_1/K_1$ and $q_2: A_2 \to A_2/K_2$ be quotients of the domain and range of f, with kernels K_1 and K_2, respectively. Show that f induces a morphism f_* which makes the following diagram commutative iff $f(K_1) \subset K_2$ (*Hint:* First prove this for the special case

$$\begin{array}{ccc} A_1 & \xrightarrow{f} & A_2 \\ \downarrow{\scriptstyle q_1} & & \downarrow{\scriptstyle q_2} \\ A_1/K_1 & \overset{f_*}{\dashrightarrow} & A_2/K_2 \end{array}$$

where $K_2 = \{e\}$, so that the requirement $f(K_1) \subset K_2$ reduces to the condition that $K_1 \subset K$. Compare Prob. I.CC.)

PROBLEMS†

AA **Automorphisms** An isomorphism $f: G \to G$ of a group G with itself is called an "automorphism" of G. It can be thought of as a "symmetry" of the group. Show that if $a \in g$, the function $K_a: G \to G$: $K_a(g) = a^{-1}ga$ is an automorphism; it is called the "inner automorphism" of G determined by a, and its value $a^{-1}ga$ is the "conjugate of g by a." The set $\mathcal{A}$ of all automorphisms of a group G is a group (under composition) of permutations of G. The subset $\mathcal{I}$ of inner automorphisms, $\mathcal{I} = \{K_a: a \in G\}$, is a subgroup. The function $K: G \to \mathcal{A}$: $K(a) = K_a$ is a morphism having as kernel the set of elements of G which commute with each element of G; $Ker\,(K) = \{a \in G: ag = ga$ for all $g \in G\}$. This normal subgroup $Ker\,(K)$ of G is called the "center" of G.

† In these and subsequent problems, verify each assertion.

The subgroup $\mathcal{I}$ is normal in $\mathcal{A}$; the quotient group $\mathcal{A}/\mathcal{I}$ is the group of "outer automorphisms" of G (some authors call the elements of $\mathcal{A} - \mathcal{I}$ the outer automorphisms).

Now apply your work to compute the centers and groups of inner and outer automorphisms for the group of integers **Z** and for the group V of permutations of $\{a,b,c\}$.

BB Categorical Matters Attempt to restate Prob. I.BB, replacing the words "set" by "group" and "function" by "morphism." (*Warning:* The characterization of epimorphisms between nonabelian groups is not reasonable without some study of free groups.)

CC Groups of Morphisms If G and H are abelian groups, then the set "$Hom\ (G,H)$" of morphisms of G into H can be given an abelian group structure by the "pointwise" multiplication of functions, $f_1f_2(g) = f_1(g)f_2(g)$, where the latter product is of two elements of H. Of course, similar statements may be made for the set of all functions from a *set* S into a group H, for example, the set of real-valued real functions.

If K is also abelian, a morphism $k: H \to K$ induces a function $k_*: Hom\ (G,H) \to Hom\ (G,K)$ by composition, $k_*(f) = k \circ f$, and k_* is a morphism. On the other side, a morphism $m: F \to G$ induces a morphism $m_*: Hom\ (G,H) \to Hom\ (F,H)$. (Do you see just why k_* "goes in the same direction" as k, while m_* goes the reverse of m?)

Furthermore, composition yields a natural morphism of

$$Hom\ (G,H) \times Hom\ (H,K)$$

into $Hom\ (G,K)$.

DD Normal Products A group G is the "normal" (or "semidirect") product of K by H if

i H and K are subgroups of G, and K is normal,
ii $H \cap K = e$,
iii $HK = G$ (this is the product of subsets).

An example is given by the affine group of Exercises B.iv and N. Can you give some nonexamples?

A group G is the normal product of K by H iff K is a normal subgroup and the quotient epimorphism $q: G \to G/K$ (which always has a right-inverse function) has a right-inverse *morphism* p, with $Im\ (p) = H$.

Also, G is the normal product of K by H iff K is a normal subgroup and H is a subgroup which is a system of coset representatives for K.

(A subset R of a group G is a "system of coset representatives" for a subgroup S of G if R contains exactly one member from each coset of S.)

Another characterization is that H and K be subgroups, with K normal, and that the multiplication function of G, when restricted to $H \times K$, be a set isomorphism—that is, that each $g \in G$ have a unique representation $g = hk$ as a product of an element of H with one of K (in that order).

EE **Cayley's Theorem** A set isomorphism of a set S with itself is called a permutation of S; if S is infinite it is more usually termed a "transformation" of S. If G is any family of permutations, or transformations, of a given set S, and G forms a group under the operation of composition of functions, then G is a "permutation group" on S (or a "transformation group" on S). Such a group is often called a "concrete" group, because it seems more substantial and intuitive than an "abstract" group, which is merely a set with an operation abstractly defined on it (that is, a group satisfying our definition).

Let G be a group, and for each $g \in G$ let $L_g\colon G \to G$ be the function which assigns to $h \in G$ the element gh [so that $L_g(h) = gh$]. Show that the set $\mathfrak{L} = \{L_g\colon g \in G\}$ is a group of permutations of the set G, and that the function $L\colon G \to \mathfrak{L}\colon L(g) = L_g$ is an isomorphism. Thus every abstract group is isomorphic to a group of permutations (with composition of functions as the group operation). Apply this conclusion to the group V of permutations of the set $\{a,b,c\}$ (see Exercise A; note also the application suggested by Exercise Q).

FF **Exact Sequences** A "sequence" of groups and morphisms is an array of the form

$$\cdots \to A_{i-1} \xrightarrow{f_i} A_i \xrightarrow{f_{i+1}} A_{i+1} \xrightarrow{f_{i+2}} A_{i+2} \to \cdots,$$

which may or may not terminate at either end. The sequence is called "exact" at A_i if the image of f_i is exactly the kernel of f_{i+1}, $Im\,(f_i) = Ker\,(f_{i+1})$. An example is

$$\{1\} \xrightarrow{f_1} S^1 \xrightarrow{i} \mathbf{C} - \{0\} \xrightarrow{abs} \{x \in \mathbf{R}\colon x > 0\} \xrightarrow{f_2} \{1\},$$

where f_1 and f_2 are the only possible morphisms from and to singleton groups, i is the inclusion of the unit circle in $\mathbf{C} - \{0\}$, and abs denotes the absolute-value function sending x to $|x|$. This sequence is exact at S^1, at $\mathbf{C} - \{0\}$, and at the positive reals. A sequence which is exact at each group along it is termed an "exact sequence." In general, prove that if

$$\{e\} \to K \xrightarrow{f_2} G \xrightarrow{f_3} H \to \{e\}$$

is exact, then f_2 is monic, f_3 is epic, and H is isomorphic to $G/Im\,(f_2)$.

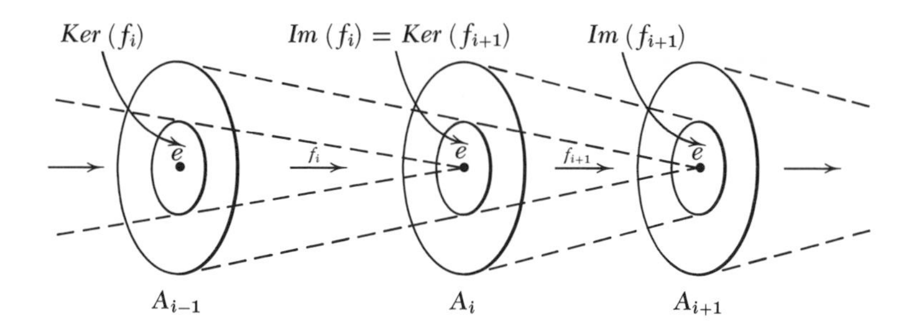

Show that if f is the function described in Exercise F, then the sequence of groups

$$\{e\} \to 2\pi\mathbf{Z} \xrightarrow{i} \mathbf{R} \xrightarrow{f} \mathbf{C} - \{0\} \xrightarrow{abs} \{x \in \mathbf{R}: x > 0\} \to \{e\}$$

is exact (here $2\pi\mathbf{Z} = \{2\pi k: k \in \mathbf{Z}\}$ and i is the inclusion function).

GG Lagrange's Theorem Let G be a group and H a subgroup of G. The "order" of G is the number of elements of G; if G is infinite it is said to have infinite order. The number of distinct cosets gH of H in G is the "index" of H in G; of course, H may have infinite index in G.

Show that there is a natural 1-1 correspondence between H and each coset gH of H, and thus that every coset of H has exactly the same number of elements as has H. Since these cosets partition G, if H has finite order m and finite index n in G, then the order of G is the product mn. Hence the order (and also the index) of each subgroup in G divides the order of G.

Prove that every group whose order is a prime integer p has no non-trivial subgroups, and that it is isomorphic to the cyclic group $\mathbf{Z}_p$ of order p. If p is a prime which divides the integer n, find a subgroup H of order p in $\mathbf{Z}_n$, and also a subgroup K of index p in $\mathbf{Z}_n$.

If g is a member of a group G, the "order" of g is the least positive integer k such that g^k, the product of g with itself taken k times, is the identity; if for no positive integer k is $g^k = e$, then g is said to have infinite order. Show that if G is finite, then the order of each element $g \in G$ divides the order of G.

HH **The Second Isomorphism Theorem** Let H and K be subgroups of a group G, with K normal in G. Observe that HK is a subgroup of G, K is normal in HK, and $H \cap K$ is normal in H; then prove that

$$\frac{H}{H \cap K} \cong \frac{HK}{K}$$

Metric Spaces

CHAPTER III

The continuity of a real-valued function of a real variable is defined early in a course in calculus. In essence, such a function f is continuous at a real number r if its values $f(s)$ are as close as you like to $f(r)$, provided only that s be close enough to r. The functions which are continuous at every point r are a special class of "nicely behaved" functions whose values do not "jump suddenly"; their graphs may be drawn without lifting the pencil.

The only property of the real numbers which is mentioned in the definition of continuity is the fact that there is defined a distance between each two real numbers; the notion of closeness depends on that of distance.

We shall now examine those facts about the reals, and functions with real domain and range, which depend only on the idea of distance between numbers. If we are careful to use only this one property of the real numbers in our proofs, each conclusion we reach should apply not only to the reals, but also to every other set where distance between members is appropriately defined. The situation will parallel that of the previous chapter: if we prove a theorem about real addition using only the group property of that operation in the proof, say the theorem that $(x + y) + (z + w) = x + [(y + z) + w]$, then that theorem applies as well to complex addition and to any other group.

But just what is the notion of distance? If r and s are real numbers, then the distance from r to s is $|r - s|$; this involves both subtraction and the absolute-value function. Since subtraction is defined in terms of the additive group structure and the absolute value $|x|$ of a real x is defined in terms of the order on $\mathbf{R}$, $|x| = x$ if $x \geq 0$ and $|x| = -x$ if $x < 0$, the recipe above for distance has not isolated the concept. That is, our intuitive understanding of distance in the three-dimensional world in which we live certainly is simpler; to state the distance from Los Angeles to Berlin does not require that either Los Angeles be less than Berlin or Berlin be less than Los Angeles. Just as for the group property, we must characterize distance in terms of its properties, not in terms of specific methods of computation. Now, distances are real numbers, and the distance from a point to itself is zero, while distances between distinct points are positive. Furthermore, the distance from x to y is the same as that from y to x. While these statements apply to distance in general, we summarize them for the real line: There exists a function $\mathfrak{d}: \mathbf{R} \times \mathbf{R} \to \mathbf{R}$ such that for all elements (r,s) of $\mathbf{R} \times \mathbf{R}$

i $\mathfrak{d}(r,s) \geq 0$,
ii $\mathfrak{d}(r,s) = 0$ iff $r = s$,
iii $\mathfrak{d}(r,s) = \mathfrak{d}(s,r)$.

Here, of course, $\mathfrak{d}(r,s) = |r - s|$. There is yet another property of distance in the world in which we live: the distance along one side of a triangle is no greater than the sum of the distances along the other two sides. On the real line this is called the "triangle inequality" for real numbers,

iv $\mathfrak{d}(r,s) \leq \mathfrak{d}(r,t) + \mathfrak{d}(t,s)$.

This has a more familiar form in the reals as

$$|x + y| \leq |x| + |y|.$$

A proof can be made by considering the four cases where $x \geq 0$ and $y \geq 0$, $x < 0$ and $y \geq 0$, etc. Now let $x = r - t$ and $y = t - s$ to get the triangle inequality:

$$|r - s| = |(r - t) + (t - s)| \leq |r - t| + |t - s|.$$

Our list of properties of distance is now complete. Notice that the distance between points on the earth's surface may be defined by an airline pilot to be "as the crow flies," that is, lengths of arcs of great circles. But an astronomer may think of the distance between surface points as being the length of the straight line, going through the body of the earth, which joins them. Each of these different definitions of distance on the earth's surface, as well as distance along the real line, has properties i through iv. Guided by these examples, let us now use these four properties as axioms to give a precise, mathematical definition of the notion of distance.

THE DEFINITION

A **metric space** (M,d) is a set M together with a function $d\colon M \times M \to \mathbf{R}$ such that for all x, y, and $z \in M$,

i $d(x,y) \geq 0$,
ii $d(x,y) = 0$ iff $x = y$,
iii $d(x,y) = d(y,x)$,
iv $d(x,z) \leq d(x,y) + d(y,z)$.

We have discussed some examples: the usual distance functions on the line, in the plane, and in real three-dimensional space. Furthermore, we have seen two different distance functions (which we shall call **metrics**) on the surface of the earth. Lest you suppose that somehow the definition of a metric space has captured *all* the qualities of our usual concept of distance, you should verify that if M is any set whatsoever, then the function d: $M \times M \to \mathbf{R}$, which assigns $d(x,y) = 1$ to each pair (x,y) with $x \neq y$ and $d(x,x) = 0$ for all $x \in M$, is a metric on M.

Just as we did in the case of groups, we shall often abuse our definition by referring to a metric space (M,d) simply as "the metric space M," when the specific metric d on the **underlying set** M is clearly enough understood.

We can now generalize the idea of the continuity of a function to metric spaces. If (M,d) and (M',d') are metric spaces and $f\colon M \to M'$ is a function from the set M to the set M', then f is **continuous at a point** $x \in M$ iff for each positive real number ε there exists a positive real δ such that for every $y \in M$, if $d(x,y) < \delta$, then $d'[f(x), f(y)] < \varepsilon$. The function f is said to be **continuous,** or **continuous from** (M,d) **to** (M',d'), if it is continuous at each point $x \in M$. Clearly, if M and M' are both the set of real numbers $\mathbf{R}$, and $d(x,y) = d'(x,y) = |x - y|$, so that d and d' are the **usual metric** for $\mathbf{R}$, then this is exactly the usual definition for the continuity of f. You should realize here that the choice of different metrics in place of d and d' for the same sets M and M' will, in general, result in a different set of continuous functions f from (M,d) to (M',d'). For instance, let $M = M' = \mathbf{R}$, as before, and let d' be the usual real metric, but define d'' to be the peculiar metric mentioned earlier, $d''(x,y) = 1$ if $x \neq y$ and $d''(x,x) = 0$ for each real x and y. We assert that *every function* $f\colon \mathbf{R} \to \mathbf{R}$ is continuous from $(\mathbf{R},d'')$ to $(\mathbf{R},d')$. For every point $x \in \mathbf{R}$ and every $\varepsilon > 0$ the choice $\delta = 1$ will suffice, since $\{y \in \mathbf{R}\colon d''(x,y) < 1\} = \{x\}$ (be sure to think this through). Can you tell which functions from $(\mathbf{R},d')$ to $(\mathbf{R},d'')$ are continuous?

Henceforth the usual metric will always be assumed present on **R** unless the contrary is made quite clear.

Exercise A

ε-BALLS

For convenience, we introduce the concept and notation of ε-balls in metric spaces: if $(M,\mathfrak{d})$ is a metric space with $x \in M$, and ε is a positive real, then the ε-**ball** $\mathfrak{b}(\varepsilon,x)$ **centered at** x is $\mathfrak{b}(\varepsilon,x) = \{y \in M\colon \mathfrak{d}(x,y) < \varepsilon\}$, the set of all points of M whose distance from x is less than ε. On the real line $\mathfrak{b}(\varepsilon,x)$ is the open interval of length 2ε and center x, in the plane $\mathfrak{b}(\varepsilon,x)$ is an edgeless disc of radius ε centered at x, and in real three-dimensional space $\mathfrak{b}(\varepsilon,x)$ is a solid skinless ball of radius ε and center x. Observe that the set M and the function $\mathfrak{d}$ both enter into the definition of $\mathfrak{b}(\varepsilon,x)$, although neither is visible in the notation; the context should make it clear just which metric space each ε-ball lies in. In this spirit, then, we phrase the definition of continuity in terms of balls: "A function $f\colon M \to M'$ from one metric space to another is continuous iff for each $x \in M$ and for each real $\varepsilon > 0$ there exists a $\delta > 0$ such that $f[\mathfrak{b}(\delta,x)] \subset \mathfrak{b}[\varepsilon,f(x)]$." Here the ball $\mathfrak{b}(\delta,x)$ is a subset of M, while $\mathfrak{b}[\varepsilon,f(x)]$ lies in M'; the requirement that $f[\mathfrak{b}(\delta,x)] \subset \mathfrak{b}[\varepsilon,f(x)]$ is exactly the requirement that if $\mathfrak{d}(x,y) < \delta$, then $\mathfrak{d}[f(x),f(y)] < \varepsilon$.

Exercises B, C, D, and E

SUBSPACES

If M is a metric space and S is a subset of M, then the restriction $\mathfrak{d}'$ of the distance function $\mathfrak{d}\colon M \times M \to \mathbf{R}$ to the subset $S \times S$ of $M \times M$ gives a function $\mathfrak{d}'\colon S \times S \to \mathbf{R}$; $\mathfrak{d}'$ is always a metric for S. It is easy to check in your head that $\mathfrak{d}'$ satisfies the axioms because $\mathfrak{d}$ does. An example is the aforementioned "astronomer's notion" of distance on the earth's surface; here M is the three-dimensional universe and S is the surface of our planet. (A nonexample in this setting is the pilot's notion of surface distance along arcs of great circles.)

With this metric, the metric space $(S,\mathfrak{d}')$ is called a **subspace** of $(M,\mathfrak{d})$. And now we have a method of manufacturing many metric subspaces in each metric space. Is it clear that the relation of being a subspace, S is related to *M* if S is a subspace of *M*, is a transitive relation between metric spaces?

One caution here: the unit interval $\mathbf{I} = [0,1]$ is a metric subspace of $\mathbf{R}$; in the metric space $\mathbf{I}$, however, if $0 < \varepsilon < 1$, then the ε-ball centered at 0 is the half-open interval $[0,\varepsilon)$, and *not* $(-\varepsilon,\varepsilon)$. The general statement is that if S is a subspace of the metric space *M*, the ball of radius ε centered at x *in* S is the intersection of S with the ball of radius ε centered at x *in M*.

A METRIC SPACE OF FUNCTIONS

The utility of this abstraction of metric spaces is best described by example. Let *C* be the set of all continuous functions from the unit interval $\mathbf{I} = [0,1]$ into the reals $\mathbf{R}$. A metric for *C* may be defined in a very intuitive way. Let the distance between two members *f* and g of *C* be the maximal

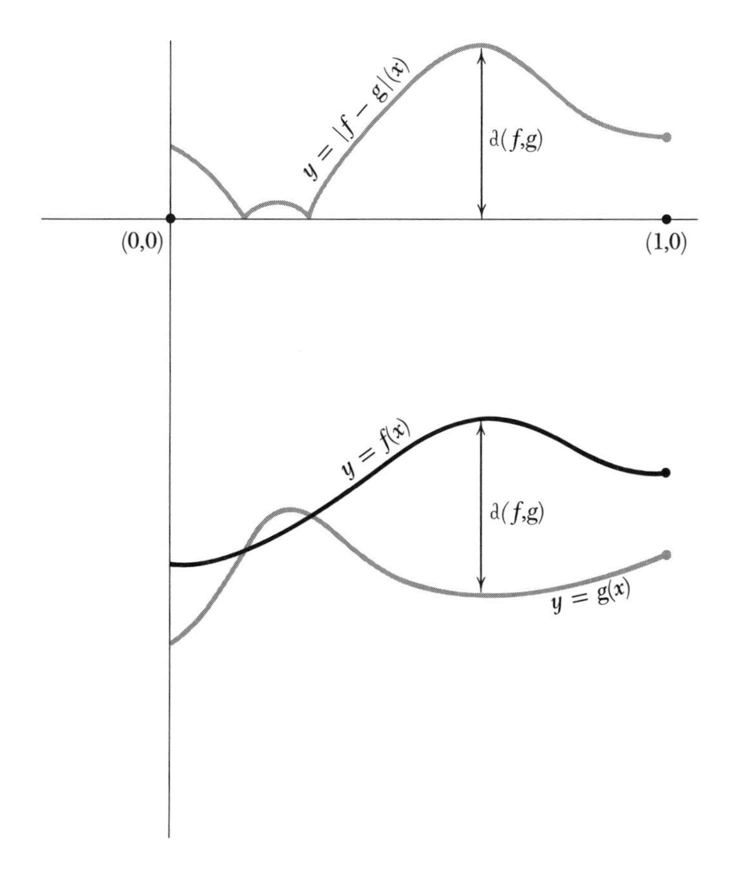

distance between their values, $\partial(f,g) = lub\,\{|f(x) - g(x)| : x \in \mathbf{I}\}$, where the least upper bound of a nonempty set S of real numbers† is denoted *lub* S. This is a reasonable definition, since the function $|f - g| : \mathbf{I} \to \mathbf{R}$: $|f - g|(x) = |f(x) - g(x)|$ is continuous whenever both f and g are continuous. Consequently, the continuous function $|f - g|$ attains a maximal value, which we take for $\partial(f,g)$, on the *closed interval* **I**. (These statements are usually proved in a calculus course. That a continuous function from **I** to **R** actually reaches a maximum value will be shown later in this chapter.) The facts, that $\partial(f,g) \geq 0$, that $\partial(f,f) = 0$, and that $\partial(f,g) = \partial(g,f)$ for all pairs (f,g), are trivial. The triangle inequality can be proved as follows. For each x (and for all f, g, and $h \in C$),

$$|f(x) - h(x)| \leq |f(x) - g(x)| + |g(x) - h(x)|;$$

this is the triangle inequality for the real numbers $f(x)$, $g(x)$, and $h(x)$. But let $|f - h|$, $|f - g|$, and $|g - h|$ be maximal at x_0, x_1, and x_2, respectively. Then, for all x,

† We assume the axiom of completeness for **R**: each nonempty subset of **R** which is bounded above has a least upper bound.

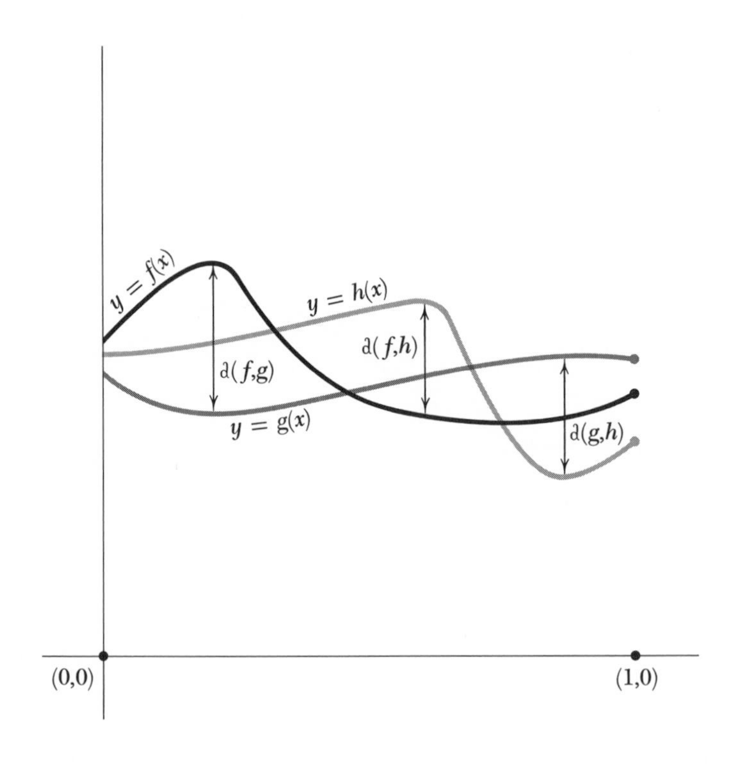

$$\begin{aligned}|f(x) - h(x)| &\leq |f(x_1) - g(x_1)| + |g(x) - h(x)| \\ &\leq |f(x_1) - g(x_1)| + |g(x_2) - h(x_2)| \\ &= \mathfrak{d}(f,g) + \mathfrak{d}(g,h).\end{aligned}$$

Hence

$$\mathfrak{d}(f,h) = |f(x_0) - h(x_0)| \leq \mathfrak{d}(f,g) + \mathfrak{d}(g,h),$$

and we do have a metric function defined for the set C. The distance $\mathfrak{d}(f,g)$ can be described as the maximal vertical distance between the graphs of f and g, or as the maximal height of the graph of $|f - g|$.

That we have a reasonable notion of distance in C, that $(C,\mathfrak{d})$ is a metric space, is itself an accomplishment. The way is now open to apply much of our intuition about distance to the set C, which is a much more complex set than real 3-space; intuitively, it is infinite dimensional. For instance, an ε-ball $\mathfrak{b}(\varepsilon,f)$ centered at a member f of C is the set of all members g of C such that, for all x, $|f(x) - g(x)| < \varepsilon$. This requires that the graph of g lie inside the vermiform strip of height 2ε whose center (in a

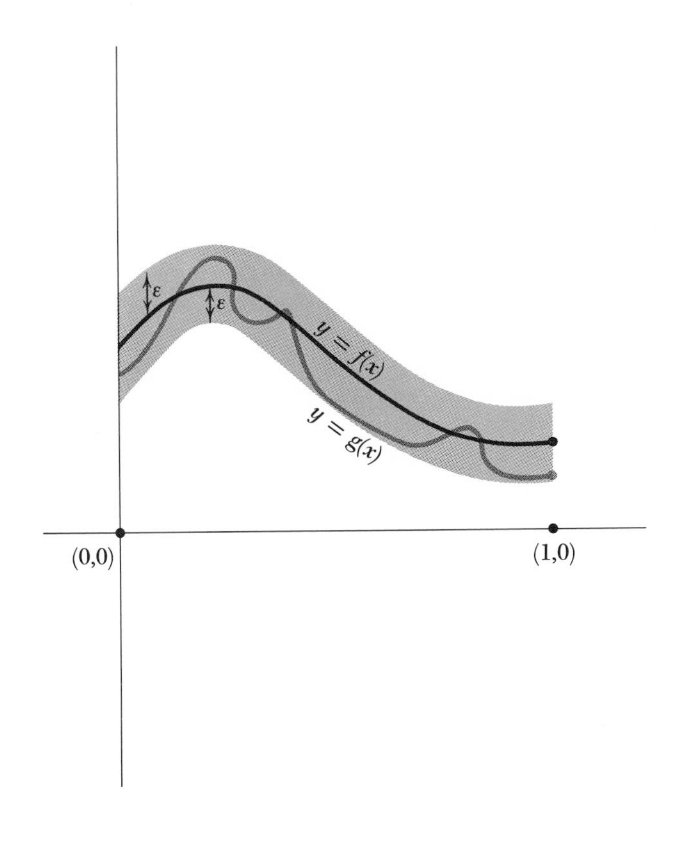

vertical sense) is the graph of f. The skin of an ε-ball in real 3-space corresponds in this example to a function whose graph lies inside this region of height 2ε, except that its graph touches the boundary lines of the region, or perhaps runs along the boundary for a while.

We can do more. With a metric defined on C, the definition of continuity applies to a function from a metric space into C or to a function from C to any metric space. For instance, there is a family of **evaluation functions** $e_t: C \to \mathbf{R}$, one for each $t \in \mathbf{I}$, each of which has values $e_t(f) = f(t)$. And it can now be shown that for each $t \in \mathbf{I}$, e_t is continuous. If $\mathfrak{b}[\varepsilon, e_t(f)]$ is an ε-ball at $e_t(f) = f(t) \in \mathbf{R}$, then the ball $\mathfrak{b}(\varepsilon, f)$ centered at f and having that same diameter has the property that $e_t[\mathfrak{b}(\varepsilon,f)] \subset \mathfrak{b}[\varepsilon, f(t)]$ [if $g \in \mathfrak{b}(\varepsilon,f)$, then for every $t \in \mathbf{I}$, $|g(t) - f(t)| < \varepsilon$]. Hence e_t is continuous at each point f of its domain.

PYTHAGORAS' THEOREM

The celebrated theorem of the Greek Pythagoras states that the distance between two points $x = (x_1, x_2)$ and $y = (y_1, y_2)$ in the plane is $\mathfrak{d}(x,y) =$

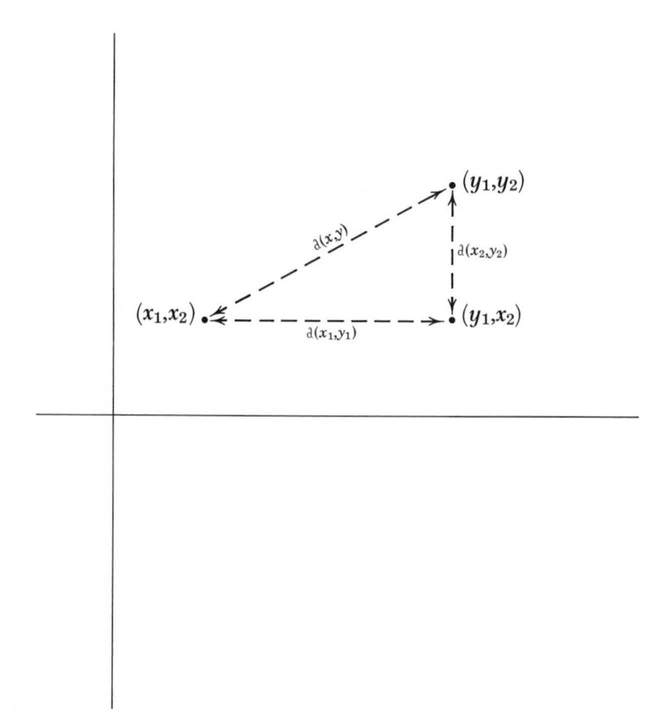

$[(x_1 - y_1)^2 + (x_2 - y_2)^2]^{1/2}$. Since the plane is the direct-product set $\mathbf{R} \times \mathbf{R}$, and each factor set $\mathbf{R}$ is a metric space with the usual metric, this may be expressed as "the distance between points x and y in the plane is the square root of the sum $|x_1 - y_1|^2 + |x_2 - y_2|^2$ of the squares of the distances between their respective (real) coordinates." Similarly, if we consider real 3-space to be $\mathbf{R}^2 \times \mathbf{R}$, the product of the (x_1,x_2)-plane and the x_3-axis, then the distance between points $x = (x_1,x_2,x_3)$ and $y = (y_1,y_2,y_3)$ is

$$\begin{aligned} d(x,y) &= (d[(x_1,x_2),(y_1,y_2)]^2 + d(x_3,y_3)^2)^{1/2} \\ &= [(x_1 - y_1)^2 + (x_2 - y_2)^2 + (x_3 - y_3)^2]^{1/2}. \end{aligned}$$

(This distance could have been calculated as well by the five other ways corresponding to diagonals of the various faces of the parallelepiped which has x and y at opposite vertices and edges parallel to coordinate axes.)

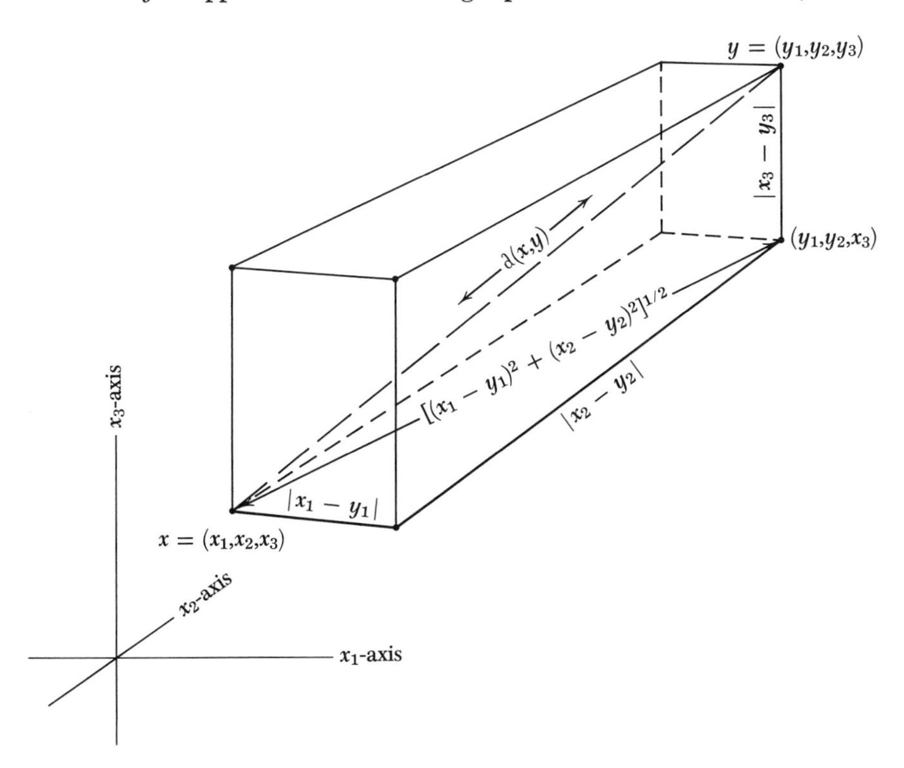

With these examples to guide us, we can easily construct a **product metric** d on the direct-product set $M_1 \times M_2$ of the underlying sets of two metric spaces (M_1,d_1) and (M_2,d_2). If $m = (m_1,m_2)$ and $n = (n_1,n_2)$ are two points of $M_1 \times M_2$, we define the distance between them to be

$$d(m,n) = [d_1(m_1,n_1)^2 + d_2(m_2,n_2)^2]^{1/2}.$$

It is easy to see that ∂ is a nonnegative real-valued function, $\partial(m,n) = 0$ iff $m = n$, and $\partial(m,n) = \partial(n,m)$. We offer a proof of the triangle inequality based on familiar geometric facts in the plane. Let $m = (m_1,m_2)$, $n = (n_1,n_2)$, and $p = (p_1,p_2)$ be three points of $M_1 \times M_2$. Use the various distances in the two coordinate spaces to define three points α, β, and γ in the plane:

$$\begin{aligned} \alpha &= [\partial_1(m_1,n_1),\partial_2(m_2,n_2)] & |\alpha| &= \partial(m,n), \\ \beta &= [\partial_1(n_1,p_1),\partial_2(n_2,p_2)] & |\beta| &= \partial(n,p), \\ \gamma &= [\partial_1(m_1,p_1),\partial_2(m_2,p_2)] & |\gamma| &= \partial(m,p). \end{aligned}$$

Since distances are always nonnegative, each of these points is in the first quadrant, as is the sum $\alpha + \beta$ (of complex numbers). Geometrically, $\alpha + \beta$ is the corner opposite the origin on a parallelogram having 0, α, and β for three corners. It is clear that $|\alpha + \beta| \leq |\alpha| + |\beta|$; the length of the diagonal is no more than the sum of the lengths of two adjacent sides. But

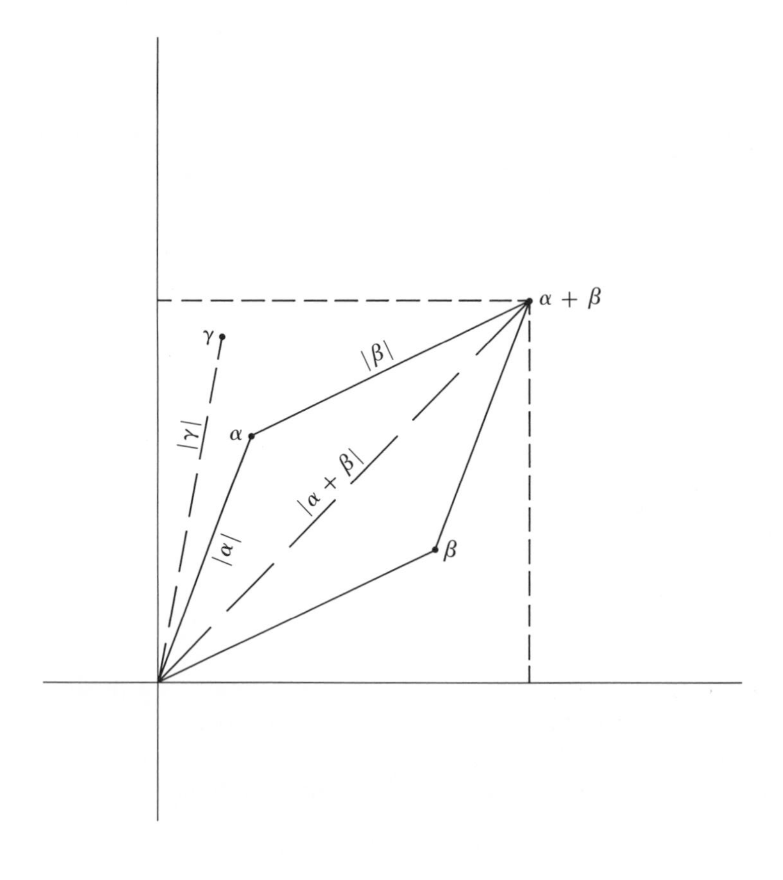

the triangle inequality in M_1 and M_2 says that $0 \leq \mathfrak{d}_1(m_1,p_1) \leq \mathfrak{d}_1(m_1,n_1) + \mathfrak{d}_1(n_1,p_1)$ and $0 \leq \mathfrak{d}_2(m_2,p_2) \leq \mathfrak{d}_2(m_2,n_2) + \mathfrak{d}_2(n_2,p_2)$; accordingly, γ lies in the rectangle bounded by the coordinate axes and having $\alpha + \beta$ for a corner. Clearly, $|\gamma| \leq |\alpha + \beta|$, and thus $|\gamma| \leq |\alpha| + |\beta|$; but this is exactly what was to be shown,

$$\mathfrak{d}(m,p) \leq \mathfrak{d}(m,n) + \mathfrak{d}(n,p).$$

If M_1, M_2, and M_3 are metric spaces, then there are product metrics on both $(M_1 \times M_2) \times M_3$ and $M_1 \times (M_2 \times M_3)$. These two direct products are set isomorphic (see Exercise I.P); such a set isomorphism can be constructed which preserves distances. (A function f preserves distances if $\mathfrak{d}[f(x),f(y)] = \mathfrak{d}(x,y)$ for every pair (x,y) in the domain of f; f is then called an "isometry.")

This should be obvious from the example earlier in this section of the product metric in real 3-space; we shall not pause to prove it. More generally, if $M_1, M_2, \ldots, M_k$ are all metric spaces, we may (by Exercise I.P) think of "the" direct product M of the M_i's as being the set of ordered k-tuples $(m_1, m_2, \ldots, m_k)$, where each $m_i \in M_i$. Further, there is a metric on $M = M_1 \times M_2 \times \cdots \times M_k$ which has as its value

$$\mathfrak{d}(m,n) = \left[\sum_{i=1}^{k} \mathfrak{d}_i(m_i,n_i)^2 \right]^{1/2}$$

where $\mathfrak{d}_i$ is the metric on M_i, and $m = (m_1, m_2, \ldots, m_k)$ and $n = (n_1, n_2, \ldots, n_k)$.

In particular, if $k = 2$ and $M_1 = M_2 = \mathbf{R}$, the real line with the usual metric, then the product-metric space $\mathbf{R} \times \mathbf{R} = \mathbf{R}^2$ is just the plane with the usual metric. Similarly, $\mathbf{R} \times \mathbf{R} \times \mathbf{R} = \mathbf{R}^3$ is the euclidean 3-space where we live, with its usual metric, and for each positive integer n we may build **euclidean** (or **real**) **n-space** $\mathbf{R} \times \mathbf{R} \times \cdots \times \mathbf{R} = \mathbf{R}^n$, the set of n-tuples of real numbers, furnished with its product metric.† Our collection of examples of metric spaces thus now includes every subspace A of real n-space. Further, if $A \subset \mathbf{R}^m$ and $B \subset \mathbf{R}^n$ are metric subspaces of euclidean spaces, we can now discuss the continuity of a function $f: A \to B$.

Exercises F and G

† Euclidean n-space is usually thought of as being furnished also with the structure of a vector space over the real field. We shall consider this additional structure, and its compatibility with the metric, later (but see Prob. EE).

PATH-CONNECTEDNESS

A fundamental intuitive property of subsets of the plane or real 3-space is that of being all in one piece, one chunk. A circle, a disc, and a straight line all have that property, but the subset $\{1,2,3\}$ of the real line has three pieces. We now set a definition based on the idea that a subset is in one piece if between each two points of the set a line (not necessarily straight) can be drawn connecting the points. A metric space M is **path connected** iff for each pair (m,n) of points of M there exists a continuous function f: $\mathbf{I} \to M$ from the unit interval $[0,1] \subset \mathbf{R}$ into M with $f(0) = m$ and $f(1) = n$. Suggestively, the path f in M begins at m and ends at n. A subset $N \subset M$ is **path connected** iff the metric subspace N of M is path connected.

Clearly, **I** itself, a metric subspace of **R**, is path connected; it is an exercise to show that the family of intervals is exactly the family of path-connected subsets of **R**. We content ourselves now with showing that **R** itself is path connected. Let r and s be real numbers and define f: $\mathbf{I} \to \mathbf{R}$ by $f(t) = r + t(s - r)$; $f(0) = r$ and $f(1) = s$, and it is easy to see that f is continuous.

A familiar theorem (the intermediate-value theorem) of the calculus states that a continuous function (real-valued, of a real variable) which has both positive and negative values must also have zero as a value. More generally, if $f(r)$ and $f(s)$ are values of a continuous function with $f(r) \leq f(s)$, then for each x in between, $f(r) \leq x \leq f(s)$, there is a t with $f(t) = x$. This theorem enables us to show that if x is a real number, then the subset $\mathbf{R} - \{x\}$, the line with x deleted, is not path connected. Indeed, $x - 1$ and $x + 1$ are both in $\mathbf{R} - \{x\}$, and if f: $\mathbf{I} \to \mathbf{R} - \{x\}$ were a path from $x - 1$ to $x + 1$, then f would have to take on the value x somewhere in its domain **I**. But then the values of f could not all lie in $\mathbf{R} - \{x\}$; hence no such function f exists, and $\mathbf{R} - \{x\}$ is not path connected.

The plane $\mathbf{R}^2$ and 3-space $\mathbf{R}^3$ are path connected. In the plane a proof goes as follows: if (x_1,x_2) and $(y_1,y_2) \in \mathbf{R}^2$, then f: $\mathbf{I} \to \mathbf{R}^2$: $f(t) = [(1 - t)x_1 + ty_1, (1 - t)x_2 + ty_2]$ is a path joining them; it is not hard to see that $\mathbf{R}^2 - \{(0,0)\}$ and $\mathbf{R}^3 - \{(0,0,0)\}$ are path connected. This construction will work for most pairs of points, unless the path goes through the origin. In that case, a new path may easily be described which dodges around the origin. Do you see how the higher dimensionality permits this for $\mathbf{R}^2$ and $\mathbf{R}^3$, yet not for **R**?

In our intuitive description of continuous functions we suggested that their graphs could be drawn "without lifting the pencil"; that is, their values do not suddenly "jump." The graph of a discontinuous function is pictured below. It is clear that if this function is restricted to a small enough open interval J centered at x_0, then the image $f(J)$ of the path-connected set J is

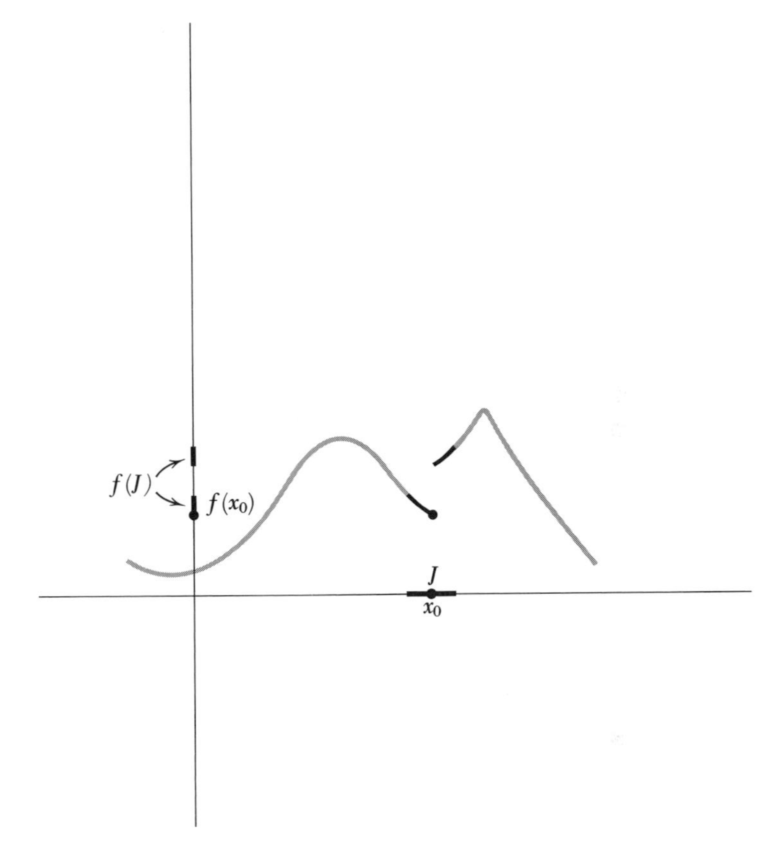

in two pieces. We shall now show that this behavior is not possible for a continuous function; *the continuous image of a path-connected set is always path connected.*

As a first step, we prove that *the composite of two continuous functions is continuous.* Assume that $f: A \to B$ and $g: B \to C$ are continuous functions, where A, B, and C are metric spaces. For each $x \in A$ and every positive real number ε we must find a positive number δ such that $(g \circ f)[\mathfrak{b}(\delta,x)] \subset \mathfrak{b}[\varepsilon, g \circ f(x)]$. Now, g is continuous, so there exists a positive number φ such that $g(\mathfrak{b}[\varphi,f(x)]) \subset \mathfrak{b}(\varepsilon,g[f(x)])$; since f is continuous, there exists a δ with $f[\mathfrak{b}(\delta,x)] \subset \mathfrak{b}[\varphi,f(x)]$. But then $(g \circ f)[\mathfrak{b}(\delta,x)] \subset g(\mathfrak{b}[\varphi,f(x)]) \subset \mathfrak{b}[\varepsilon, g \circ f(x)]$, which was to be shown; $g \circ f$ is continuous. ■

Now let $f: \mathbf{I} \to B$ be a path in B, and let $g: B \to C$ be continuous; by the preceding result, $g \circ f: \mathbf{I} \to C$ is a (continuous) path in C. Suppose B to be path connected, and let $g(x)$ and $g(y)$ be any two points in $g(B)$, the image of B under g. There is a path $f: \mathbf{I} \to B$ beginning at x and ending at y; hence $g \circ f$ is a path beginning at $g(x)$ and ending at $g(y)$. This shows that $g(B)$ is a path-connected set. ∎ Do you see how this is a natural generalization of the intermediate-value theorem of calculus?

Exercises H, J, and K

COMPACTNESS

Here we need to enlarge our vocabulary a bit. If S is a subset of a metric space M, and $x \in S$, then S is a **neighborhood** of x iff there is a positive ε such that the ε-ball at x *in* M lies in S, $\mathfrak{b}(\varepsilon,x) \subset S$. For instance, a closed interval $[a,b] \subset \mathbf{R}$ is a neighborhood of each of its points except for the end points a and b. The subset S of M is called **open** if it is a neighborhood of each of its points; thus, in the real line, an open interval (a,b) is an open set, while a closed interval $[a,b]$ is not an open set.

More generally, in a metric space M each ε-ball $\mathfrak{b}(\varepsilon,x)$ is an open set. This is easily proved. If $y \in \mathfrak{b}(\varepsilon,x)$, then $\mathfrak{d}(x,y) < \varepsilon$, so the number $\delta = \varepsilon - \mathfrak{d}(x,y)$ is positive. The triangle inequality says that if $z \in \mathfrak{b}(\delta,y)$, so that $\mathfrak{d}(y,z) < \delta$, then

$$\mathfrak{d}(x,z) \leq \mathfrak{d}(x,y) + \mathfrak{d}(y,z) < \mathfrak{d}(x,y) + [\varepsilon - \mathfrak{d}(x,y)] = \varepsilon.$$

Hence $\mathfrak{b}(\delta,y) \subset \mathfrak{b}(\varepsilon,x)$, which latter set must be open.

The complement of an open set in M is called **closed;** closed intervals of $\mathbf{R}$ are closed sets in $\mathbf{R}$ (why?). Of course, a subset of $\mathbf{R}$ need be neither open nor closed; $(0,1]$ is such a set.

One more definition: a subset S of a metric space M is **bounded** iff it lies inside some ε-ball; that is, S is bounded iff there exists an $x \in M$ and a positive ε such that $S \subset \mathfrak{b}(\varepsilon,x)$. An interval is a bounded set in $\mathbf{R}$; the integers $\mathbf{Z}$ are not a bounded set. Now to the point: a subset of $\mathbf{R}$ (or $\mathbf{R}^n$) which is both closed and bounded is defined in the calculus to be **compact;** $[0,1]$ is compact, while $(0,1)$ and $\mathbf{Z}$ are not compact. This notion of compactness is an extremely important tool of topology and of the applications of topology to analysis. For instance, a continuous real-valued function whose domain is a compact subset of $\mathbf{R}^n$ is **bounded** (its set of values is bounded) and also is uniformly continuous. We begin the study of this notion with a classic characterization of the compact subsets of the real line.

HEINE-BOREL THEOREM *A subset S of* **R** *is compact iff, whenever $\mathcal{O}$ is a family of open sets of* **R** *and $\bigcup\mathcal{O} \supset S$, there exists a finite subfamily $\mathcal{F} = \{O_1, O_2, \ldots, O_k\}$ of members of $\mathcal{O}$ with $S \subset \bigcup\mathcal{F}$.*

Proof Assume that S is not compact; we shall exhibit a family $\mathcal{O}$ of open sets, with $S \subset \bigcup\mathcal{O}$, such that for no finite subfamily $\mathcal{F}$ of $\mathcal{O}$ is $S \subset \bigcup\mathcal{F}$. There are two cases to consider; either S is not closed or S is not bounded. If S is not bounded, let $\mathcal{O}$ be the family of all ε-balls centered at the origin $\mathcal{O} = \{\mathcal{b}(\varepsilon,0): \varepsilon > 0\}$. $\bigcup\mathcal{O} = \mathbf{R} \supset S$, yet the union $\bigcup\mathcal{F}$ of a finite subfamily $\mathcal{F}$ of $\mathcal{O}$ is just the largest ball in $\mathcal{F}$; since S is not bounded, $\bigcup\mathcal{F} \not\supset S$. In case S is not closed, its complement S' is not open, so there is a point $x \in S'$ for which no ball $\mathcal{b}(\varepsilon,x)$ is contained in S'. That is, for each positive ε, there is a point y of S with $y \in \mathcal{b}(\varepsilon,x)$; y is less than ε distant from x. It is clear that for each positive ε the set $O_\varepsilon = (-\infty, x - \varepsilon) \cup (x + \varepsilon, \infty)$ is open, and that if $\mathcal{O} = (O_\varepsilon: \varepsilon > 0)$, then $\bigcup\mathcal{O} = \mathbf{R} - \{x\} \supset S$. Again, however, the union of any finite subfamily of $\mathcal{O}$ is just the largest member of that subfamily; by the comments above, no member of $\mathcal{O}$ could contain S.

For the second part of the proof we assume that S is compact, so S is closed and is contained in some bounded interval (a,b); also, $S \subset [a,b]$, and the set $U = S' \cap (a - 1, b + 1)$ is open in **R**. Now observe that it will suffice to prove the theorem for the compact set $[a,b]$. Whenever $\mathcal{O}$ is a family of open sets with $S \subset \bigcup\mathcal{O}$, we may construct the family $\mathcal{O}' = \mathcal{O} \cup \{U\}$ of open sets such that $[a,b] \subset \bigcup\mathcal{O}'$. If $\mathcal{F}'$ is a finite subfamily of $\mathcal{O}'$ with $[a,b] \subset \bigcup\mathcal{F}'$, then $\mathcal{F} = \mathcal{F}' - \{U\}$ is the desired finite subfamily of $\mathcal{O}$, $S \subset \bigcup\mathcal{F}$. Hence we suppose $S = [a,b]$ and form the set $T \subset [a,b]$ by this rule: if $a \le x \le b$, then $x \in T$ iff $[a,x]$ lies in the union of some finite subfamily of $\mathcal{O}$. Certainly $a \in T$ and b is an upper bound for T; let $l = lub\ T$, and let L be a member of $\mathcal{O}$ with $l \in L$. Since L is open, there exists $\varepsilon > 0$ such that $(l - \varepsilon, l + \varepsilon) \subset L$, and since $l = lub\ T$, there is a $t \in T$ with $t > l - \varepsilon$. But when $\mathcal{F}$ is a finite subfamily of $\mathcal{O}$ with $[a,t] \subset \bigcup\mathcal{F}$, then $\mathcal{F} \cup \{L\}$ is a finite subfamily of $\mathcal{O}$ whose union contains $[a, l + \delta]$ for every $\delta \in [0,\varepsilon)$. Since l is an upper bound for T, $l + \delta \notin T$, and thus $l + \delta > b$ for all $\delta > 0$, or $l \ge b$. By definition, $l \le b$; this shows that $l = b$ and concludes the proof. ■ The Heine-Borel theorem is true in each real n-space; a proof is sketched in Prob. FF, and a complete proof appears in Chap. VI.

Exercises L and P

n-SPHERES

We shall now construct a family of useful examples of metric spaces which are both compact and path connected. If $x \in \mathbf{R}^{n+1}$, let $\|x\| = \mathfrak{d}(x,0)$ be the distance from x to the origin ("the length of the vector x"), $\|x\| = \left(\sum_{i=1}^{n+1} x_i^2\right)^{1/2}$, if $x = (x_1, x_2, \ldots, x_{n+1})$. The set $S^n = \{x \in \mathbf{R}^{n+1}: \|x\| = 1\}$ of elements of $\mathbf{R}^{n+1}$ at distance exactly 1 from the origin is called the **unit n-sphere in $\mathbf{R}^{n+1}$**. It is, in suggestive language, the bounding skin of the ball $\mathfrak{b}(1,0)$ of radius 1 centered at 0. The 0-sphere in $\mathbf{R}^1 = \mathbf{R}$ is thus the two-point set $\{-1,1\}$, the 1-sphere is the unit circle in the plane, and the 2-sphere is the surface of a balloon of radius 1 in real 3-space.

The n-sphere is a closed subset of real $(n + 1)$-space; if $\|y\| \neq 1$, then it is an exercise (see Exercise Q) to show that a point z whose coordinates are sufficiently close to those of y will have $\|z\| \neq 1$. Consequently, there is an ε-ball centered at y all of whose elements lie off the n-sphere; the complement of S^n is open. S^n is clearly contained in the ball $\mathfrak{b}(2,0)$; it is closed and bounded, and thus compact.

The n-sphere is also path connected if $n > 0$. To see this, recall that the set $\mathbf{R}^{n+1} - \{0\}$ is path connected for $n > 0$. Further, the function g: $\mathbf{R}^{n+1} - \{0\} \to S^n$ which sends each $x \in \mathbf{R}^{n+1}$, $\|x\| \neq 0$, to $g(x) = x/\|x\|$ is continuous; more explicitly, $g(x) = (x_1/\|x\|, x_2/\|x\|, \ldots, x_{n+1}/\|x\|)$. A detailed proof of this fact would observe that the values $g(y)$ can be made as close as you please to $g(x)$ by merely requiring that the coordinates of y be close enough to those of x. The values of g lie on the n-sphere, since for all x,

$$\|g(x)\| = \left[\sum_{i=1}^{n+1}\left(\frac{x_i}{\|x\|}\right)^2\right]^{1/2} = \left(\frac{1}{\|x\|^2}\sum_{i=1}^{n+1} x_i^2\right)^{1/2} = \left(\frac{\|x\|^2}{\|x\|^2}\right)^{1/2} = 1.$$

Also, g is an onto function, since $x \in S^n$ implies $g(x) = x$. But then, S^n is the continuous image of the path-connected space $\mathbf{R}^{n+1} - \{0\}$; S^n itself is path connected.

MORE ABOUT CONTINUITY

An open subset of a metric space was defined as a subset which was a neighborhood of each of its points. That is, a subset S is open in the metric space M iff for each member x of S there is an $\varepsilon > 0$ such that $\mathcal{b}(\varepsilon,x) \subset S$. The notion of open set can be used to characterize continuous functions.

THEOREM *A function $f: M \to N$ from one metric space to another is continuous iff the set $f^{-1}(S)$ is open in M whenever S is open in N.*

Proof We first assume that f is continuous and that S is open in N; we must show that if $x \in f^{-1}(S)$, then there is a $\delta > 0$ such that $\mathcal{b}(\delta,x)$ lies inside $f^{-1}(S)$. But $f(x) \in S$, and S is open; therefore there is an $\varepsilon > 0$ with $\mathcal{b}[\varepsilon,f(x)] \subset S$. Since f is continuous, there exists a $\delta > 0$ such that if the distance $\mathcal{d}(x,y)$ (in M) is less than δ, then $\mathcal{d}[f(x), f(y)] < \varepsilon$ (in N); that is, $f(y) \in \mathcal{b}[\varepsilon,f(x)]$. This says clearly that f carries each member of $\mathcal{b}(\delta,x)$ into S, or $\mathcal{b}(\delta,x) \subset f^{-1}(S)$. Since such a number δ exists for each point x of $f^{-1}(S)$, that set is open.

For the second half of the proof we assume that for each open set S of N, $f^{-1}(S)$ is open in M. Let x be an arbitrary point of M and let ε be a positive number; we wish to find a δ with $f[\mathcal{b}(\delta,x)] \subset \mathcal{b}[\varepsilon,f(x)]$. But balls are open sets, so $f^{-1}(\mathcal{b}[\varepsilon,f(x)])$ is open and x is one of its points. Hence there is a ball $\mathcal{b}(\delta,x)$ centered at x and lying entirely inside $f^{-1}(\mathcal{b}[\varepsilon,f(x)])$, which was to be shown; f must be continuous. ∎

An easily remembered, if imprecise, statement of this result is that "a function is continuous iff, under it, inverse images of open sets are open." Notice that, even when this theorem is applied to a function whose domain and range are metric subspaces of the real line, continuity is determined by the open subsets of its domain and range as metric spaces, without reference to open subsets of the whole line. This is not deep. An ε-ball in the subspace A of $\mathbf{R}$ is merely the intersection of A with the ball in $\mathbf{R}$ of the same radius; hence the open subsets of A are just the intersections of A with the open subsets of $\mathbf{R}$. This observation can be used to characterize, by the Heine-Borel theorem, the compact subsets of $\mathbf{R}$ in terms of their own internal structure: *a subset A of* $\mathbf{R}$ *is compact iff every cover of the metric space A by a family of its open subsets has a finite subcover.* ∎ Let us immediately use this idea to establish that continuous real functions preserve compactness.

THEOREM *If $f: A \to B$ is a continuous real-valued function whose domain A is a compact subset of* $\mathbf{R}$, *then its image $f(A)$ is compact.*

Proof The Heine-Borel theorem will be used here. Assume that $\mathcal{O}$ is a family of open sets with $f(A) \subset \bigcup\mathcal{O}$. Then the family $\mathcal{P} = \{f^{-1}(O): O \in \mathcal{O}\}$ of inverse images of members of $\mathcal{O}$ is a family of open subsets of A. Further, $A \subset \bigcup\mathcal{P}$, since $x \in A$ implies that there is a member O of $\mathcal{O}$ with $f(x) \in O$, and then $x \in f^{-1}(O)$. But A is compact, so there is a finite subfamily $\mathcal{F}$ of $\mathcal{P}$

with $A \subset \bigcup \mathcal{F}$; thus the family $\mathcal{G} = \{O: f^{-1}(O) \in \mathcal{F}\}$ has a finite number of members, and $f(A) \subset \bigcup \mathcal{G}$. We have shown that each family $\mathcal{O}$ contains a finite subfamily $\mathcal{G}$; by the Heine-Borel theorem, $f(A)$ is compact. ∎ You should realize that a continuous function need preserve neither closedness nor boundedness by itself. For instance, the function $f(x) = 1/x$ interchanges the closed but unbounded set $[1,\infty]$ with the bounded but nonclosed set $(0,1]$.

COROLLARY *The function f of the theorem has a maximum value if $A \neq \varnothing$; that is, there is a point x of A for which $f(x) \geq f(y)$ for every $y \in A$.*

Proof Since $f(A)$ is nonempty and bounded, there is a least upper bound $l = lub\, f(A)$. Suppose $l \notin f(A)$; since $f(A)$ is closed, its complement is open. Thus there is an interval $(l - \varepsilon, l + \varepsilon)$ lying outside $f(A)$. But then $l - \varepsilon$ is an upper bound for $f(A)$, a clear contradiction. Our supposition is thus impossible, and $l \in f(A)$, which is exactly what was to be proved. ∎ Is it implicit that f has a minimal value too?

The function $f: \mathbf{Z} \to \mathbf{R}$ whose value at an integer n is $f(n) = n$ is one example which does not attain a maximum value; the function $f: (0,1) \to \mathbf{R}$: $f(x) = 1/x$ is another. In neither case is the domain compact.

The proof of this theorem depended crucially on the preceding result that a function is continuous iff all inverse images of open sets are open. This simple statement deserves comparison with our earlier *definition* of continuity. That definition evolved only after hundreds of years; many great mathematicians did their work without benefit of our set-theoretic precision, relying instead on what we now call intuitive notions of what a continuous function is. One reason for the long delay was that this was an exceedingly difficult problem; the logic of our definition was more complex than any but the most contorted of sentences in everyday language. If $f: M \to N$ is a function from one metric space to another, "f is continuous iff *for each* $x \in M$ and *for each* real $\varepsilon > 0$ *there exists* a real $\delta > 0$ such that *for all* $y \in M$, if $\mathrm{d}(x,y) < \delta$, then $\mathrm{d}[f(x),f(y)] < \varepsilon$." Essential logical phrases are italicized here; can you construct an English sentence with logic this complicated? In contrast, we might as well have said, "f is continuous iff *for all* $S \subset N$, if S is open, then $f^{-1}(S)$ is open." Again, in this equivalent definition essential logic is italicized. An enormous simplification is afforded by the use of the notion of an open set. You might reasonably respond, "But the complexity has merely been swept under the definition of *open set*." True enough, but the use of this device has far-reaching consequences, some of which we shall explore, beginning with the next chapter.

Exercises M and N

REFERENCES AND FURTHER TOPICS

A treatment of the continuity of functions on real n-space can be found in any textbook of advanced calculus. Much of the material presented here is covered in elementary calculus texts as well. To find the discussion you want in a given book, look up key words (for example, "compact") in the index.

Our treatment of metric spaces is not complete, in any sense of the word. It is designed to provide motivation, introductory information, and examples for the study of topology to come. Two discussions in a similar spirit are

B. Mendelson, *Introduction to Topology,* chap. 2 (Boston: Allyn and Bacon, 1962).

G. F. Simmons, *Topology and Modern Analysis*, chap. 2 (New York: McGraw-Hill, 1963).

A more thoroughgoing treatment of metric spaces, including many further topics as well as an introduction, is given by

M. H. A. Newman, *Topology of Plane Sets of Points* (New York: Cambridge University Press, 1939).

EXERCISES

A Show that if $(M,\mathfrak{d})$ is a metric space, then the function $\mathfrak{d}': M \times M \to \mathbf{R}$ is also a metric for M, where $\mathfrak{d}'(x,y) = \mathfrak{d}(x,y)$ whenever $\mathfrak{d}(x,y) \leq 1$, and $\mathfrak{d}'(x,y) = 1$ for those pairs (x,y) where $\mathfrak{d}(x,y) \geq 1$. A metric such as $\mathfrak{d}'$, all of whose values form a bounded subset of $\mathbf{R}$, is called a **bounded metric,** and $(M,\mathfrak{d}')$ is a **bounded metric space.**

B Define a function $\mathfrak{d}: \mathbf{C} \times \mathbf{C} \to \mathbf{R}$ by letting the value of $\mathfrak{d}$ at a pair $(z,w) \in \mathbf{C} \times \mathbf{C}$, $z = (z_1,z_2)$ and $w = (w_1,w_2)$, be their maximal coordinate separation,

$$\mathfrak{d}(z,w) = max\ \{|z_1 - w_1|,\ |z_2 - w_2|\}.$$

Show that $\mathfrak{d}$ is a metric for $\mathbf{C}$. What do ε-balls look like in $(\mathbf{C},\mathfrak{d})$?

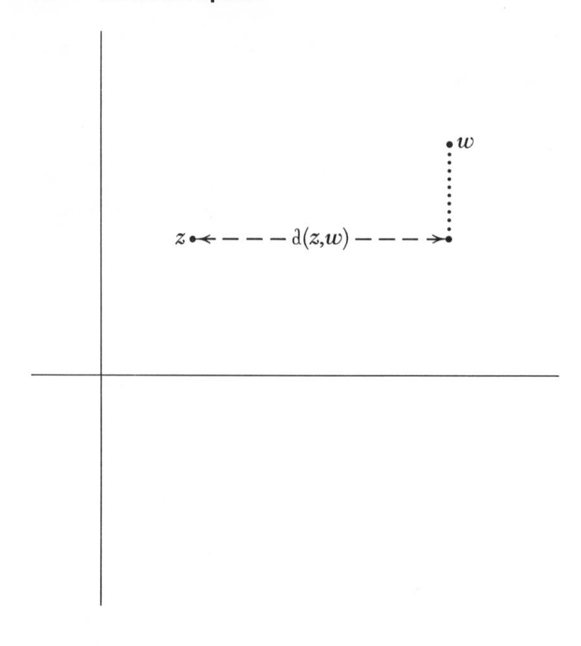

C Show that the projections p_1 and p_2 of the plane **C** onto the real and imaginary axes, $p_1(z_1,z_2) = z_1$ and $p_2(z_1,z_2) = z_2$, are continuous.

D A function $f: M \to N$ between metric spaces is a **homeomorphism** iff it has a two-sided inverse function f^{-1}, and both f and f^{-1} are continuous. When such a homeomorphism exists, the spaces M and N are said to be homeomorphic to one another. Show (by constructing a homeomorphism) that any two open intervals, considered as subspaces of **R** (with the usual metric), are homeomorphic. Can you prove that $(0,1)$ and $[0,1]$ are *not* homeomorphic?

E If ∂ and ∂' are two metrics for the same set M, ∂ is **equivalent** to ∂' iff the identity map 1_M is a homeomorphism of (M,∂) with (M,∂'). Show that the metrics ∂ and ∂' of Exercise A are equivalent (and hence that each metric space is homeomorphic to a bounded metric space).

F Prove that the metric ∂ defined for **C** in Exercise B is equivalent to the usual product metric ∂' for **C**. Can you generalize this example to arbitrary product spaces? (The sort of metric constructed in Exercise B is often defined to *be* the product metric.)

G Show that if $M \times N$ is the product of two metric spaces, with the product metric, then the projection $p: M \times N \to M: p(m,n) = m$ of the product space onto one of its factors is continuous. Show also that the diagonal function Δ from M into the direct product $M \times M$ of M with itself, $\Delta(m) = (m,m)$, is continuous.

H Show that a subset of $\mathbf{R}$ is path connected iff it is an interval (perhaps infinite). Here an interval means a set of the form (a,b), $[a,b)$, $(a,b]$, or $[a,b]$, with the usual conventions if a or b is $\pm\infty$ (for example, $[a,\infty) = [a,\infty] = \{x: a \leq x\}$).

J A **closed ε-ball centered at** x in a metric space M is a set of the form $\{y \in M: \mathfrak{d}(x,y) \leq \varepsilon\}$, that is, an ε-ball with its skin added to it. Prove first that a subset S of M is path connected iff every point of S is the end of some path beginning at a fixed point x of S. Then show that if S is a subset of $\mathbf{R}^n$ which contains $\mathfrak{b}(\varepsilon,x)$ and is contained in the closed ε-ball at x, then S is path connected.

K Let S be the subset of the plane which is the union of the segment $A = \{(x,y): x = 0 \text{ and } -1 \leq y \leq 1\}$ of the y-axis with the part $B = \{(x,y): 0 < x \leq 1 \text{ and } y = \sin(1/x)\}$ of the graph of $\sin(1/x)$ which lies above $(0,1]$. Prove that S is not path connected.

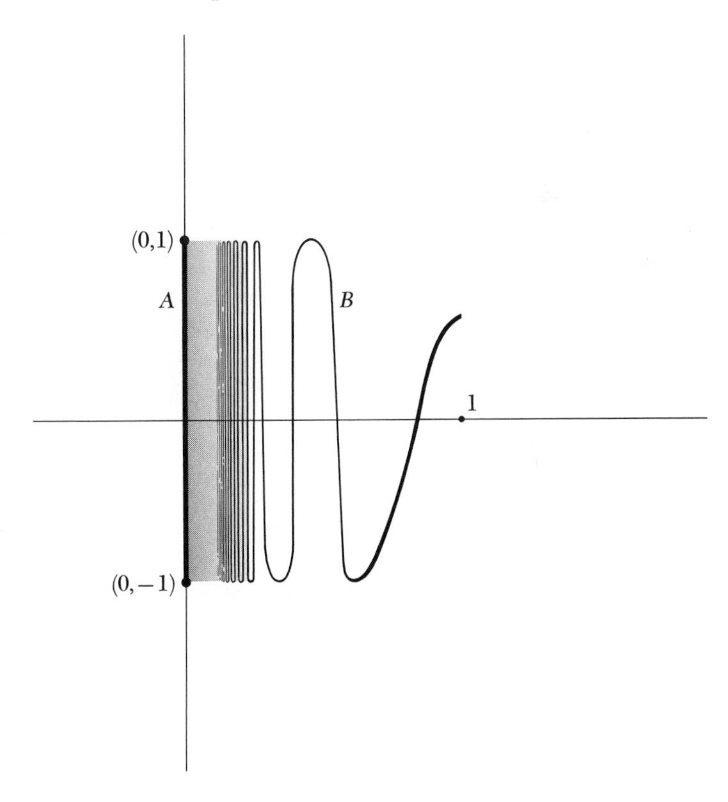

L Show that the subset S described in Exercise K is closed in $\mathbf{R}^2$, and is thus compact.

M Show that two metrics $\mathscr{d}$ and $\mathscr{d}'$ on a set M are equivalent (see Exercise E) iff the two classes of open subsets of M which they define are exactly the same, that is, iff $\{S\text{: S is open in } (M,\mathscr{d})\} = \{S\text{: S is open in } (M,\mathscr{d}')\}$.

N Show that if $\mathcal{S}$ is a family of open subsets of a metric space M, then $\bigcup\mathcal{S}$ is open. Further, if $\mathcal{S}$ has a finite number of members, then $\bigcap\mathcal{S}$ is open. Give an example of a family $\mathcal{S}$ (infinite, of course) of open sets for which $\bigcap\mathcal{S}$ is not open. Observe also that M is open. Can you see that M is closed as well, that is, that $\varnothing$ is open?

P A family $\mathcal{S}$ of sets has the "finite intersection property" iff every finite subset $\mathcal{F}$ of $\mathcal{S}$ has a nonempty intersection; $\bigcap\mathcal{F} \neq \varnothing$. Translate the Heine-Borel theorem into the language of closed sets by proving that a subset S of $\mathbf{R}$ is compact iff, whenever $\mathcal{S}$ is a family of closed subsets of S with the finite intersection property, then $\bigcap\mathcal{S} \neq \varnothing$.

Q Assume the results of Exercise N and use the De Morgan formulas (Theorem II.2) to prove that if $\mathcal{S}$ is a family of closed sets, then $\bigcap\mathcal{S}$ is closed, and if $\mathcal{S}$ is a finite family, then $\bigcup\mathcal{S}$ is also closed. Show that a closed ε-ball (see Exercise J) is a closed set; then show that S^n is an intersection of closed subsets of $\mathbf{R}^{n+1}$, and so is closed.

R Let $f\colon M \to N$ be a continuous function between metric spaces, and let $f|_S$ be the restriction of the function f to the subspace S of M. Show that $f|_S$ is continuous. Then give an example of a function $f\colon \mathbf{R} \to \mathbf{R}$ which is not continuous at 0, but whose restriction to $[0,1]$ is continuous. (The restriction of your function f to an *open* interval containing 0 is discontinuous.)

S Prove that each open subset of $\mathbf{R}$ is the union of a mutually disjoint family of open intervals.

T Let S be a compact subset of $\mathbf{R}$, and let $\mathcal{O}$ be a family of open sets of $\mathbf{R}$ with $S \subset \bigcup\mathcal{O}$. Show that there exists a positive number ε (called a "Lebesgue number") such that for every member x of S there is some member O_x of $\mathcal{O}$ which contains the interval $(x - \varepsilon, x + \varepsilon)$. It will be helpful in your proof to show that the function $g\colon S \to \mathbf{R}$ is continuous, where $g(x) = lub\ \{\varepsilon\text{: there exists an } O \in \mathcal{O} \text{ with } \mathscr{b}(\varepsilon,x) \subset O\}$ when that least upper bound exists, or otherwise $g(x) = 1$.

U Show that if (M,d) is a metric space, then $\text{d}\colon M \times M \to \mathbf{R}$ is continuous when $M \times M$ is given the product metric.

W Show that the set of all homeomorphisms (see Exercise D) of a fixed metric space with itself forms a group under the operation of composition.

PROBLEMS

AA Sequences An "infinite sequence" a in a set S is a function a from the set of positive integers to S. Customarily the value $a(i)$ of the function at an integer i is denoted a_i and called a "term" of the sequence. A sequence $a_1, a_2, a_3, \ldots, a_i, \ldots$ of real numbers "converges" to b, or $\lim_{i\to\infty} a_i = b$, if for every $\varepsilon > 0$ there is an integer N such that if $i \geq N$, then $|a_i - b| < \varepsilon$. More generally, a sequence $a_1, a_2, \ldots$ of points in a metric space A converges to $b \in A$, $\lim_{i\to\infty} a_i = b$, if for each real $\varepsilon > 0$ there is an N such that if $i \geq N$, then $\text{d}(a_i,b) < \varepsilon$. The sequence is said to "converge," or to "be convergent," if it converges to some b; b is called the "limit" of the sequence. Show that if $a_1, a_2, \ldots$ is a sequence converging to b in $\mathbf{R}^n$, then, and only then, for each integer j between 1 and n the sequence of j-th coordinates of $a_1, a_2, \ldots$ (which is a sequence of real numbers) converges to the j-th coordinate of b.

If $a_1, a_2, \ldots$ is a sequence in $\mathbf{R}^n$, then a "subsequence" of that sequence is a sequence $a_{i_1}, a_{i_2}, a_{i_3}, \ldots$ whose terms are among those of the original sequence, and for which $i_1 < i_2 < \cdots < i_j < i_{j+1} < \cdots$. Use the Heine-Borel theorem to show that every sequence in a compact subset of $\mathbf{R}$ has some convergent subsequence. Use that result to prove that a subset A of $\mathbf{R}^n$ is compact iff every sequence in A has a convergent subsequence.

Characterize the continuous functions $f\colon A \to B$, with A and B metric spaces, as exactly the set of functions $f\colon A \to B$ such that for every convergent sequence a in A, $\lim_{i\to\infty} a_i = b \in A$ implies $\lim_{i\to\infty} f(a_i) = f(b)$. Notice that this characterization differs from those previously offered in that it involves only properties of f, not those of f^{-1}.

BB Fixed Points If f is a function from some set S to S itself, then a "fixed point" of f is a point $s \in S$ such that $f(s) = s$. Clearly, some functions have fixed points and some do not, but *every* continuous function f:

$\mathbf{I} \to \mathbf{I}$ has a fixed point. Prove this by contradiction: assume f has no fixed point, and define a new function $g: \mathbf{I} \to \{0,1\}$ by having $g(x) = 0$ if $f(x) < x$ and $g(x) = 1$ if $f(x) > x$. The new function is well defined, since there is no x for which $f(x) = x$. Show that g is continuous and argue that there exists no continuous onto function $g: \mathbf{I} \to \{0,1\}$. Thus either $g(\mathbf{I}) = 0$ or $g(\mathbf{I}) = 1$, and in either case a new contradiction arises.

Now let S be the half-open interval $[0,1)$, and construct a continuous function from S into itself which has no fixed point.

CC Compact Connected Functions Each characterization offered except that of Prob. AA for the continuity of a function involves a property of the inverse of the function. But a continuous function always carries compact sets to compact sets and path-connected sets to path-connected sets. Does this property characterize continuous real functions of a real variable? That is, is it true that if $f: A \to B$, where A and B lie in $\mathbf{R}$, if $f(K)$ is compact for each compact set $K \subset A$, and also if $f(C)$ is path-connected for each path-connected set $C \subset A$, then f must be continuous?

DD Composite Continuity Let f and g be functions whose domains and ranges are metric spaces, with $f \circ g$ defined and continuous. Does the continuity of either f or g imply continuity of the other?

A function $h: A \to B$ is called "open" if for each open set S of A, $h(S)$ is an open set of B. Now assume that $f \circ g$ is continuous and that one of the functions, f or g, is both continuous and open. Does that imply the continuity of the other?

EE The Hilbert Space $\mathbf{R}^n$ We shall call the members of $\mathbf{R}^n$ "vectors"; if x is a vector, $x = (x_1, x_2, \ldots, x_n)$, then x_i is the "i-th coordinate of x." If x and y are vectors, the sum $x + y = z$ is the vector whose i-th coordinate is $z_i = x_i + y_i$; this sum is an abelian group operation for $\mathbf{R}^n$ (the n-fold direct product of the additive reals). This group addition is continuous from $\mathbf{R}^n \times \mathbf{R}^n$ to $\mathbf{R}^n$, and negation (the group inversion function) is continuous from $\mathbf{R}^n$ to $\mathbf{R}^n$.

Real numbers will also be called "scalars"; if s is a scalar and x a vector, then the product $sx = (sx_1, sx_2, \ldots, sx_n)$ is defined; this yields a continuous function from $\mathbf{R} \times \mathbf{R}^n$ to $\mathbf{R}^n$. Further, this multiplication of vectors by a fixed scalar $s \neq 0$ defines an "automorphism" A_s of the group $\mathbf{R}^n$, and the assignment of A_s to s gives a monomorphism of the multiplicative group of nonzero reals into the automorphism group of $\mathbf{R}^n$ (see Prob. II.AA).

For each pair x,y of vectors there is an "inner product" $(x,y) = \sum_{i=1}^{n} x_i y_i$, a scalar. Clearly, $(x,y) = (y,x)$; if in (x,y) we fix x, the resulting function of y is a morphism to $\mathbf{R}$. The inner products define a continuous function from $\mathbf{R}^n \times \mathbf{R}^n$ to $\mathbf{R}$; algebraically, $s(x,y) = (sx,y) = (x,sy)$. The "norm" of a vector x is the scalar $\|x\| = (x,x)^{1/2}$; it is a continuous function of x and $\|sx\| = |s|\,\|x\|$. We could have used this "norm on $\mathbf{R}^n$" to define our metric in the first place: $\mathfrak{d}(x,y) = \|x - y\|$.

FF The Heine-Borel Theorem in the Plane Prove that subset S of $\mathbf{R}^2$ is compact iff whenever $\mathcal{O}$ is a family of open sets of $\mathbf{R}^2$ and $\bigcup\mathcal{O} \supset S$ there exists a finite subfamily $\mathcal{F} = \{O_1, O_2, \ldots, O_k\}$ of members of $\mathcal{O}$ with $S \subset \bigcup\mathcal{F}$. Do this by imitating the proof given in the text for the real case. In the first half of the proof, the only alteration you will need to make is the definition of the open sets O_ε; here note that O_ε is just the complement of the closed ε-ball in $\mathbf{R}$ (see Exercises J and Q).

The second half of the proof needs more work. First observe that it will suffice to prove the theorem for sets S which are rectangles, $S = \{x,y): a_1 \leq x \leq a_2$ and $b_1 \leq y \leq b_2\}$. Then form the subset T of $\mathbf{I}$ by the rule that $t \in T$ iff there is a finite subfamily $\mathcal{F}$ for the set $S_t = \{(x,y): a_1 \leq x \leq a_1 + t(a_2 - a_1)$ and $b_1 \leq y \leq b_1 + t(b_2 - b_1)\}$. Then $\ell = lub\ T$ exists as before. The next step is hard; a hint is provided by the diagram. Suppose every dotted rectangle inside the small rectangle corresponding to $\ell \in \mathbf{I}$ has a finite subfamily $\mathcal{F}$ of $\mathcal{O}$, so that it lies inside $\bigcup\mathcal{F}$. The union of the upper and right edges of the rectangle

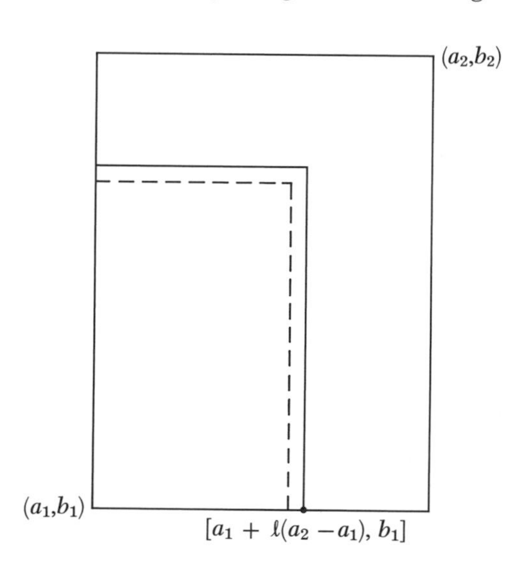

corresponding to ℓ is homeomorphic to a compact line segment, and intersections of members of $\mathcal{O}$ with that union correspond under the homeomorphism to open subsets of the line segment. Choose a finite family of these, and use that choice to find a strip of positive width centered along the union of upper and right edges of the ℓ-rectangle, which strip lies in the union of a finite number of members of $\mathcal{O}$. Then choose an appropriate dotted rectangle corresponding to a member of T less than ℓ and arrive at a contradiction (unless $\ell = 1$).

Do you now see how to alter your proof further for the case $\mathbf{R}^2$ so as to get a proof of the theorem for $\mathbf{R}^n$ under the assumption that it is true for $\mathbf{R}^{n-1}$? This is the inductive step in a proof of the theorem for every euclidean n-space. It depends crucially on the fact that half the boundary of an n-dimensional rectangle is homeomorphic to an $(n - 1)$-dimensional rectangle. And also, of course, an implicit proof is needed of the fact that compactness is invariant under homeomorphisms, that two homeomorphic metric subspaces of $\mathbf{R}^n$ are either both compact or both not compact.

GG Integrals and Derivatives Let C be the metric space of all continuous functions from $\mathbf{I}$ into $\mathbf{R}$ which was described in this chapter. Is the function continuous from C to $\mathbf{R}$ which has as its value for each function f the Reimann integral $\int_0^1 f(x)\,dx$? How about the function from C to C which assigns to f the (continuous?) function $\int_0^x f(t)\,dt$? Ask yourself a similar question concerning the differentiation operation (furnished with a suitable domain and range).

HH R is Homeomorphic to an Open Interval Show that the function $f\colon \mathbf{R} \to (-1,1)\colon f(x) = x/(1 + |x|)$ is a homeomorphism. It is immediate, then, that if (M,∂) is a metric space, then (M,∂') is an equivalent bounded metric space, where

$$\partial'(m,n) = \frac{\partial(m,n)}{1 + \partial(m,n)}.$$

JJ Continuous Group Operations Let C be the set of all continuous functions from $\mathbf{I}$ into $\mathbf{R}$; there is an abelian group structure on C induced by the addition of $\mathbf{R}$. The sum of two elements f and g of C is defined by pointwise addition: $f + g$ is the member of C which has the value $(f + g)(x) = f(x) + g(x)$ for each $x \in \mathbf{I}$ (the sum is known to be continuous from the calculus). If C is given the metric described in the text, and $C \times C$ has the product metric, then the addition function is continuous from $C \times C$ into C, and the negation (or inversion) function is continuous from C to C. In fact, negation is a homeomorphism of C. Does the pointwise product, where $(fg)(x) = f(x)g(x)$, also define a continuous function from $C \times C$ to C?

Topologies

CHAPTER IV

The study of those geometric problems originating in the concepts of limit and continuity is nearly as old as mathematics itself. The ancient Greeks worried about Zeno's paradox and the existence of irrational numbers. But only during the past few hundred years, as the number and complexity of such problems increased, has a clarification of these concepts evolved. The thoughts of many men culminated in the definitions by Frechet of metric spaces and by Hausdorff of topological spaces, both early in this century.

Metric spaces are a broad generalization of their examples, the euclidean n-spaces, and most of the arguments mathematicians make today about limits and continuity could be expressed in terms of metric spaces. However, there are important cases where this is not so, and frequent cases where the definition of a metric for a space would be laborious and unnecessary. For many discussions the exact values of the metric function, the distances, are not needed; for instance, we have seen that a function from one metric space to another is continuous iff the inverse image of each open set is open. Similarly, a sequence of real numbers converges to a limit number iff that sequence eventually becomes and remains inside each open set containing the limit (this statement readily generalizes to metric spaces; see Prob. III.AA). Hence, to discuss continuity and limits in metric spaces, it suffices to know the family of open sets in each metric space. That is, suppose you are told that M and N are metric spaces, and that f is a function from M to N. Suppose you are not told what the metrics are for M and N, but are merely informed just which subsets of M and of N are the open sets. You can then decide whether or not f is continuous.

This suggests a further question: given a set M and a family $\mathfrak{M}$ of subsets of M, is there any metric whatsoever for M in which $\mathfrak{M}$ is exactly the family of open sets? We shall not answer this question entirely; however, it was an exercise (Exercise III.N) to show that the family of open subsets of a metric space is closed under unions and finite intersections. That is, if there is a subset $\mathcal{S}$ of $\mathfrak{M}$ with $\bigcup \mathcal{S} \notin \mathfrak{M}$, or if there is a finite subset $\mathcal{F}$ of $\mathfrak{M}$ with $\bigcap \mathcal{F} \notin \mathfrak{M}$, then there could exist no metric on M such that $\mathfrak{M}$ is the class of open sets.

There are other properties, as we shall see, which the class $\mathfrak{M}$ must have in order that it be the class of open sets under some metric, but we begin here by posing the following question: how much of the theory of metric spaces developed in Chap. III can be discussed, and how much is true, if we are given that $\mathfrak{M}$ is closed under unions and finite intersections, even if we are not sure there is any metric at all which gives rise to the family $\mathfrak{M}$?

THE DEFINITION

Let T be a set; a topology τ on T is a collection τ of subsets of T which satisfies certain axioms. A family τ of subsets of T is a **topology on** (or **for**) T if

- **i** if $\mathcal{S} \subset \tau$, then $\bigcup \mathcal{S} \in \tau$,
- **ii** If $\mathcal{S} \subset \tau$ and $\mathcal{S}$ is finite, then $\bigcap \mathcal{S} \in \tau$,
- **iii** $\varnothing \in \tau$ and $T \in \tau$.

Then (T,τ) is a **topological space,** or just a **space,** T is its **underlying set,** and the members of τ are called the **open sets** of (T,τ) or the **τ-open sets** of T. A subset $N \subset T$ is a **neighborhood,** or **τ-neighborhood,** of a point $t \in T$ if there is a member S of τ with $t \in S \subset N$. Can you see that a set S is τ-open iff it is a τ-neighborhood of each of its points?

An example is, of course, given by the collection μ_n of all the subsets of real n-space $\mathbf{R}^n$ which are usually called "open." Exercise III.N says that μ_n is indeed a topology. It is termed the **usual** (or "euclidean") **topology** for $\mathbf{R}^n$. Perhaps it should be pointed out right here that, in the definition of a topology, property iii is completely unneeded. Since $\varnothing$ is a *finite* family of members of any class τ of subsets of a set T, when properties i and ii are satisfied it is necessary that $\bigcup \varnothing = \varnothing$ and $\bigcap \varnothing = T$ be members of τ. Nonetheless, it is a good idea to check these special cases separately; one's mind sometimes plays the trick of ignoring the void case. With this danger understood, the definition of a topology can be rephrased as "a family of subsets closed under unions and finite intersections."

In particular, the usual open subsets of the real line **R** form a topology μ for **R**. [Note here that $\varnothing$ is a μ-open set of **R**; it is the open interval (0,0), if you like.] But there are other topologies possible on **R**. One is the discrete topology δ which consists of every subset of **R**; clearly, δ is closed under unions and finite intersections. This topology is exactly the class of open subsets if **R** is given the peculiar metric mentioned in Chap. III, $\mathfrak{d}(x,x) = 0$, and if $x \neq y$, then $\mathfrak{d}(x,y) = 1$. To check this, observe that the ball $\mathfrak{b}(1,x) = \{x\}$ for each $x \in R$; hence $\{x\}$ is an open subset under this metric, and so is every union of singleton sets, that is, every subset of **R**.

Every topology for **R** is contained in δ (that is, is a subfamily of δ); this is expressed by saying that δ is the largest topology for **R**. The other extreme, $\varepsilon = \{\varnothing, \mathbf{R}\}$, is the smallest topology for **R**. Yet a fourth topology φ for **R** is the family of all subsets S of **R** for which either the complement S′ is finite or else $S = \varnothing$; check φ against the axioms to be sure it is a topology. Neither ε nor φ could be the class of open sets under some metric for **R**; in the case of ε we need merely note that $0 \neq 1$, so the distance d from 0 to 1 is positive in any metric. Hence the open subsets of **R** under a metric always include an open set $\mathfrak{b}(d,0)$ which contains 0 and does not contain 1, whereas the only open set of ε which contains 0 is **R**. A topology which is just the family of open sets for some metric is called **metrizable;** thus μ and δ are metrizable, but ε is not; it is an exercise to show that φ is not metrizable.

Evidently, $\varepsilon \subset \varphi \subset \mu \subset \delta$. If σ and τ are two topologies with $\sigma \subset \tau$, we shall say that σ is the **smaller topology** and τ the **larger.** (Other nomenclatures are "coarser-finer" and "weaker-stronger." Be warned that there is some confusion in their use; we shall stick to "smaller" and "larger.") However, such a linear ordering is not always possible. If $x \in \mathbf{R}$, let $\nu_x = \{S \subset R\colon x \in S \in \mu \text{ or else } S = \varnothing\}$; this is yet another topology for **R**, and $\varepsilon \subset \nu_x \subset \mu$, but φ neither contains nor is contained in ν_x (why?). This gives us five different topologies for the underlying set **R**—or, really, an infinite number, since $\nu_x \neq \nu_y$ if $x \neq y$.

For each set T the collection δ of all subsets of T is a topology for T, and it contains each other topology on T; this largest topology δ is called the **discrete topology** for T. Also, the collection $\varepsilon = \{\varnothing, T\}$ is always a smallest topology for T; it is called the **indiscrete topology** on T. It is an amusing combinatorial exercise to count the total number of topologies possible on a given finite set T (see Exercise C).

Exercises A, B, and C

METRIZABLE SPACES AND CONTINUITY

We have pointed out that when M is a metric space the family of all those subsets of M which are open in the metric (that is, contain an ε-ball about each of their points) is a topology μ, called the **metric topology,** for M. It may happen that two different metrics on M define the same topology μ for M; in that case the two metrics are called equivalent. An example of this was given in Chap. III, where two different bounded metrics were described for **R**, each of which defined the usual metric topology for **R.**

Though it was defined in terms of the metrics, the continuity of a function between metric spaces can be described solely in terms of the metric topologies of its domain and range spaces. We now use this theorem to suggest our *definition* of the continuity of a function from one *space* to another: a **continuous function** f from a topological space (S,σ) into a space (T,τ) is a function f: $S \to T$ such that $f^{-1}(U)$ is an σ-open set whenever U is τ-open. We now adopt the word **map** as a synonym for "continuous function." That is, f is a map iff $U \in \tau$ implies $f^{-1}(U) \in \sigma$. By design, if σ and τ are metric topologies, this new and broader notion of continuity agrees with the metric definition. Recall that our definition of the product metric merely broadened the pythagorean theorem. Here again we have been led to a definition in a general setting from a theorem in a particular case.

Each choice of topologies σ and τ for the fixed sets S and T selects a subset of the set of functions from S to T, the continuous ones, as those which are, in some sense, "nicely behaved." In general, as the domain topology σ "gets larger" (is replaced by a larger one), more and more functions on S become continuous, and as τ gets smaller, more and more functions into T become continuous. It is easy to see that if σ is the discrete topology for S, then every function on (S,σ) is a map, since inverse images of sets must always be open. Conversely, if the range topology τ for T is chosen to be the indiscrete one, $\tau = \{T, \varnothing\}$, then every function into (T,τ) is continuous, since $f^{-1}(T) = S \in \sigma$ and $f^{-1}(\varnothing) = \varnothing \in \sigma$, no matter which topology σ is chosen for S. Perhaps it is worth noting here that, for every choice of σ and τ, each constant function from S to T is continuous, since $f^{-1}(U)$ is either $\varnothing$ or S, depending on whether or not the only value of f lies in U.

Exercise D

CLOSURE, INTERIOR, AND BOUNDARY

Throughout this section let (T,τ) be a fixed but arbitrary topological space. Since there is only one topology on T under consideration, we shall speak simply of the space T [instead of (T,τ)], and "open" will mean "τ-open," etc. In fact, when confusion will not result, we shall henceforth use the name of the underlying set for the space as well.

In real 3-space, the closure of a subset A is A together with all its skin, whether that skin is part of A or not. Alternatively, the closure is the set of points arbitrarily close to the points of A. The interior of A is the set of all points of A for which A is a neighborhood, that is, A minus any part of its skin which it contains. The boundary of A is its skin, the set of points arbitrarily close to both A and its complement A'. Each of these ideas may readily be extended to topological spaces in general; we have only to phrase their definitions in the topological language of open sets.

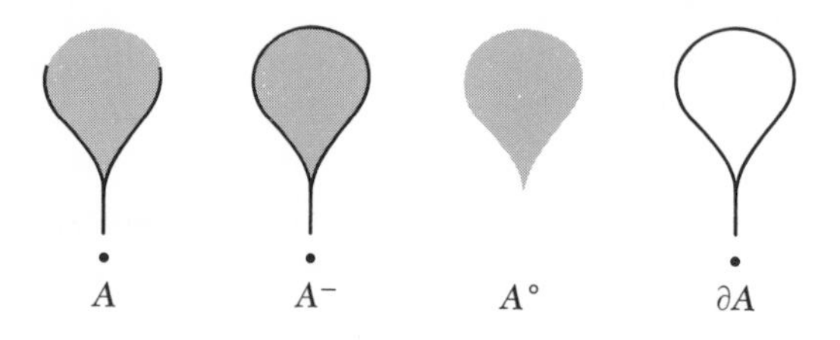

The complement of an open set is called **closed.** The **closure** A^- of a subset A of T is the intersection of the family of all closed sets which contain A. The **interior** A° of a set A is the union of the family of all open sets which are contained in A. The **boundary** of A is $\partial A = A^- \cap (A')^-$.

1 THEOREM

i *Finite unions and arbitrary intersections of (families of) closed sets are again closed.*

ii *For each set A, A^- is closed and contains A and $(A^-)^- = A^-$. If A is closed, then $A^- = A$.*

iii *For each set A, A° is open and is contained in A and $(A^\circ)^\circ = A^\circ$. If A is open, then $A^\circ = A$.*

iv *If $A \subset B$, then $A^\circ \subset B^\circ$ and $A^- \subset B^-$.*

v *For each set A, ∂A is closed and $\partial A = A^- - A^\circ = \partial(A')$. Further, $A^- = A^\circ \cup \partial A$.*

vi *A set A is open iff $A \cap \partial A = \varnothing$ and closed iff $A \supset \partial A$.*

vii *$A^\circ = [(A')^-]'$ and $A^- = [(A')^\circ]'$ for all sets A.* ■

The proofs of these statements are short and straightforward, and an understanding of them is essential. The proofs are asked for formally in Exercise E, but a wise reader will construct them at once.

In the examples of the metric spaces $\mathbf{R}^n$, the above definitions specialize to the usual notions of the closure, boundary, and interior of a set. In any space an intuitive feeling for the boundary of a set is given by the contention that a point of the space is in the boundary of a set iff every neighborhood of the point intersects both the set and its complement. This is true, since the point could then be in the interior of neither the set nor its complement. Similarly, a point is in the closure of the set iff every neighborhood of the point intersects the set. A point is in the interior of the set iff the set itself is a neighborhood of the point. The closure of a set can also be characterized as the minimal (or smallest) closed set containing the given set; this makes sense, since intersections of closed sets are closed. Correspondingly, the interior of a set is just the maximal (or largest) open set contained in the given set.

The continuity of a function, a property defined in terms of open sets, can be described by the behavior of the function with respect to closed sets and by the closure operation as well.

2 THEOREM *The following statements about a function f from a topological space (S,σ) to another space (T,τ) are equivalent:*

i *f is continuous,*
ii *$f^{-1}(B)$ is σ-closed for each τ-closed set B,*
iii *$f(A^-) \subset f(A)^-$ for every subset A of S,*
iv *$f^{-1}(B)^- \subset f^{-1}(B^-)$ for every subset B of T.*

Proof The proof proceeds by a circular sort of logic; with "$\Longrightarrow$" used to abbreviate the word "implies," our plan of attack is

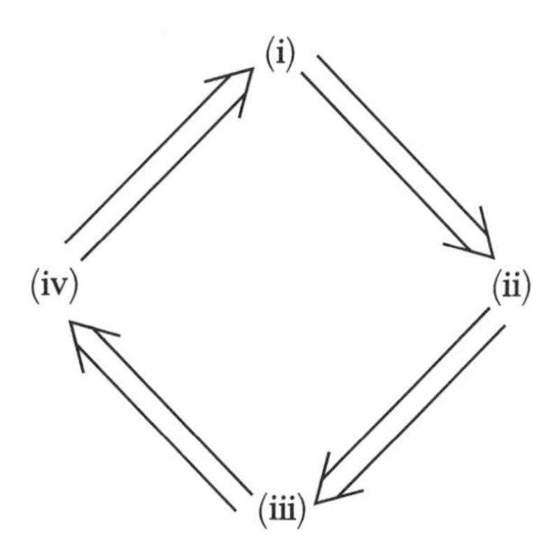

Statement i implies statement ii: If B is τ-closed, then B' is τ-open; therefore $f^{-1}(B')$ is σ-open, since f is continuous. But $f^{-1}(B') = [f^{-1}(B)]'$, so $f^{-1}(B)$ is closed.

Statement ii implies statement iii: If $A \subset S$, then $f^{-1}[f(A)^-]$ is closed (by ii) and contains A, and so $A^- \subset f^{-1}[f(A)^-]$. Thus $f(A^-) \subset f(f^{-1}[f(A)^-]) \subset f(A)^-$. The latter set inclusion is a particular case of a more general statement: if C is a subset of the range of an arbitrary function f, then $f[f^{-1}(C)] \subset C$.

Statement iii implies statement iv: Let $A = f^{-1}(B)$; then (by iii) $f[f^{-1}(B)^-] \subset (f[f^{-1}(B)])^- \subset B^-$. But for all subsets C of the domain of f, $f^{-1}[f(C)] \supset C$, so $f^{-1}(B)^- \subset f^{-1}(f[f^{-1}(B)^-]) \subset f^{-1}(B^-)$.

Statement iv implies statement i: If C is open in T, then C' is closed, and (by iv) $f^{-1}(C') \supset f^{-1}(C')^- \supset f^{-1}(C')$, so $f^{-1}(C')$ is closed. But $f^{-1}(C') = [f^{-1}(C)]'$, so $f^{-1}(C)$ is open. ■

Exercises E and F

SUBSPACES

If A is a subset of T and τ is a topology on T, then $\{A \cap S: S \in \tau\} = \alpha$ is a topology on A. The proof that α is closed under unions is an application of the distributive law (Theorem I.2.i); if $\{A_\lambda: \lambda \in L\}$ is a subfamily of α, so that for each index $\lambda \in L$, $A_\lambda = A \cap S_\lambda$ is a member of α, then

$$\bigcup\{A_\lambda: \lambda \in L\} = \bigcup\{A \cap S_\lambda: \lambda \in L\} = A \cap \bigcup\{S_\lambda: \lambda \in L\}.$$

The closure of α under finite intersections is similarly proved. The space (A,α) is called a **subspace** of the space (T,τ); we shall often say simply that "A is a subspace of T." The topology α is the **relative topology** on A. It is an exercise to show that if T is a metric space with metric topology τ, the metric topology on a metric subspace A is the same as the relative topology on A. This will show that the definition of subspace of a topological space suitably generalizes the earlier notion of a metric subspace.

As another example, if $A \subset T$ and τ is the discrete topology on T, then the relative topology on A is the discrete topology on A. If the topology τ on T is indiscrete (so that $\tau = \{\varnothing, T\}$), then the relative topology $\{\varnothing, A\}$ is also indiscrete.

Notice that the inclusion $A \subset T$, when regarded as a function $i: A \to T$: $i(a) = a$, is continuous on α to τ. This is obvious, since for each open set $S \in \tau$, $i^{-1}(S) = S \cap A$ is, by definition, a member of α. Furthermore, all the members of α are obtained in just this fashion, so α is the smallest topology for A with which i is continuous. Now suppose $A \subset B \subset T$ and that α and β are the relative topologies on A and B, respectively, induced by a topology τ on T. Since $A \subset B$, there is another relative topology α' for A induced by β; is it true always that $\alpha = \alpha'$? Well, if $i: A \subset T$, $j: A \subset B$, and $k: B \subset T$ are the inclusion functions, then $i = k \circ j$ is continuous from α' to τ, since j is continuous on α' and k is continuous on β. Thus $\alpha' \supset \alpha$ (be sure you agree). But if $U \in \alpha'$, then there exists $V \in \beta$ such that $U = A \cap V$, and there is $W \in \tau$ such that $V = B \cap W$, or

$$U = A \cap (B \cap W) = (A \cap B) \cap W = A \cap W \in \alpha.$$

Hence $\alpha' = \alpha$. This may be rephrased as *every subspace of a subspace of T is a subspace of T.* ■ An example is provided by the discrete relative topology of $\mathbf{Z} \subset \mathbf{R} \subset \mathbf{C}$; another example is $\mathbf{R} \subset \mathbf{R}^2 \subset \mathbf{R}^3$.

Exercises G and H

BASES AND SUBBASES

The process that was used to define the metric topology on a metric space is useful in many other situations as well. Remember that the family of all unions of ε-balls is a topology. This fact motivates our next definition, which is rephrased in the language of arbitrary topological spaces: Let (T,τ) be a space, and let $\beta \subset \tau$; then β is a **base** (or "basis") for τ iff each $S \in \tau$ is the union of a subset of β, that is, iff $S \in \tau$ implies that there exists $\gamma \subset \beta$ such that $\bigcup \gamma = S$. Clearly, if μ is a topology for T and $\beta \subset \mu$, then $\tau \subset \mu$, so τ is the (unique) smallest topology containing β. Not every subfamily β of a topology τ is a base for τ; two sorts of trouble may appear. As an example, let δ be the discrete topology on $S = \{a,b,c\}$ and define $\beta = \{\{a,b\},\{a,c\}\} \subset \delta$. The set of unions of subsets of β does not contain $\{a\}$ and so is not closed under finite intersections; this set of unions is not a topology, and hence β is not a base for δ. On the other hand, the set $\varepsilon = \{\varnothing, S\} \subset \delta$ is itself a topology on S, but it also is not a base for δ.

3 THEOREM *A collection β of sets is a base for some topology τ on $T = \bigcup \beta$ iff for each pair S,T of elements of β, and for each $t \in S \cap T$, there is an element U of β with $t \in U \subset S \cap T$. In that case, $\tau = \{\bigcup \gamma: \gamma \subset \beta\}$.*

Proof Suppose $\beta \subset \tau$ is a base for τ, and let S, $T \in \beta$. Since $\beta \subset \tau$, S and T, and thus $S \cap T$, are members of τ. Hence $S \cap T$ is a union of members of β, $S \cap T = \bigcup\{U: U \in \beta$ and $U \subset S \cap T\}$, and thus each member t of $S \cap T$ lies in some member $U \in \beta$ which is a subset of $S \cap T$.

Conversely, if the intersection of each pair of members of β is a union of members of β, then it is easy to see that $\tau = \{\bigcup\gamma: \gamma \subset \beta\}$ is a topology for $\bigcup\beta$; we check only that τ is closed under pairwise intersections. Let $t \in \bigcup\gamma_1 \cap \bigcup\gamma_2$; each $\gamma_i \subset \beta$, so there is an $S_i \in \beta$ such that $t \in S_i \in \gamma_i$ for $i = 1$ and 2. By assumption, there exists $U \in \beta$ such that $t \in U \subset S_1 \cap S_2$; hence $U \subset \bigcup\gamma_1 \cap \bigcup\gamma_2$; therefore that latter set is a union of members of β. ∎

Theorem 3 states clearly that if σ is an arbitrary family of sets, then the family β of all finite intersections of members of σ is a base for a topology τ on $\bigcup\sigma$. Then σ is called a **subbase** (or "subbasis") for τ. We may claim now that a family σ is a subbase for a topology τ iff $\sigma \subset \tau$ and each member of τ is the union of finite intersections of elements of σ. It is obvious that if μ is any topology and $\sigma \subset \mu$, then $\tau \subset \mu$; τ is the smallest topology containing σ, and it is unique. An example of a subbase for the usual real topology is the family σ of all open intervals of length 1. No member of σ lies inside $(0,1) \cap (\frac{1}{2},\frac{3}{2})$, but every open interval (a,b) of length less than 1 is $(a,b) = (a, a + 1) \cap (b - 1, b)$; the open intervals of length less than 1 form a base for the usual topology. This latter example should be amplified; every base for a topology is also a subbase for that topology.

We have already used bases to construct topologies and will do so frequently. However, for some purposes it is not necessary ever to mention the topology which a base or a subbase "generates." For instance, the metric definition of a continuous function involves ε-balls and δ-balls, that is, members of the base only. Perhaps the most important fact about subbases is the following.

4 THEOREM *Let f be a function from (A,α) to (B,β), and let σ be a subbase for β. Then f is continuous iff for each member $S \in \sigma$, $f^{-1}(S)$ is open in A.*

Proof If f is continuous and $S \in \sigma \subset \beta$, then certainly $f^{-1}(S) \in \alpha$. To show the converse, we note that every member $T \in \beta$ is the union of finite intersections of members of σ. We assume $f^{-1}(S)$ is open for each $S \in \sigma$; if we can prove that f^{-1} preserves all unions and intersections, then we shall have shown $f^{-1}(T)$ to be a union of finite intersections of open sets, and thus open. This proof is consequently completed by the following lemma, which is stated separately, since it applies to every function and holds interest independent of the proposition at hand.

5 LEMMA *If $f: A \to B$ is a function, then f^{-1} preserves differences, intersections, and unions of sets in B. Precisely, if S, $T \subset B$, then $f^{-1}(S - T) = f^{-1}(S) - f^{-1}(T)$, and if $\mathfrak{T}$ is a family of subsets of B, then $f^{-1}(\bigcap \mathfrak{T}) = \bigcap\{f^{-1}(T): T \in \mathfrak{T}\}$ and $f^{-1}(\bigcup \mathfrak{T}) = \bigcup\{f^{-1}(T): T \in \mathfrak{T}\}$.*

Proof First assume $x \in f^{-1}(S - T)$, so that $f(x) \in S - T$, $f(x) \in S$, and $f(x) \notin T$. Clearly, $x \in f^{-1}(S)$ and $x \notin f^{-1}(T)$; hence $x \in f^{-1}(S) - f^{-1}(T)$. This shows that $f^{-1}(S - T) \subset f^{-1}(S) - f^{-1}(T)$. Conversely, if $x \in f^{-1}(S) - f^{-1}(T)$, then $f(x) \in S$; and if $f(x) \in T$, then $x \in f^{-1}(T)$, a contradiction. Therefore $f^{-1}(S - T) = f^{-1}(S) - f^{-1}(T)$.

To show that $f^{-1}(\bigcap \mathfrak{T}) \subset \bigcap\{f^{-1}(T): T \in \mathfrak{T}\}$, let $x \in f^{-1}(\bigcap \mathfrak{T})$, so that $f(x) \in \bigcap \mathfrak{T}$; that is, $f(x) \in T$ for each $T \in \mathfrak{T}$. Thus $x \in f^{-1}(T)$ for each $T \in \mathfrak{T}$, and $x \in \bigcap\{f^{-1}(T): T \in \mathfrak{T}\}$. To prove the other inclusion, assume $x \in \bigcap\{f^{-1}(T): T \in \mathfrak{T}\}$, so that $T \in \mathfrak{T}$ implies $x \in f^{-1}(T)$ or $f(x) \in T$. This says $f(x) \in \bigcap \mathfrak{T}$, so $x \in f^{-1}(\bigcap \mathfrak{T})$.

The proof that f^{-1} preserves unions is similar; it is left for an exercise. ∎

Note that the above result does *not* suggest that the domain topology α may be replaced by a subbase(or even base) τ so that continuity of f may be verified merely by checking inverse images $f^{-1}(S)$ for membership in τ, no matter which class (of subsets of B) S is allowed to range over. Inspection of a random example will probably convince you of this.

Exercise L

PRODUCT SPACES

If (A,α) and (B,β) are spaces, then there is a natural way of defining a topology on $A \times B$ in terms of α and β. Specifically, the **product space** $(A \times B, \alpha \times \beta)$ is the cartesian-product set $A \times B$ furnished with the **product topology** $\alpha \times \beta$ which has for a base the family $\{S \times T: S \in \alpha$ and $T \in \beta\}$. By way of example, we investigate the case where $A = B = \mathbf{R}$, with α and β both the usual topology μ. A subset N of the plane $\mathbf{R} \times \mathbf{R} = \mathbf{R}^2$ is open in the product topology $\mu \times \mu$ iff for each point $(x,y) \in N$ there is a pair of open sets S,T in μ with $x \in S$, $y \in T$, and $S \times T \subset N$. If, for instance, $N = \mathfrak{b}[\varepsilon_1,(x,y)]$, the ε_1-disc at (x,y), then S and T may be taken to be the open intervals of length $\sqrt{2}\varepsilon_1$ centered at x and y, respectively. Since each set M which is open in the metric topology for $\mathbf{R}^2$ contains some ε-disc about each of its points, each such set M must be in $\mu \times \mu$. On the other

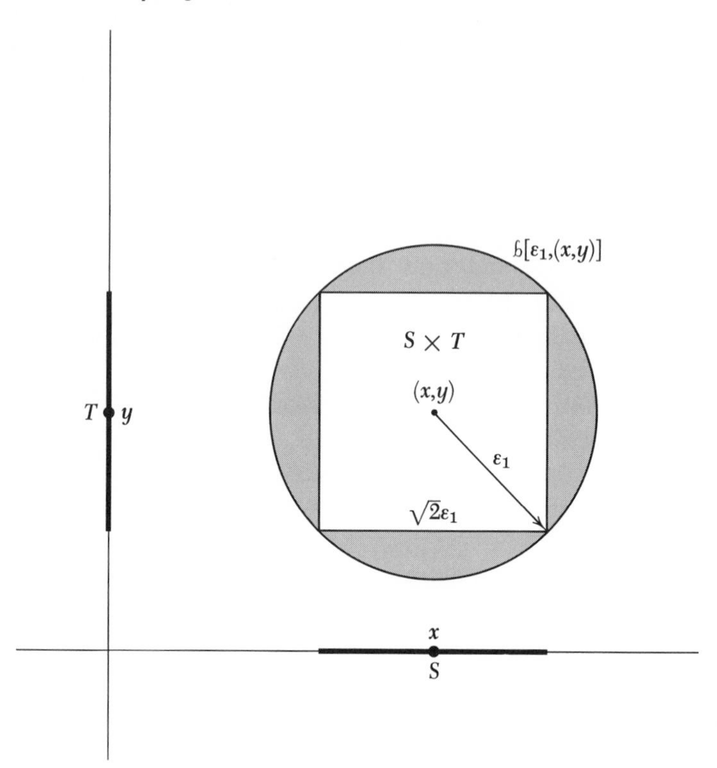

hand, each base element $S \times T$ of $\mu \times \mu$ must contain, for each of its points, a rectangular base element (that is, the product $I_1 \times I_2$ of two open intervals) centered at that point, and of course that rectangle contains a concentric ε-disc. Thus each such $S \times T$ is open in the usual topology of the product metric; hence $\mu \times \mu$ equals the metric topology.

It is comforting to know that the product topology is exactly the usual metric topology on the most familiar of cartesian products. Another route to an understanding of product spaces goes by way of the two projections $p: A \times B \to A$ and $q: A \times B \to B$, where $p(a,b) = a$ and $q(a,b) = b$. If $S \in \alpha$, then $p^{-1}(S) = S \times B$ is a base element in $\alpha \times \beta$, and thus p is continuous. Similarly, $T \in \beta$ implies $q^{-1}(T) = A \times T$ is in $\alpha \times \beta$; thus q also is continuous. Furthermore, $p^{-1}(S) \cap q^{-1}(T) = (S \times B) \cap (A \times T) = S \times T$. Hence, if γ is a topology on $A \times B$ for which p and q are continuous, then the base $\{S \times T: S \in \alpha \text{ and } T \in \beta\} \subset \gamma$, and so $\alpha \times \beta \subset \gamma$. We may now characterize the product topology as the smallest topology on the product space for which the projections are continuous.

Exercises J and K

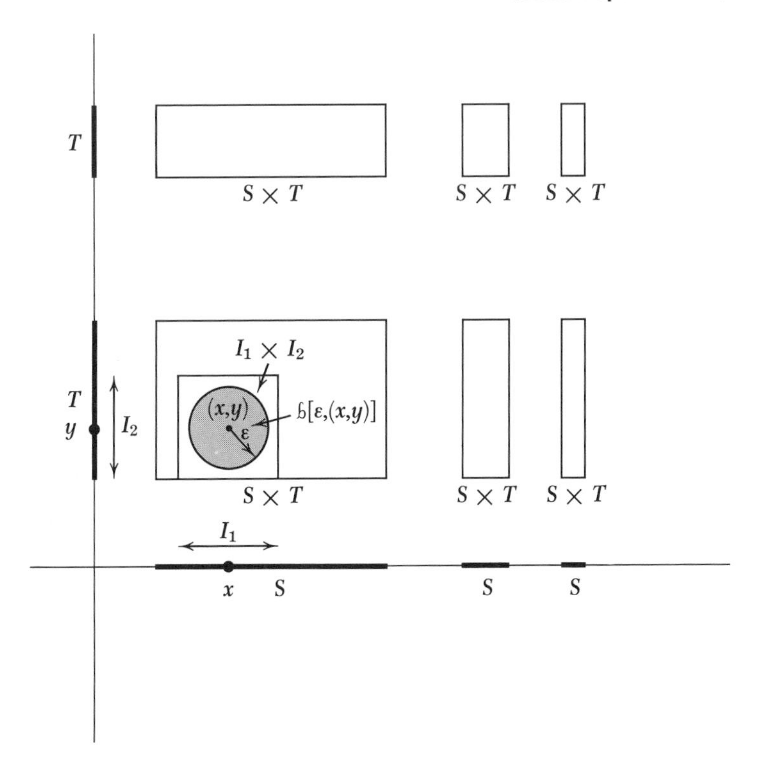

QUOTIENT SPACES

Recall from Chap. I that if R is an equivalence relation on a set S, then the family $\mathfrak{T}$ of equivalence classes under R is a partition of S. The quotient function q: $S \to \mathfrak{T}$ assigns to each $x \in S$ the class $[x] = q(x)$ of all elements of S which are R-related to x. Now, if S has a topology σ on it, we would like to construct a topology τ for the quotient set $\mathfrak{T}$ such that q is continuous from σ to τ. That much is easy, of course; the indiscrete topology on $\mathfrak{T}$ is guaranteed to make q continuous, regardless of which topology σ is chosen for S. But for a given σ there is a unique largest topology to which q is continuous; the **quotient topology** for $\mathfrak{T}$ is defined to be

$$\tau = \{\mathfrak{U} \subset \mathfrak{T}: q^{-1}(\mathfrak{U}) \in \sigma\}.$$

Lemma 5 assures us that τ is indeed a topology for $\mathfrak{T}$; the inverse image of a union (or intersection) of τ-members is just the union (or intersection) of their individual inverse images, since q is a function. Since σ is closed under unions and finite intersections, so is τ. Clearly, q is continuous to τ, since the τ-open sets have been chosen so that their inverse images are open. Equally clearly, τ is the largest topology to which q is continuous, for if a topology τ' for $\mathfrak{T}$ contains a subset $\mathcal{V}$ of $\mathfrak{T}$ with $\mathcal{V} \notin \tau$, then $q^{-1}(\mathcal{V})$ is not open in S and thus q is not continuous to τ'.

An example is in order here. Let S be the unit interval $\mathbf{I} = [0,1]$, and define the relation R on $\mathbf{I}$ by $R = \{(x,x): x \in \mathbf{I}\} \cup \{(0,1),(1,0)\}$, that is, xRx for all x, and $0R1$:

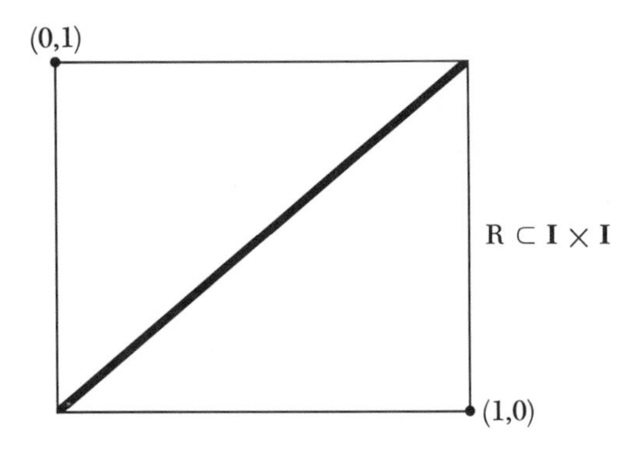

The equivalence class for $x \in (0,1)$ is $[x] = \{x\}$, a singleton. The other class is $[0] = [1] = \{0,1\}$. The quotient set $\mathfrak{T}$ may be visualized as an interval which has been bent so that its two end points are glued together to become a single point of $\mathfrak{T}$; that is, a circle:

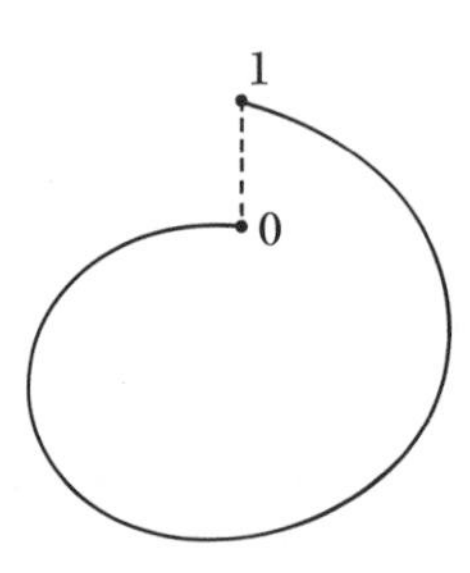

The quotient topology τ for $\mathfrak{T}$ contains a subset $\mathcal{U}$ of $\mathfrak{T}$ iff $q^{-1}(\mathcal{U})$ is open in $\mathbf{I}$. Now

$$q^{-1}(\mathcal{U}) = \mathbf{U}\mathcal{U} = \{x: [x] \in \mathcal{U}\} = \{x: q(x) \in \mathcal{U}\},$$

and it is not hard to decide that in case $\{0,1\} \notin \mathcal{U}$, $\mathcal{U} \in \tau$ iff $\mathcal{U} = q(U)$ for some open subset U of $(0,1)$. However, if $\{0,1\} \in \mathcal{U}$, then $\mathbf{U}\mathcal{U}$ can be open only if it contains a set $[0,\varepsilon) \cup (1 - \varepsilon, 1]$ for some $\varepsilon > 0$, that is to say, only if $\bigcup\mathcal{U}$ is a neighborhood in **I** of both 0 and 1. [To be open, $\bigcup\mathcal{U}$ must also contain a neighborhood in (0,1) of each of its other points, as well.] All this should convince you that $(\mathcal{T},\tau)$ is very much like S^1, the unit circle in the plane.

The general operation we have performed, the manufacture of a topology on a quotient set, has fundamental importance and frequent use in topology. The space $(\mathcal{T},\tau)$ is said to be a **quotient** ("identification," "factor," or "decomposition") **space** of (S,σ), and τ may be called the "identification topology." In our example the quotient map q "collapsed (or identified) the subset $\{0,1\}$ of **I** to a point." Some light was shed on the corresponding operation in the cases of sets (Chap. I) and groups (Chap. II) by a factorization of functions. This is possible in the topological case as well.

6 THE QUOTIENT THEOREM FOR SPACES *If $f\colon S \to T$ is a map, there exist unique maps q, r, and i such that $f = i \circ r \circ q$, where q is a quotient, r is a* 1-1 *correspondence, and i is an inclusion:*

$$\begin{array}{ccc} S & \xrightarrow{f} & T \\ {\scriptstyle q}\downarrow & & \uparrow{\scriptstyle i} \\ \mathcal{T} & \xrightarrow{r} & Im\,(f) \end{array}$$

Proof The diagram is familiar by now. The relation R on S has the property that xRy iff $f(x) = f(y)$. Our functions q, r, and i are defined by their values: $q(x) = [x]$, $r([x]) = f(x) \in Im\ (f)$, and $i[f(x)] = f(x) \in T$. These functions are known (from Chap. I) to be unique; hence we need only decide that they are continuous. But i is the inclusion of a subspace, and so is continuous. And q is a quotient map, and so is continuous. We need to show that if U is open in $Im\,(f)$, then $r^{-1}(U)$ is open in $\mathcal{T}$; that is, that r is continuous. But an open set U of $Im\,(f)$ must be the intersection $U = Im\,(f) \cap V$ of that subspace with an open set V of T. Furthermore, $f^{-1}(U) = f^{-1}(V)$, since $x \in f^{-1}(V)$ implies $f(x) \in U$; hence $f^{-1}(U)$ is open in S. We assert that $q^{-1}[r^{-1}(U)] = f^{-1}(U)$ [which makes $r^{-1}(U)$ open in the quotient topology for $\mathcal{T}$]; clearly, for all $x \in S$, $f(x) = r[q(x)]$, so $f(x) \in U$ iff $q(x) \in r^{-1}(U)$. ■
The quotient space $\mathcal{T}$ will sometimes be denoted S/f.

As an example of the Quotient Theorem, consider the map $f\colon \mathbf{I} \to S^1$ which wraps the interval once around the unit circle in the plane, $f(x) = (\cos 2\pi x, \sin 2\pi x)$. The functions *cos* and *sin* are known from the calculus to be continuous functions; if you assume that, it is not difficult to prove that f is continuous onto the circle (see Exercise Q for an easy proof). The factorization $f = i \circ r \circ q$ yields the quotient q of **I** which we considered earlier; since f is onto, i is the identity map on S^1, and r is a continuous 1-1 correspondence of the quotient space with the circle.

Exercise S

HOMEOMORPHISMS

Just as in the metric case, a map of spaces having a two-sided inverse which is also a map is called a **homeomorphism.** In the above example, the 1-1 correspondence r is a homeomorphism; this statement is implied by the fact that $r(\mathfrak{U})$ is open in S^1 for each open set $\mathfrak{U}$ of the quotient space $\mathbf{I}/f$ (why?), since $(r^{-1})^{-1}(\mathfrak{U}) = r(\mathfrak{U})$. If $f: S \to T$ is a homeomorphism, then S is **homeomorphic** to T, and S and T are **homeomorphs** of one another. This is an equivalence relation on the class of all topological spaces, because identity maps, inverses of homeomorphisms, and composites of homeomorphisms are always homeomorphisms. Since a homeomorphism f from S to T sets up a 1-1 correspondence between the topologies on S and T, the study of topology may be fairly said to be the study of these equivalence classes, rather than the study of individual spaces. We shall return to this matter later.

A map f: $S \to T$ is defined to be **open** (or "interior") iff the image $f(U)$ of each open set U of S is open in T. Similarly, a **closed map** is one which carries each closed set to a closed set. Clearly, a homeomorphism is just an open (or closed) map which is a 1-1 correspondence. Composites of open (or closed) maps are again open (or closed) maps. The function f which wraps the interval around the circle is not open; $[0,½)$ is open in $\mathbf{I}$, but $f[0,½)$ is not a neighborhood of $f(0) = (1,0)$ in the circle. Since the factors r and i assigned to f by the Quotient Theorem are (in this particular case) both open maps, the quotient q defined by f must fail to be open, since $f = i \circ r \circ q$ cannot be the composite of open functions. The existence of a map (in fact, a quotient map) which is not closed is the subject of an exercise. Nevertheless, we can add the following to the Quotient Theorem.

7 COROLLARY *If f is open (or closed), then each of the factors q, r, and i is open (or closed). Hence r is a homeomorphism if f is either open or closed.*

Proof If U is open in S, then each of the sets $f(U)$, $i^{-1}[f(U)]$ and $r^{-1}(i^{-1}[f(U)]) = q(U)$ is open, so q is open. Similarly, if $\mathfrak{U}$ is open in S/f, then $q^{-1}(\mathfrak{U})$, $f[q^{-1}(\mathfrak{U})]$, and in turn, $i^{-1}(f[q^{-1}(\mathfrak{U})]) = r(\mathfrak{U})$ are all open, so r is open. To vary the argument, i is open since $Im\,(f) = f(S)$ is an open subset of T.

The proof of the assertion for a closed map f can be had by replacing each instance of the word "open" in the proof above by the word "closed." And a 1-1 correspondence r is open iff it is closed. ■ The topological analog of an isomorphism of sets or groups is evidently a homeomorphism. However, the Quotient Theorem for Spaces is not strictly analogous to those similar theorems in the cases of sets and of groups, since r is not always a homeomorphism (see Chap. V). The corollary partially mends this defective analogy.

Whenever r is a homeomorphism we shall say that $Im\,(f)$ has the **quotient topology.**

Now consider our example, the map f which wraps $\mathbf{I}$ around S^1. If V is closed in $\mathbf{I}$, then V is compact, as is $f(V)$, which is therefore a closed (and bounded) subset of the plane. But then $f(V) = f(V) \cap S^1$ is a closed subset of S^1. Hence f is closed, though we have shown it not to be open, and the quotient space $\mathbf{I}/f$ is homeomorphic to the circle S^1. We shall express this fact by saying that "$\mathbf{I}/f$ is a circle."

We now use the notion of direct product to manufacture a further example from the above. The map $f\colon \mathbf{I} \to S^1$ defines a function $f \times f\colon \mathbf{I} \times \mathbf{I} \to S^1 \times S^1$ by $(f \times f)(x,y) = [f(x),f(y)]$; we may use the diagram of continuous functions,

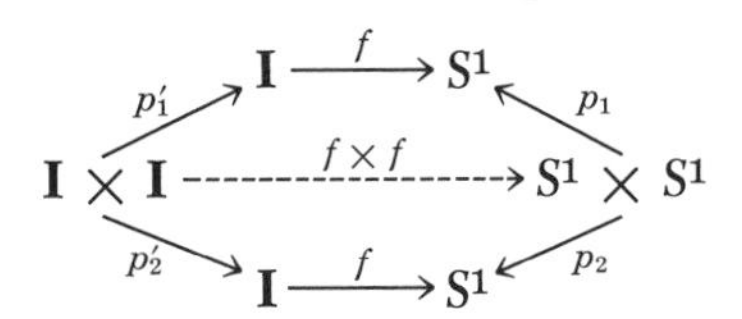

to argue that $f \times f$ is continuous (see Exercise Q). Geometrically, $\mathbf{I} \times \mathbf{I}$ may be thought of as its homeomorph, the square $\{(x,y)\colon 0 \leq x \leq 1$ and $0 \leq y \leq 1\}$ in the plane; similarly, the torus $S^1 \times S^1$ is homeomorphic to

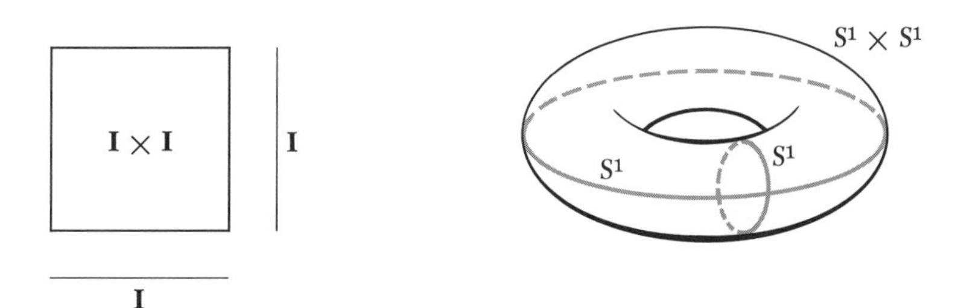

the idealized surface of an innertube in 3-space (it is a "circle of circles"). Consequently, we may think of the map $f \times f$ as having as domain the square in the plane and as range the torus in 3-space. It is not hard to show that $f \times f$ is a closed map in this case. Therefore the torus has the quotient topology; it may be regarded as a quotient of the square. The identification occurring in this quotient may be described geometrically as: sew the left and right edges of the square together (to get a cylinder), and then sew the top and bottom edges together.

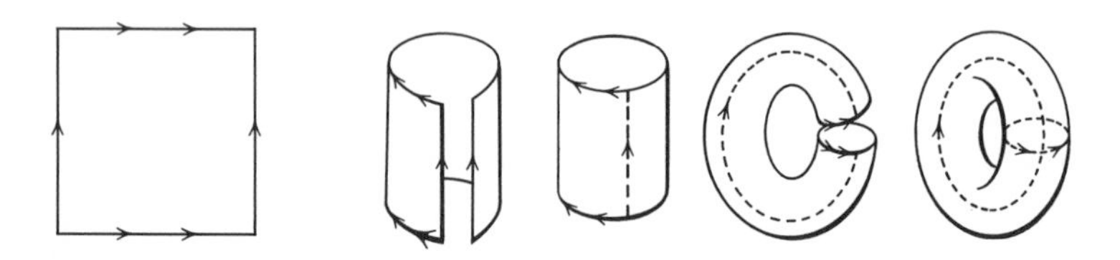

The torus and the 2-sphere have in common the property that sufficiently small patches of each look just like a disc in the plane; physically, a disc-shaped tire patch could be applied to repair any puncture in either. A space with this property is called a 2-manifold. Precisely, an **n-manifold** (or just **manifold**) M is a topological space M which, about each of its points m, contains an open set N, $m \in N$, that is homeomorphic to euclidean n-space (the same space $\mathbf{R}^n$ must work for every $m \in M$). Obviously $\mathbf{R}^n$ itself is an n-manifold (with $N = \mathbf{R}^n$ for every m). The circle S^1 is a 1-manifold. Furthermore, any space X homeomorphic to an n-manifold must also be an n-manifold. We shall not study manifolds as such, but they constitute a large class of interesting topological spaces which will frequently provide us with examples.

Exercises Q and U

FACTORING THROUGH QUOTIENTS

Let $p\colon \mathbf{I} \to X$ be a path in a space X with $p(0) = p(1)$; that is, p is a "closed path," which begins and ends at the same place, a "loop" in X. Then p is constant on each equivalence class of the quotient map of the interval $\mathbf{I}$ into the circle S^1 which identifies the end points of $\mathbf{I}$, and there obviously exists a unique *function* $g\colon S^1 \to X$ such that $g \circ q = p$;

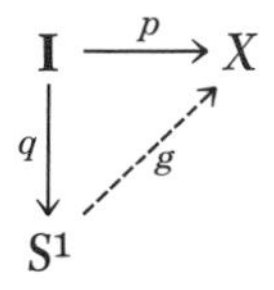

It is natural to hope that g is continuous, that a closed path in X corresponds uniquely to a map of S^1 into X. It does. In order to see this, we rephrase the problem in the greatest possible generality.

Suppose q: S $\to$ S$/R$ is the quotient *function* on a *set* S modulo a relation R on S and f: S $\to T$ is a function. Can f be factored through S$/R$? That is, is there a function g: S$/R \to T$ such that $f = g \circ q$? Clearly, if there exist members x and y of S with xRy, so that $q(x) = q(y)$, and yet $f(x) \neq f(y)$, then $g \circ q(x) = g \circ q(y)$ for every such function g. This is contradictory, and we conclude that no g exists with $g \circ q = f$. On the other hand, if for every x, $y \in$ S, xRy implies $f(x) = f(y)$ (that is, if the relation induced by f on S contains R), then g may be uniquely defined: if $[x]$ is the R-class of x, then $g([x]) = f(x)$ (compare Prob. I.CC). If f is a *morphism* of the *group* S into

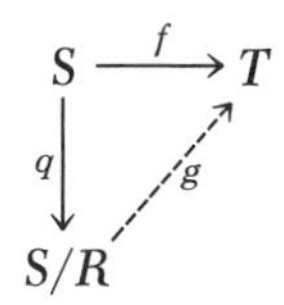

the *group* T, it is easy to see that this requirement reduces to the demand that $Ker\,(f) \supset Ker\,(q)$ (compare Exercise II.Z). How about the topological case, where f: S $\to T$ is a *map* from the *space* S to the *space* T? This question is settled by a proof of the fact that *a function* g: S$/R \to T$ *on a quotient space is continuous if and only if* $g \circ q$: S $\to T$ *is continuous.* It is clear that $g \circ q$ is continuous when g is. In the other direction, assume that $g \circ q$ is continuous, and let U be an open subset of T. The subset $g^{-1}(U)$ of S$/R$ is necessarily open in the quotient topology, since $q^{-1}[g^{-1}(U)] = (g \circ q)^{-1}(U)$ is open in S. ∎

Now it is evident that a map f: S $\to T$ has a factoring $f = g \circ q$ through S$/R$, where the factor g is a *map*, iff f can be factored through S$/R$ as a *function*, that is, iff the relation induced by f contains R. This result means, for example, that there is a 1-1 correspondence between the set of all maps of S^1 into a space X and the set of those maps of **I** into X which begin and end at the same point, the set of closed paths in X. Can you phrase a similar correspondence between the set of maps from a torus into X and a certain subset of the set of maps of $\mathbf{I}^2$ into X?

Exercises W and Z

REFERENCES AND FURTHER TOPICS

Although this chapter began by proposing a discussion of "limits and continuity," it delivered no comments on the definition of "limit" in a topological space (Prob. III.AA examines sequences in metric spaces). The definition is easy enough: a sequence $s_1, s_2, s_3, \ldots$ of points of a space T converges to a limit s, $\lim_{n\to\infty} s_n = s$, iff for each neighborhood U of s there is an integer N such that $s_n \in U$ if $n \geq N$. Among metric spaces sequences suffice to determine the topologies and the classes of continuous functions. For example, a subset S of a metric space M is open iff for each point s of S and each sequence $s_1, s_2, s_3, \ldots$ with limit s there exists an integer N such that $s_n \in S$ if $n \geq N$. However, this statement is not necessarily true for a nonmetrizable space M. This problem is solved by use of "nets," which are a generalization of sequences; every topology can be completely described in terms of its convergent nets. A good account of this is given in

J. Dugundji, *Topology*, chap. X (Boston: Allyn and Bacon, 1966).

J. L. Kelley, *General Topology*, chap. 2 (Princeton, N. J.: Van Nostrand, 1955).

Another question we have not answered is, "Which topological spaces are metrizable?" While we have exhibited nonmetrizable spaces, and we shall later observe conditions on spaces which are necessary for metrizability, for a full answer we suggest chap. 4 of Kelley, cited above. A partial answer, which will satisfy our needs, is that subspaces of $\mathbf{R}^n$ are metrizable, and that each manifold we shall see (and in fact every manifold) is homeomorphic to a subset of some real n-space.

The objects we defined to be manifolds are often called "topological manifolds." It is possible to build into a manifold a differential structure; the study of such structures and their interrelations constitutes the subjects of differential geometry and differential topology. Good introductions to these matters may be found in

Auslander and MacKenzie, *Introduction to Differentiable Manifolds* (New York: McGraw-Hill, 1963).

J. R. Munkres, *Elementary Differential Topology*, rev. ed. (Princeton, N.J.: Princeton University Press, 1966).

The material presented in this chapter is quite standard; parallel discussions may be found in a number of recently published textbooks possessing such titles as "Introduction to (or Elementary or Elements of) Point Set (or General) Topology." Find the discussion you want by looking up key words (for example, "base") in the index.

One excellent text, on a somewhat higher level than our discussion, is

N. Bourbaki, *General Topology,* parts I and II (Reading, Mass.: Addison-Wesley, 1967).

This is a translation of N. Bourbaki, *Topologie Générale* (Paris: Hermann, 1953); if you do not yet read math in French, here is an excellent time and place to begin. Try it; using the translation as a pony, you will find it possible even if you have never studied that language.

EXERCISES

A Let $\varphi = \{S \subset \mathbf{R}: S = \varnothing$ or S' is finite$\}$; show in detail that φ is a topology for $\mathbf{R}$, and also that it is not metrizable.

B Show that if $\mathfrak{U}$ is a family of sets, each of which is a union of ε-balls in a metric space M, then $\bigcup \mathfrak{U}$ is itself a union of ε-balls.

C Count the number of distinct topologies there are for $\varnothing$, $\{a\}$, $\{a,b\}$, $\{a,b,c\}$, and $\{a,b,c,d\}$. What may be said about the family $\sigma = \{S: S' \in \tau\}$ of τ-closed sets when τ is a topology for a finite set?

D Is the usual metric topology on the set of rational numbers discrete?

E Give a detailed proof of each statement of Theorem 1.

F If $A \subset B$, is $\partial A \subset \partial B$? Give examples of a nonopen intersection of open sets and a nonclosed union of closed sets.

G A point x in a space T is an "accumulation," "cluster," or "limit" point of a subset A of T iff x is in the closure of $A - \{x\}$. A point $x \in A$ is an "isolated point" of A iff $\{x\}$ is open in the subspace A. Show that x is an isolated point of a subset A iff $x \in A^-$ yet x is not an accumulation point of A.

H Prove that if A is a subset of a metric space M, then the relative topology on A is exactly the same as the metric topology for the metric subspace A.

J Prove that if T_1 and T_2 are spaces having bases β_1 and β_2, then the family $\beta = \{B_1 \times B_2 : B_1 \in \beta_1 \text{ and } B_2 \in \beta_2\}$ is a base for the product topology of $T_1 \times T_2$.

K Generalize the example in the text (for $\mathbf{R} \times \mathbf{R}$) to show that the topology of the product metric is equal to the product topology defined by the metric topologies for the direct product of two metric spaces. (*Hint*: Use Exercise J.)

L Show that the set of balls with rational radii whose centers have all coordinates rational forms a basis for the usual topology on $\mathbf{R}^n$. A topology with a countable† (or denumerable) basis is said to be "second countable" or to satisfy the "second axiom of countability."

M Let τ_λ be a topology on T for each index λ in L, and show that $\tau = \bigcap\{\tau_\lambda : \lambda \in L\}$ is a topology on T. Also show that $\bigcup\{\tau_\lambda : \lambda \in L\}$ need not be a topology; is it a base for some topology?

N If $S \subset A$ and $T \subset B$, then is the product of the relative topologies the same as the relative topology on $S \times T$ in the product topology on $A \times B$? What if S is a singleton?

P Prove that the projections of a product space are open functions (and hence that each factor space has the quotient topology).

Q Show that *a function into a product space is continuous iff its composition with each projection is continuous.* ■ Then use this important result to establish that, if $a \in A$, then the subspace $a \times B$ of $A \times B$ is homeomorphic to B.

A subset A of a space T is "dense" in T if $A^- = T$. The space T is "separable" if there exists a countable† dense subset A of T. Prove that every second-countable space (see Exercise L) is separable (and therefore that $\mathbf{R}^n$ is separable).

S Let p be the projection of the plane $\mathbf{R} \times \mathbf{R}$ onto the real axis $\mathbf{R}$. Show that the range $\mathbf{R}$ has the quotient topology, but that p (and the quotient function it defines) is not closed. [*Hint*: Consider the subset $\{(x,1/x) : x > 0\}$ of $\mathbf{R} \times \mathbf{R}$.]

T Prove that every homeomorph of a metric space is metrizable.

† A set is "countable" if it may be put in 1-1 correspondence with a subset of the natural numbers $N = \{1,2,3,\ldots\}$.

U Separate the following alphabet into classes of homeomorphic letters:

A B C D E F G H I J K L M N O P Q R S T U V W X Y Z

W Produce examples showing that in the factorization $f = i \circ r \circ q$ of a function any pair of the factors may be open, yet the third factor may not be open.

Y Show that every ε-ball in $\mathbf{R}^n$ is homeomorphic to $\mathbf{R}^n$ itself. A homeomorph of $\mathbf{R}^n$ is called an "open n-cell"; show that if $I = (0,1)$, then I^n is also an n-cell. (Homeomorphs of closed ε-balls in $\mathbf{R}^n$ are called "closed n-cells.")

Z Show that the quotient space of the n-**cube** $\mathbf{I}^n$ resulting from identification of all boundary points (here think of $\mathbf{I}^n \subset \mathbf{R}^n$) is homeomorphic to S^n. Then assert that the set of all maps from S^n into a space X is in 1-1 correspondence with the set of all those maps from $\mathbf{I}^n$ into X which are constant on the boundary of $\mathbf{I}^n$. (*Hint*: Use the corollary to the Quotient Theorem, and draw pictures for the cases $n = 1$, 2, and maybe 3.)

PROBLEMS

AA **Quotients of a Function Space** Let C be the metric space of continuous functions from $\mathbf{I}$ into $\mathbf{R}$ which was described in Chap. III. The distance between two functions f and g is

$$\mathfrak{d}(f,g) = lub\ \{|f(x) - g(x)| : x \in \mathbf{I}\},$$

and there is a continuous evaluation function $e_t\colon C \to \mathbf{R}$ defined for each $t \in \mathbf{I}$: $e_t(f) = f(t)$. Show that the quotient space C/e_t is homeomorphic to $\mathbf{R}$ for each $t \in \mathbf{I}$. [If you are familiar with ring theory, you may prove that the family $\{e_t^{-1}(0) : t \in \mathbf{I}\}$ is exactly the family of maximal ideals in the ring C, when the operations in C are the usual pointwise addition and multiplication of functions. Hence these quotients of the space C are just the ring quotients of C which are fields.]

BB **Quotients of the Plane** From Exercise Z you may easily construct a map which exhibits the 2-sphere as a quotient of the plane. The dis-

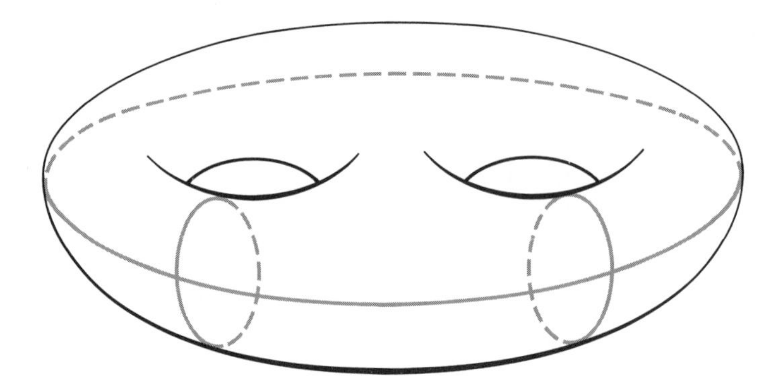

cussion in the text easily yields the torus as a quotient of the plane; another more regular construction is suggested by group theory (see Chap. II for hints). Armed with these examples, attempt to construct a map which shows the "double torus" (see illustration) to be a quotient of the plane. The drawing of many sketches may help you wrap the plane around this object.

CC A New Function Space Topology Let C be the *set* of maps from $\mathbf{I}$ into $\mathbf{R}$ (see Prob. AA), and for each $x \in \mathbf{I}$ and each open (in the usual topology) set S of $\mathbf{R}$, let $S_x = \{f \in C: f(x) \in S\}$. Let the topology τ be generated by the subbase σ consisting of all the subsets S_x of C. Is τ the same as the metric topology μ for C? Does either contain the other? Is the abelian group operation of pointwise addition continuous on $\tau \times \tau$ to τ? How about the inversion function (compare Prob. III.JJ)?

DD Due to Kuratowski If T is a set, then the power set $\mathcal{P}(T)$ of T is the set $\{S: S \subset T\}$ of all subsets of T. If T is a space, the closure and complementation operations define two functions from $\mathcal{P}(T)$ into itself. Show that the set of all possible compositions of these two functions with themselves or each other, in any order, has at most 14 members (one member, for example, of this set will send S to $S'''^{-'--}$). Demonstrate that there are exactly 14 of these functions if the space T is $\mathbf{R}$ with the usual topology by exhibiting a single subset S of $\mathbf{R}$ for which each of your 14 composite functions has a different value. You might begin by observing that always $S'' = S$ and $S^{--} = S^{-}$, and then show that composites of sufficiently many of these functions must be the same as some composite of fewer of them.

EE **Categorical Matters** Let $p_1: B \to A_1$ and $p_2: B \to A_2$ be continuous functions with the property that, given any space C and continuous functions $f_1: C \to A_1$ and $f_2: C \to A_2$, there exists a unique continuous function $f: C \to B$ such that $f_1 = p_1 \circ f$ and $f_2 = p_2 \circ f$. Use Prob. I.BB to show that such functions p_1 and p_2 always exist (for each pair of spaces A_1 and A_2) and that every such space B is homeomorphic to the space $A_1 \times A_2$.

FF **More Categorical Matters** The "dual" problem to the preceding one is gotten by reversing the directions of all arrows. Let $p_1: A_1 \to B$ and $p_2: A_2 \to B$ be continuous functions with the property that, given any space C and continuous functions $f_1: A_1 \to C$ and $f_2: A_2 \to C$, there exists a unique continuous function $f: B \to C$ such that $f \circ p_1 = f_1$ and $f \circ p_2 = f_2$:

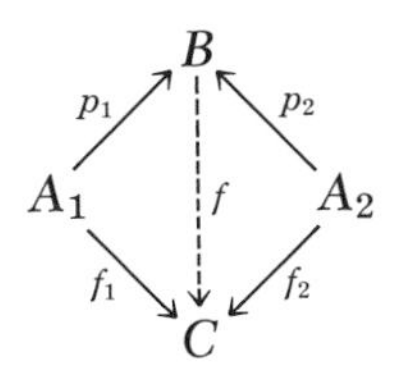

Show that if two spaces B and B' have this property for the same pair A_1, A_2 and certain given functions p_1, p_2 and p_1', p_2', then there is a homeomorphism from B to B'. Find one such space B, assuming $A_1 \cap A_2 = \varnothing$.

GG **Interior and Closure Operators** Let $\mathcal{P}$ be the power set of a set X (that is, $\mathcal{P}$ is the set of all subsets of X; see Prob. DD), and let $f: \mathcal{P} \to \mathcal{P}$ be a function. Show that there exists a topology on X for which f is the interior operator, $f(A) = A°$ for every subset A of X, iff f has the following properties:

i $f(X) = X$,
ii $f(A) \subset A$ for all $A \in \mathcal{P}$,
iii $f[f(A)] = f(A)$ for all $A \in \mathcal{P}$,
iv $f(A \cap B) = f(A) \cap f(B)$ for all $A, B \in \mathcal{P}$.

Can you phrase similar axioms for closure operators?

Topological Groups

CHAPTER V

You may have already been struck by the realization that many of the familiar geometric objects which we have used as examples of metric and topological spaces were also used as examples of groups. The list includes the real line $\mathbf{R}$ and $\mathbf{R} - \{0\}$, the plane $\mathbf{C}$ and $\mathbf{C} - \{0\}$, the circle S^1, and the torus $S^1 \times S^1$, and we should add to the list real n-space $\mathbf{R}^n$, which has the group structure of a direct product of n copies of the real numbers. In each case the underlying set supports a group structure and also a metric topological structure (we intend here the usual topology for each space). Furthermore, there are some familiar interrelations of these structures; for instance, it was pointed out in Chap. III that real addition is a continuous function from $\mathbf{R} \times \mathbf{R}$ to $\mathbf{R}$ [a pair (r,s) goes to $r + s$], and negation is continuous from $\mathbf{R}$ to $\mathbf{R}$ (a real r goes to $-r$). One might say that the group structure and the topology of $\mathbf{R}$ are "compatible"; the group operations are continuous. We shall see that a similar statement is true for each of the other groups-with-topology which are listed above.

The properties of addition in these examples led us naturally to the concept of group; the properties of these metric examples were generalized by our definition of metric spaces, and this led onward to the notion of topological space. Each of these processes of abstraction has given us a broadened understanding of the motivating special instances by supplying us with many new examples which have the same abstract properties. And with each instance of abstraction (or generalization) has come a new concise language in which to discuss mathematics.

We now wish to abstract a fact of a higher order, the interrelation of the topological and group structures on the above examples. In this process we shall certainly be introducing new theory; however, in another sense we now embark on the exploration of an example of nearly all that has gone before it.

THE DEFINITION

A **topological group** (or "continuous group") is a triple (G,m,τ), where (G,m) is a group and (G,τ) is a topological space (so that G is a set, m is a product on G, and τ is a topology on G). Further, the multiplication m: $G \times G \to G$ must be continuous from the product topology $\tau \times \tau$ to τ, and the inversion function ι: $G \to G$ of the group (G,m) must be continuous from τ to τ. A final requirement (which is not made by some authors) is that the singleton subset $\{e\}$ be closed in G; here e is the identity of the group G, and we are requiring that $G - \{e\} \in \tau$.

The **underlying set** of (G,m,τ) is G, the **underlying group** is(G,m), and the **underlying space** is (G,τ). Of course, when the choice of multiplication and topology is clearly enough understood in a particular case, we shall refer to (G,m,τ) simply as "the topological group G."

We have seen one example of a topological group, the additive reals. Another example of such compatibility is the group $(\mathbf{R} - \{0\}, \cdot)$ of nonzero reals under real multiplication furnished with the usual (or relative) topology. Here again, the group multiplication, which sends an ordered pair (x,y) to their product xy (as real numbers), is continuous, as is the inversion, $\iota(x) = 1/x$. We shall check these last assertions.

Let N be an open set of $\mathbf{R} - \{0\}$, and let the product xy be an element of N. We shall find open sets S and T of $\mathbf{R} - \{0\}$, with $(x,y) \in S \times T$, such that $ST = \{st\text{: } (s,t) \in S \times T\} \subset N$, and thus prove that the set $\{(u,v)\text{: } uv \in N\}$ is open in $(\mathbf{R} - \{0\}) \times (\mathbf{R} - \{0\})$. This will show that multiplication is continuous. Now, N is a relative neighborhood of xy, so $N \cup \{0\}$ is a neighborhood of xy in $\mathbf{R}$; therefore there is an ε such that the interval $(xy - \varepsilon,\ xy + \varepsilon) \subset N \cup \{0\}$, and if $\eta = min\ \{\varepsilon, |xy|\}$, that is if η is the smaller of the two numbers ε and $|xy|$, then $(xy - \eta,\ xy + \eta) \subset N$. Choose $\delta = min\ \{\eta/3x, \eta/3y, (\eta/3)^{1/2}\}$ and define the open intervals $S = (x - \delta,\ x + \delta)$ and $T = (y - \delta,\ y + \delta)$. Let a and b be two numbers of absolute value less than δ, and let $z = x + a$ and $w = y + b$; (z,w) is an element of $S \times T$, and we compute that

$$|zw - xy| = |ay + bx + ab| \leq |ay| + |bx| + |ab| < \frac{\eta}{3} + \frac{\eta}{3} + \frac{\eta}{3} \leq \varepsilon.$$

Hence $zw \in N$, and consequently, $ST \subset N$.

To construct a proof that inversion is continuous, note first that it suffices to show this separately for the positive and for the negative half of the group $\mathbf{R} - \{0\}$. But inversion (on each half) is order reversing, $x < y$ iff $1/x > 1/y$. Thus the inverse of an open interval (a,b) is just the open interval $(1/b,1/a)$, so inversion carries open sets to open sets. But the *functional inverse* for the inversion function is the inversion function itself, $\iota \circ \iota = 1$; hence it is continuous and, in fact, a homeomorphism.

Thus the real numbers under addition and the nonzero reals under multiplication form topological groups. Other examples of topological groups are provided by the sets $\mathbf{R}^m$ of real m-tuples, with the usual metric topology of Chap. III and the addition of *real m-vectors*: If $p = (p_1, p_2, \ldots, p_m)$ and $q = (q_1, q_2, \ldots, q_m)$ are m-tuples, then $p + q = (p_1 + q_1, p_2 + q_2, \ldots, p_m + q_m)$ gives the group "product" (written additively). The identity is $(0, 0, \ldots, 0) = 0$ and

$$-p = (-p_1, -p_2, \ldots, -p_m)$$

is the group inverse of p. It is a straightforward chore to verify that the addition and negation of real m-vectors are continuous operations (see Exercise B). The addition of complex numbers, regarded as real pairs, is just the case $m = 2$.

We now establish that the nonzero complex numbers, with the usual topology and complex multiplication, form a topological group. [Here the "complex number zero" means (0,0), which we shall usually denote "0." It is the *additive* identity for $\mathbf{C}$.] In terms of real coordinates, the product on $\mathbf{C} = \mathbf{R}^2$ is given by $(x,y)(z,w) = (xz - yw, yz + xw)$. The (*multiplicative*) identity for this product is (1,0); it will usually be denoted by "1." Since the usual topology for $\mathbf{C} = \mathbf{R} \times \mathbf{R}$ is the product topology, and the product topology on $\mathbf{C} \times \mathbf{C}$ is just that of $\mathbf{R}^4$, this multiplication will be continuous iff its compositions with the two projections (of its range) are continuous (see Exercise IV.T). Hence we examine the function $f: \mathbf{R}^4 \to \mathbf{R}$: $f(x,y,z,w) = xz - yw$; since real multiplication is continuous, as is real subtraction, f is continuous:

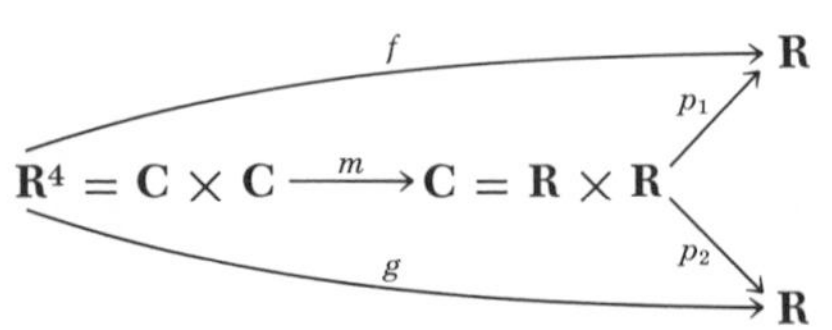

A similar argument shows the continuity of g: $\mathbf{R}^4 \to \mathbf{R}$: $g(x,y,z,w) = yz + xw$. The same treatment may be made for the inversion function $\iota(x,y) = [x/(x^2 + y^2), -y/(x^2 + y^2)]$ when one remembers that the quotient of continuous real-valued functions is continuous if the denominator is never zero (that is, if the quotient is defined at all).

Another interesting topological group is the **circle group,** the subgroup $\{z: |x| = 1\}$ of the multiplicative nonzero complex numbers with the relative topology. This is an instance of the following generality.

THEOREM *If H is, in the algebraic sense, a subgroup of the group G, and G is a topological group, then H is a topological group with its relative topology as a subset of G.*

The proof is immediate to the following useful topological fact.

LEMMA *If $f: X \to Y$ is continuous and $A \subset X$, $f(A) \subset B \subset Y$, then $f|_{A \times B}$ is continuous from relative topology to relative topology.*

Proof Let T be B-open, so there is an open set $U \subset Y$ with $U \cap B = T$. Now, $f^{-1}(U)$ is open in X; hence $f^{-1}(U) \cap A = A \cap f^{-1}(U) \cap f^{-1}(B) = A \cap f^{-1}(U \cap B) = A \cap f^{-1}(T) = (f|_{A \times B})^{-1}(T)$ is an A-open set. ■

It is worth noting that the statement of the above theorem is made imprecise by the confusion of the sets G and H with groups, and also with topological spaces, having those underlying sets. You should understand how to rewrite the statement to obtain precision, though perhaps even more confusion is introduced by a multiplicity of symbols.

Exercises A, B, and C

HOMOGENEITY

Each element g of a topological group G determines a **left-multiplication,** or **left-translation, function** L_g on G to G; its value at each $h \in G$ is $L_g(h) = gh$. If m: $G \times G \to G$ is the product on G, and σ_g: $G \to G \times G$ is the function assigning $\sigma(h) = (g,h)$, then $L_g = m \circ \sigma_g$. But σ_g is readily seen to be continuous (you need only check its inverse image of a "rectangular"-basis element of $G \times G$; see Exercises IV.N, P, and Q); hence L_g is continuous for each $g \in G$. Another way to say this is that $g \times G$ is a subspace of $G \times G$ which is homeomorphic to G, and L_g may be regarded as the restriction of m to $g \times G$; since restrictions of continuous functions are continuous, L_g is a map. But the corresponding function $L_{g^{-1}}$, which left-multiplies elements of G by g^{-1}, is also continuous, and $L_gL_{g^{-1}}(h) = L_g(g^{-1}h) = gg^{-1}h = h$, so $L_gL_{g^{-1}} = 1_G$. Similarly, $L_{g^{-1}}$ is a left-inverse function for L_g; hence $L_g{}^{-1} = L_{g^{-1}}$, and L_g is a set isomorphism which is continuous and which has a continuous inverse; that is, *L_g is a homeomorphism of G with G itself.* ■ Similar statements hold for the right translation R_g determined by $g \in G$, $R_g(h) = hg$; each R_g is a homeomorphism of G onto G. These differ from the left translations only if G is nonabelian.

As an example, if the topological group above were the additive real numbers and g were 17, then L_{17} would be the real-valued real function which slides or "translates" the whole real line to the right by 17 units; $L_{17}(x) = 17 + x$ for each real x, and $L_{17}{}^{-1} = L_{-17}$, of course.

A topological space T is said to be **homogeneous** if it has the property that for each pair t,u of points in T there exists a homeomorphism f of T onto itself such that $f(t) = u$. A homogeneous space is one which "looks the same when viewed from any one of its points." It is clear that each topological group is homogeneous (considered as a topological space), because if g and h are members of the group, $L_h \circ L_{g^{-1}} = L_{hg^{-1}}$ is a homeomorphism carrying g to h. This provides a fund of examples of homogeneous spaces, $\mathbf{R}^n$ for all n, etc. A nonexample is the closed unit interval $\mathbf{I}$; it looks different when viewed from an end point than when viewed from an interior point. Precisely, inside every neighborhood of 0 *in* $\mathbf{I}$ there is a neighborhood of 0 which has just one point on its boundary; this is not true for the interior point ½, since every neighborhood N of ½ lying inside (¼,¾) must have the two points *lub* N and *glb* N in ∂N. Can you see why no homeomorphism could take 0 to ½? The space $\mathbf{I}$ is thus not homogeneous; it could not be a topological group under any multiplication whatsoever.

A topological group whose underlying space is an n-manifold is called a **group manifold.** Since a topological group G is homogeneous, it will suffice to show that G is a group manifold if only the point e of G has a neighborhood N which is homeomorphic to $\mathbf{R}^n$. This is enough to assume, since then each point $g \in G$ has a neighborhood $gN = L_g(N)$ homeomorphic to $\mathbf{R}^n$. Do you see that all our examples thus far of topological groups are group manifolds?

Neighborhoods of the identity of a topological group are important enough to deserve a special name; they are called **nuclei.** If, in any topological space X, $\mathfrak{N}$ is a family of neighborhoods of a fixed point x such that every neighborhood of x contains some member of $\mathfrak{N}$, then $\mathfrak{N}$ is called a **local base** (or "neighborhood base") at x. It is evident that if X is homogeneous, then a local base at a single point x determines a local base at every other point of X and so determines the entire topology of X. That is, if y is an arbitrary point of X and g is a homeomorphism of X with itself that takes y to x, then we know that a set N is a neighborhood of y iff $g(N)$ contains a member of the local base at x. In particular, if $\mathfrak{N}$ is the family of nuclei of the topological group G, and $g \in G$, then the family of all neighborhoods of g is the family of translates of members of $\mathfrak{N}$, $\{L_g(N): N \text{ is a nucleus of } G\}$. Of course, a nucleus is not necessarily open; however, a subset S of G is open iff for each point $s \in S$ there is a nucleus N of G with $L_s(N) \subset S$, that is, iff S is a neighborhood of each of its points. We may turn this around, too, to assert that S is open iff $L_{s^{-1}}(S)$ is a nucleus for each $s \in S$ (how is this equivalent?).

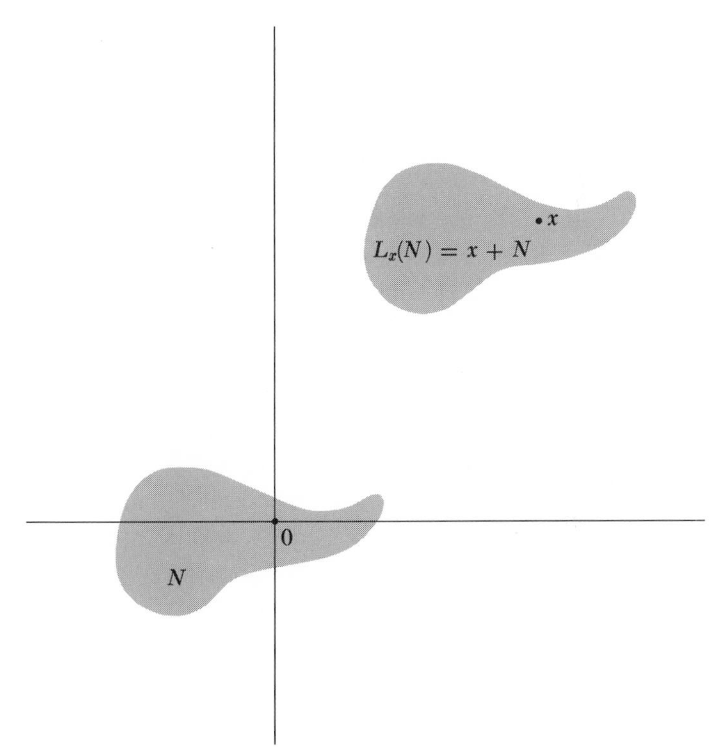

A translated nucleus in the additive plane.

SEPARATION

Since a topological group is homogeneous and has $\{e\}$ as a closed set, each singleton subset $\{g\}$ of the group must be a closed set (homeomorphisms carry closed sets to closed sets). A space which has the property that all its singleton subsets are closed ("points are closed") is called a T_1-**space** (or "Frechet space"). The requirement on a topological group (G,m,τ) that $\{e\}$ be closed could thus be replaced by the seemingly stronger requirement that (G,τ) be a T_1-space.

There is another definition of T_1-ness (or the property of being a T_1-space): *a space* (T,τ) *is* T_1 *iff, given two distinct points t and u of T, there exists an open neighborhood N of t with* $u \notin N$. That each T_1 space has this property is trivial, since $N = T - \{u\}$ will do nicely. On the other hand, if we fix u in mind and assume we can find, for each $t \neq u$, a neighborhood N of t with $u \notin N$, then $T - \{u\}$ contains a neighborhood of each of its points; it is open, so $\{u\}$ is closed. ∎ This characterization of T_1-spaces suggests the definition of a weaker property of spaces: (T,τ) is a T_0-**space** iff, given a pair t,u of distinct points of T, there exists an open set N such that $N \cap \{t,u\}$ is a singleton; that is, there exists a neighborhood of one of the points which does not contain the other. Another way to put it is that either $T - \{u\}$ is a neighborhood of t, or else $T - \{t\}$ is a neighborhood of u. Obviously, if (T,τ) is T_1, then it is T_0. The following example of a T_0-space which is not T_1 shows that this new property is really weaker than T_1-ness. Let $T = \{1,2\}$ and $\tau = \{\varnothing,\{1\},\{1,2\}\}$. The singleton $\{1\}$ is not closed, but $\{2\}$ is, so T is T_0 but not T_1. Clearly T could not be the space of any topological group.

However, if the requirement that $\{e\}$ be closed in the definition of a topological group is replaced by the demand that (G,τ) be T_0 (and the other axioms are left unchanged), it must follow that (G,τ) is T_1! This will be clear if $G - \{e\}$ is a neighborhood of each point $g \neq e$. But let $g \neq e$; since G is T_0, either $G - \{e\}$ is a neighborhood of g (and we are done), or else $G - \{g\}$ is a neighborhood of e. In the second case, the homeomorphism $L_{g^{-1}}$ carries $G - \{g\}$ to $G - \{e\}$ and shows the latter set to be a neighborhood of g^{-1}. But the inversion homeomorphism ι carries $G - \{e\}$ to itself and shows it to be a neighborhood of g; therefore $\{e\}$ is closed. To recapitulate this proof, if $G - \{g\}$ is a neighborhood of e, then $\iota \circ L_{g^{-1}}(G - \{g\}) = G - \{e\}$ is a neighborhood of $\iota \circ L_{g^{-1}}(e) = g$. Hence, *in the presence of the other axioms for a topological group,* T_0 *implies* T_1. ∎

If (T,τ) is T_0 (or T_1) and $\tau \subset \mu$, then (T,μ) is also T_0 (or T_1). These properties of a topology τ say that there are enough open sets in τ to "distinguish" points, in one or the other sense. They are often called **separation axioms.** A third such axiom for topological spaces (T,τ), and an even stronger one, is the requirement that for each pair t,u of distinct points of T there exist neighborhoods M and N of t and u, respectively, with $M \cap N = \varnothing$. A space satisfying this axiom is said to be a T_2-**space;** more often it is called a **Hausdorff space,** in honor of Felix Hausdorff, a pioneer of topology.

Clearly, T_2-spaces are T_1. An example of a T_1-space which is not T_2 has already been given: the topology on the set of real numbers which has as its members $\emptyset$ and all complements of finite sets. If $x \neq y$, then the complement of $\{x\}$ is an open set containing y but not x, so this space is T_1. However, every nonempty open set has a finite complement, and thus each pair of nonempty open sets must have a nonempty intersection (is this clear?). Hence this space is not T_2; no two of its points possess disjoint neighborhoods.

The space above is another example of one which cannot support a topological group structure; since, in the presence of the other axioms for a topological group, T_1 implies T_2. To show this we need to collect a few facts about topological groups. First, let a neighborhood V of e be called **symmetric** if $V = V^{-1}$. Certainly not all neighborhoods of e are symmetric, but each contains a symmetric neighborhood, $V \cap V^{-1} \subset V$.

LEMMA *If U is a neighborhood of e in a topological group, then there is a symmetric neighborhood V of e with $VV = VV^{-1} \subset U$.*

Proof The multiplication function m is continuous on $G \times G$ to G, so $m^{-1}(U)$ contains a neighborhood N (in the product topology) of (e,e). Now, N contains a base element $S \times T$, where S and T are open at e, and therefore a "square" base element $R \times R$, where $R = S \cap T$. Choose $V = R \cap R^{-1}$. Obviously, V is symmetric, and if g and h are elements of V, then $(g,h) \subset V \times V \subset S \times T \subset N$, so $gh \in U$; thus $VV \subset U$. ∎ This proof uses the fact that m is continuous; do you see where the continuity of inversion was needed? As an example of the lemma, let U be a neighborhood of the origin in the plane; then U contains some ε-ball at 0. Since the absolute-value function is a morphism of the multiplicative nonzero complex numbers to the positive reals (moduli multiply; $|zw| = |z|\,|w|$), V may be taken to be the ball at 0 with radius $\delta = min\ \{1,\varepsilon\}$.

We now argue that *each topological group is Hausdorff.* Let x and y be distinct elements of the group, so that $x^{-1}y \neq e$. The neighborhood $G - \{x^{-1}y\}$ of e contains a symmetric neighborhood V of e such that $x^{-1}y \notin VV$. Surely xV and yV are neighborhoods of x and y, respectively, but if $xV \cap yV$ is not empty, then there exist v_1 and v_2 in V with $xv_1 = yv_2$, or $x^{-1}y = v_1v_2{}^{-1} \in VV^{-1} = VV$. This is contradictory, so $xV \cap yV = \emptyset$; G is a Hausdorff space. ∎

Exercises E and F

TOPOLOGICAL PROPERTIES

A property of topological spaces, such as T_1-ness or homogeneity, is a **topological property** iff whenever a space X has the property and Y is homeomorphic to X, Y also has the property. For example, the property of being T_1 is topological. If f: $X \to Y$ is a homeomorphism and each singleton subset of X is closed, then each singleton subset of Y, being the image under f of a singleton of X, is also closed. Topological properties are sometimes called **topological invariants;** they may be described as those properties of spaces which are "preserved by homeomorphisms."

We have seen several examples of topological invariants: metrizability (see Exercise IV.T) and homogeneity are readily shown to be such, as are the separation axioms T_0, T_1, and T_2. We have shown this to be so for T_1; it is an exercise (see Exercise J) to establish it for T_0 and T_2. We shall only prove that homogeneity is a topological property. Let X be homogeneous, and let f: $X \to Y$ be a homeomorphism. If y and z are points of Y, choose a homeomorphism h: $X \to X$ which carries $f^{-1}(y)$ to $f^{-1}(z)$; is it not clear that the homeomorphism $f \circ h \circ f^{-1}$: $Y \to Y$ carries y to z?

Other topological properties (trivial ones) are those of having an infinite number of points or of having exactly n points and of being discrete or of having an infinite number of open sets. A nonexample is the property of being a bounded metric space. We have seen that there is a bounded metric for **R**, equivalent to the usual metric, which induces the usual topology; hence a bounded metric space may be homeomorphic to an unbounded metric space.

Exercises D, J, and T

COSET SPACES

Let (G,m,τ) be a topological group; we adopt the adjective **algebraic** to signify that our discussion is, for the moment, about only the group (G,m) and does not involve the topology on G. If H is an algebraic subgroup of G (but not necessarily normal), then the family G/H of cosets of H is defined; this family is a quotient set of the space (G,τ), and there is a quotient topology defined for it. The quotient space G/H is called a **coset space** of G. Since the inverse image of each point of the coset space G/H is a coset of H, and each coset $gH = L_g(H)$ is closed iff H is closed, we find immediately that *G/H is T_1 iff H is closed.*

A more startling fact is that *the quotient map $q: G \to G/H$ is always open!* To prove this we need only observe that if U is an open set of G, then $q(U) = \{uH: u \in U\}$ and $q^{-1}[q(U)] = \bigcup\{uH: u \in U\} = UH$, the product of the sets U and H in G. But we claim that *the product of an open set U with any subset S of G must be open,* for $US = \bigcup\{Us: s \in S\}$ is the union of the right translates $Us = R_s(U)$, and each of these sets is open in G. Hence for every subgroup H and every open set U, $q(U)$ is a set whose inverse image under q is open; $q(U)$ is thus open in the quotient topology on G/H. ∎

We know that G/H has an algebraic group structure when H is normal in G; if H is closed in G as well, then G/H has a T_1 quotient topology on it. Is G/H then a topological group? To see that it is, it will suffice to prove that for each pair f_0H, f_1H of elements of G/H and for each neighborhood $\mathfrak{U}$ of their product $(f_0H)(f_1H) = f_0f_1H$ there exist neighborhoods $\mathfrak{F}_0$ and $\mathfrak{F}_1$ of f_0H and f_1H, respectively, with the product $\mathfrak{F}_0\mathfrak{F}_1$ contained in $\mathfrak{U}$. But the multiplication of G is continuous, so there are neighborhoods F_0 and F_1 of f_0 and f_1 in G such that the product set F_0F_1 lies inside the neighborhood $q^{-1}(\mathfrak{U})$ of f_0f_1; take $\mathfrak{F}_0 = q(F_0)$ and $\mathfrak{F}_1 = q(F_1)$ (remember that q is open). A similar computation shows that inversion is continuous (or rather, open); $\mathfrak{U}^{-1} = (q[q^{-1}(\mathfrak{U})])^{-1} = q([q^{-1}(\mathfrak{U})]^{-1})$ is open if $\mathfrak{U}$ is open, since $q^{-1}(\mathfrak{U})$ and its inverse are open in G, and q is open. To recapitulate, *the quotient of a topological group modulo a closed normal subgroup is a topological group under the quotient topology and product.* ∎

Exercise H

MORPHISMS

It is natural to single out those functions from one topological group to another which suitably preserve both the topological and the group structures. If G and G' are topological groups and f is a function from G to G', then f is called a **morphism (of topological groups)** iff f is both a map (on the underlying spaces) and an algebraic morphism (on the underlying groups). Each inclusion function on a subgroup G of G' is an example, as is each quotient function from a group G to a quotient G' of G. A more specific example is given by the absolute-value function $abs: \mathbf{C} - \{0\} \to \mathbf{R} - \{0\}$ from the multiplicative group of nonzero complex numbers to the group of nonzero reals; it is continuous and preserves products.

The definitions of **epimorphisms** and **monomorphisms** are just as in the algebraic case; they are, respectively, the onto and the 1-1 morphisms. An **isomorphism** of topological groups is required to be an algebraic isomorphism and also a homeomorphism.† A word of caution: if the domain and range of a function are both topological groups, then that function must be continuous to be a "morphism"; however, when the domain and range are thought of as having only the algebraic structure of groups, that function need only preserve products to be called a "morphism." The context should make it clear which is meant; we shall henceforth use the words "group" and "morphism" in both cases, occasionally inserting the words "topological" or "algebraic" when confusion might otherwise occur.

The topological requirement that a morphism of topological groups be continuous and the algebraic requirement that it preserve products affect each other in interesting ways. We illustrate this by showing next that *if* $f\colon G \to G'$ *is an algebraic morphism from one topological group to another which is continuous at the single point* e *of its domain, then* f *is continuous at every point of* G, and so is a topological morphism. We are assuming, then, that f preserves products and that $f^{-1}(N)$ is a nucleus (or neighborhood of e) of G whenever N is a nucleus of G'. Suppose S is open in G'; we wish to show that $f^{-1}(S)$ is open; that is, that $f^{-1}(S)$ is a neighborhood of each of its points. This will be so iff for each $y \in S$ and each $x \in G$ with $f(x) = y$, $x^{-1}[f^{-1}(S)]$ is a nucleus in G. But $y^{-1}S$ is a nucleus in G', and $f(x^{-1}S) = f(x)^{-1}S = y^{-1}S$, so $x^{-1}S \subset f^{-1}(y^{-1}S)$. To complete our proof we need only show $x^{-1}S \supset f^{-1}(y^{-1}S)$. Assume $f(z) = y^{-1}s$ for some $s \in S$; then $f(xz) = f(x)y^{-1}s = yy^{-1}s = s$, and thus $xz \in f^{-1}(S)$, or $z \in x^{-1}[f^{-1}(S)]$. ■ The plan of this proof is that to find the inverse image of S we may translate S back to the identity in G', find the inverse image of this translate of S, and then translate forward in G to get $f^{-1}(S)$. The translations in G and G' are homeomorphisms, so continuity at e suffices to guarantee the continuity of f over all of G.

This theorem is quite handy; it allows us to check the continuity of a morphism by considering only nuclei, and this often simplifies arithmetic considerably.

Exercise Q

† Some authors require epimorphisms to be open maps; in their language (but not ours) a monic epimorphism is an isomorphism. Still others define all morphisms to be open; in that case our merely continuous morphisms are called "representations."

FACTORING MORPHISMS

Now suppose that $f: G \to G'$ is a morphism of topological groups. Since f is an algebraic morphism, there exists a unique factorization $f = i \circ r \circ q$ of f through the (algebraic) quotient group $G/Ker\,(f)$ of G and the subgroup $Im\,(f)$ of G':

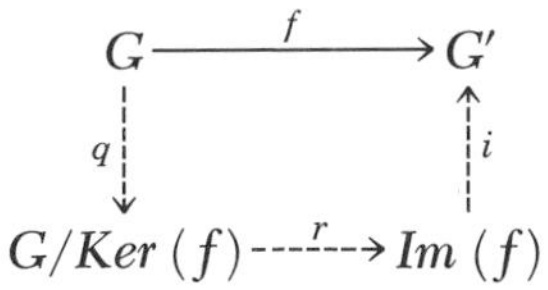

$$\begin{array}{ccc} G & \xrightarrow{f} & G' \\ {\scriptstyle q}\downarrow & & \uparrow{\scriptstyle i} \\ G/Ker\,(f) & \xrightarrow{r} & Im\,(f) \end{array}$$

Furthermore, if we consider f as a map, with domain and range the underlying spaces of G and G', there exist a unique factorization of f into the composite of a quotient, a 1-1 correspondence, and a subspace inclusion. In the case at hand, the quotient group $G/Ker\,(f)$ and the subgroup $Im\,(f)$ are both topological groups, and q and i are topological morphisms. Hence these two factorizations, the algebraic and the topological, are the same, and each of the factors q, r, and i of f is a topological morphism. This proves the first statement of the following theorem. The remainder of the conclusion has been previously established.

THE QUOTIENT THEOREM FOR TOPOLOGICAL GROUPS *Each morphism f: $G \to G'$ of topological groups possesses a unique factorization $f = i \circ r \circ q$ into topological morphisms, where q is a quotient, r is an algebraic isomorphism, and i is an inclusion. The quotient q is always open, i is open iff $Im\,(f)$ is open in G', and r is a topological isomorphism iff f is open onto its image.* ■

COROLLARY *The morphism f factors through a quotient q: $G \to G/K$ of its domain,*

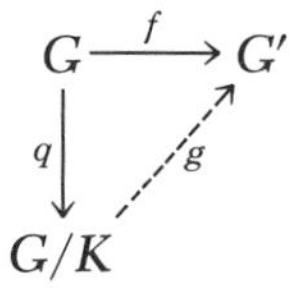

$$\begin{array}{ccc} G & \xrightarrow{f} & G' \\ {\scriptstyle q}\downarrow & \nearrow{\scriptstyle g} & \\ G/K & & \end{array}$$

iff $Ker\,(f) \supset K$.

Proof Though this statement is called a corollary to the quotient theorem, its proof merely pastes together the relevant facts about factoring f through a quotient when f is regarded first as an algebraic morphism (see Chap. II) and then as a map (see Chap. IV). ■

Exercise K

A QUOTIENT EXAMPLE

One example of the quotient theorem is offered by the morphism *abs*: $\mathbf{C} - \{0\} \rightarrow \mathbf{R} - \{0\}$, which has the value $|z|$ for each nonzero complex number z. The factorization $i \circ r \circ q$ of *abs* yields the quotient q: $\mathbf{C} - \{0\} \rightarrow (\mathbf{C} - \{0\})/S^1$, whose cosets are circles centered at the origin in the plane; r is an isomorphism (why is it open?) of the quotient group with the positive reals, and i includes that image set in the range.

A less intuitive example is afforded by the group C of all real-valued continuous functions on **I**. We studied the metric structure of C in Chap. III (see also Prob. IV.AA); now we claim that C is a topological group when addition is defined by $(f + g)(t) = f(t) + g(t)$ and negation is defined by $(-f)(t) = -f(t)$. It is trivial to verify that this addition gives C a group structure. We first check that negation is continuous. If $\mathfrak{b}(\varepsilon,f)$ is a ball at f in C, then $-\mathfrak{b}(\varepsilon,f) = \mathfrak{b}(\varepsilon,-f)$, since $|f(x) - g(x)| < \varepsilon$ iff $|(-f)(x) - (-g)(x)| < \varepsilon$. Therefore the negatives of open sets, which are unions of balls, are themselves unions of balls; negation is thus an open function, and since it is its own functional inverse, it is continuous.

As for the continuity of addition, let $\mathfrak{b}(\varepsilon, f + g)$ be given; we must find two positive numbers δ_1 and δ_2 such that the set of sums $\mathfrak{b}(\delta_1,f) + \mathfrak{b}(\delta_2,g) \subset \mathfrak{b}(\varepsilon, f + g)$. Take $\delta_1 = \delta_2 = \varepsilon/2$, and let $(f',g') \in \mathfrak{b}(\varepsilon/2,f) \times \mathfrak{b}(\varepsilon/2,g)$, a basic open set in the product-metric topology of $C \times C$. Now, for each $t \in \mathbf{I}$, $|f'(t) - f(t)| < \varepsilon/2$ and $|g'(t) - g(t)| < \varepsilon/2$; hence

$$\begin{aligned} |(f' + g')(t) - (f + g)(t)| &= |f'(t) - f(t) + g'(t) - g(t)| \\ &\leq |f'(t) - f(t)| + |g'(t) - g(t)| \\ &< \frac{\varepsilon}{2} + \frac{\varepsilon}{2}, \end{aligned}$$

so $\mathfrak{d}(f' + g', f + g) < \varepsilon$. This completes the proof that C is a topological group (Exercise G shows it to be Hausdorff). Since the evaluation e_t: $C \rightarrow \mathbf{R}$: $e_t(f) = f(t)$ is continuous, $S_t = e_t^{-1}(0)$ is a closed set in C. But S_t is a subgroup of the abelian group C; hence S_t is a closed normal subgroup for each $t \in \mathbf{I}$. In fact, e_t is a morphism, and its kernel is S_t; thus the image group **R** of additive reals will be isomorphic to C/S_t for every $t \in \mathbf{I}$ if only e_t is an open map. To argue this, we define, for each real u, the constant map $k_u: \mathbf{I} \to \mathbf{R}$: $k_u(t) = u$ for each t. Now consider the image under e_t of $\mathfrak{b}(\varepsilon,f)$. It contains $(f + k_u)(t) = f(t) + u$ for each u for which $|u| < \varepsilon$, and thus $e_t[\mathfrak{b}(\varepsilon,f)]$ contains the interval $(f(t) - \varepsilon, f(t) + \varepsilon)$ (equality holds, in fact) and e_t is open.

Thus we behold an uncountably infinite collection of quotient groups C/S_t, one for each $t \in \mathbf{I}$. All are distinct, since the subgroups S_t are different for differing values of t, yet they are all (topologically) isomorphic to the additive reals, and thus to each other.

DIRECT PRODUCTS

We have noted (in Chap. II) with satisfaction that the additive plane $\mathbf{C} = \mathbf{R} \times \mathbf{R}$ was the (algebraic) direct product of two copies of the additive reals $\mathbf{R}$. Later we saw (in Chap. IV) that the plane was the (topological) direct product of the space $\mathbf{R}$ with itself. But $\mathbf{R}$ is a topological group; the above facts may be summed up by saying that the direct-product set $\mathbf{R} \times \mathbf{R}$, furnished with the product group structure and the product topology, is a topological group. This is a completely general phenomenon.

If G and H are topological groups, then $G \times H$ has a product topology and also a product group structure. We shall show that with these structures *the direct product of two topological groups is a topological group, and the projections onto the factors are open topological epimorphisms.* Let m, n, and p be the multiplications on G, H, and $G \times H$, respectively, and define a "shuffling map" $s: (G \times H) \times (G \times H) \to (G \times G) \times (H \times H)$ by $s[(g,h),(g',h')] = [(g,g'),(h,h')]$. Then p factors through $m \times n$, $p = (m \times n) \circ s$:

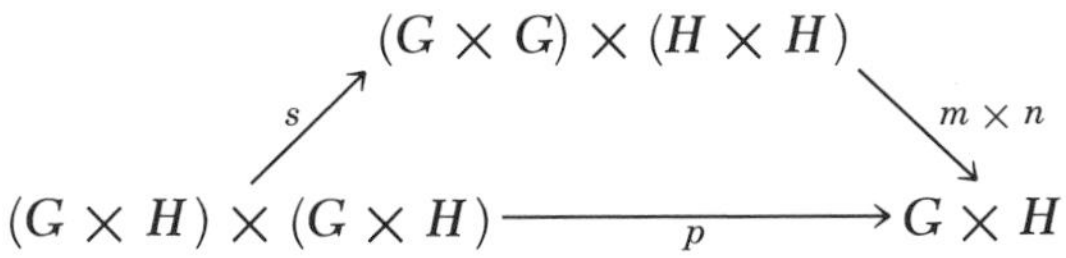

$$\begin{array}{ccc} & (G \times G) \times (H \times H) & \\ {}^{s}\nearrow & & \searrow^{m \times n} \\ (G \times H) \times (G \times H) & \xrightarrow{\quad p \quad} & G \times H \end{array}$$

It is now routine to check the continuity of s; that of $m \times n$ is guaranteed. Do you see an even simpler statement about the inversion of $G \times H$? ∎

Since the kernel of the open epimorphism which projects $G \times H$ onto its factor H is the closed normal subgroup $G \times e$, the quotient $G \times H/G \times e$ is isomorphic to H; similarly, $G \times H/e \times H \cong G$. This gives us a number of examples of quotient groups: the circle is the torus modulo one of its factors, the real line is the plane modulo the pure imaginaries, etc. A slightly more general situation is suggested by the absolute-value morphism *abs* on $\mathbf{C} - \{0\}$, which we regard here as having range the multiplicative group $\mathbf{R}_+$ of positive reals, $abs\ (a,b) = (a^2 + b^2)^{1/2}$, with its kernel $S^1 = \{(a,b) \in \mathbf{C}: (a^2 + b^2)^{1/2} = 1\}$. Now, $\mathbf{C} - \{0\}$ is not the direct product $\mathbf{R}_+ \times S^1$. Its elements are ordered pairs of reals, but every element z of $\mathbf{C} - \{0\}$ has a unique factorization $z = |z|(z/|z|)$ into the product of an element $|z| \in \mathbf{R}_+$ with an element $z/|z|$ whose absolute value is $|z/|z|| = |z|/|z| = 1$. This factorization is called the "polar form" of z; the element $z/|z|$ of S^1 is called the "argument" of z, $z/|z| = arg\ (z)$. Evidently, the function h: $\mathbf{C} - \{0\} \to \mathbf{R}_+ \times S^1$, which has values $h(z) = [|z|, arg\ (z)]$, is a homeomorphism and a morphism of groups, since *abs* is an open morphism. Hence, if p is the projection of the product $\mathbf{R}_+ \times S^1$ onto its first factor, the following diagram is commutative:

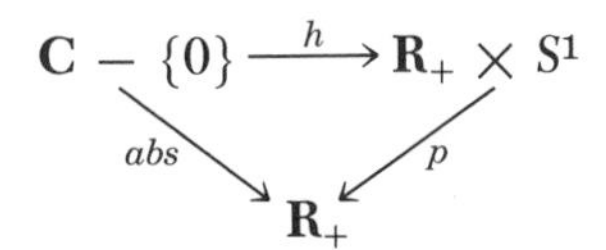

All this may be expressed by saying that "*abs* is homeomorphic to the projection of a direct product." However, mathematicians customarily abbreviate their description of such a situation by saying something like "*abs* is a direct-product projection."

So that you are not led to believe that, at least in the above sense, every morphism of topological groups is the projection of a direct product, we announce that this is not the case for the morphism *exp* of $\mathbf{R}$ onto S^1 which wraps the line around the circle, $exp\ (r) = (cos\ 2\pi r, sin\ 2\pi r)$. Each factor of a direct product is isomorphic to a subgroup of the direct product; $G \times e$ is a subgroup of $G \times H$ which is isomorphic (via the projection onto G) to G, and $e \times H \cong H$. However, there is a purely algebraic proof that S^1 is not isomorphic to a subgroup of $\mathbf{R}$; we need only observe that the square of -1 in S^1 is the identity 1, yet no nonzero element of $\mathbf{R}$ exists which when added to itself gives zero. A topological proof is requested in Exercise R.

Exercises P, R, and U

REFERENCES AND FURTHER TOPICS

We shall continue the study of topological groups in subsequent chapters; numerous facets of the subject will be explored, and we shall develop in detail the interplay of covering groups and the fundamental, or Poincaré, group associated with a topological group. However, many questions will remain untouched; one such is the matter of differentiability. In the previous chapter we gave reference to the notion of a differential structure on a manifold. It turns out that every group manifold can be given a differential structure; furnished with such a structure, it is called a "Lie (or analytic) group." You will find a complete introduction to the large and important theory of Lie groups in

Montgomery and Zippen, *Topological Transformation Groups* (New York: Interscience, 1955).

Another question concerns integration. On the real line, the fact that the length of the interval (a,b) is the same as the length of its translate $(r + a, r + b)$ for each real r and its negation $(-b, -a)$ may be expressed by saying that the measure of length of **R** is invariant under the group operations. This is responsible for the equality, for each $r \in \mathbf{R}$,

$$\int_{-\infty}^{\infty} f(x)\, dx = \int_{-\infty}^{\infty} f(x + r)\, dx = \int_{-\infty}^{\infty} f(-x)\, dx.$$

That is, there is an "invariant integration" on **R**. An appropriate notion of abstract integration which is invariant in the above sense may be defined for a large class of topological groups. An exposition of this is offered in

L. Pontrjagin, *Topological Groups*, chap. IV (Princeton, N.J.: Princeton University Press, 1939).

This reference also contains an excellent discussion of the possible "representations" (or morphisms) of a topological group into a topological group of matrices. In a particular case of this, the group of all morphisms of a topological group into S^1 is called the "character group" of the group.

Three other references touching the above topics, the first of which also contains a detailed discussion of the classical groups of matrices, are

C. Chevalley, *Theory of Lie Groups* (Princeton, N.J.: Princeton University Press, 1946).

P. M. Cohn, *Lie Groups* (New York: Cambridge University Press, 1957).

T. Husain, *Introduction to Topological Groups* (Philadelphia: Saunders, 1966).

Each of the references mentioned provides an introductory treatment which parallels this chapter. Another such is

N. Bourbaki, *General Topology*, part I, chap. 3 (Reading, Mass.: Addison-Wesley, 1967).

There is also a set of lecture notes, entitled *Topological Groups*, by D. Montgomery (Haverford College).

Of course, for the purely topological notions of this chapter, consult the references of Chap. IV, several of which contain some facts about topological groups as well.

An understanding of topological groups is essential to a study of several topics of modern physical theory. For an introduction addressed to physicists see

M. Hamermesh, *Group Theory and Its Application to Physical Problems*, chap. 8 (Reading, Mass.: Addison-Wesley, 1962).

This text presumes that the reader is familiar with quantum mechanics. It offers applications of the theory to the splitting of atomic levels in crystalline fields, to atomic spectra, and to nuclear structure.

Two other references for physical applications, the first of which covers the elements of quantum mechanics as well as its relation to group theory, are

H. Weyl, *The Theory of Groups and Quantum Mechanics*, 2d ed. (New York: Dover, undated).

E. Wigner, *Group Theory and Its Application to the Quantum Mechanics of Atomic Spectra* (New York: Academic Press, 1959).

EXERCISES

A Give a detailed proof that the real numbers, with the usual topology and the additive group structure, form a topological group. That is, show that addition and negation are continuous functions and that $\{0\}$ is a closed set.

B Verify that real n-space, with the usual euclidean metric topology and the coordinatewise addition of real n-vectors, is a topological group. (*Hint:* Use the projections onto coordinate spaces.)

C Show that quotients of continuous real-valued functions are continuous if they are defined at all. That is, given two continuous functions f and g from some space X to $\mathbf{R}$ (with the usual topology), where $g(x)$ is never zero, show that the function

$$\frac{f}{g}: X \to \mathbf{R}: \frac{f}{g}(x) = \frac{f(x)}{g(x)}$$

is continuous.

D Prove in detail that the union of the two axes is a nonhomogeneous subspace of the plane, and hence that this space, just as $\mathbf{I}$, could be the underlying space of no topological group. In performing this exercise be careful not to use a concept (such as boundedness) which is not a *topological* property of the set in question, and be sure you understand why you must take this care.

E Prove that every finite T_1-space is also T_2.

F Show that if $n > 0$ is an integer and U is a neighborhood of e in a topological group, there is a symmetric neighborhood W of e such that $W^n \subset U$ (here W^n means $WW \cdots W$, n factors).

G Euclidean n-space is Hausdorff. Show that, in fact, every metric space is Hausdorff.

H Show that an open subgroup of a topological group must be closed as well.

J Give a proof that T_0-ness and T_2-ness are topological properties. Also, exhibit a homeomorphism of the unit circle in the plane with its circumscribed square; this proves that roundness and straightness are not topological invariants. Be sure you make precise your function and show that it and its inverse are continuous.

K Prove that the group $\mathbf{R}/\mathbf{Z}$ of additive reals modulo the integers is (topologically) isomorphic to S^1.

L To define the topological group G of affine transformations of the line (see Exercise II.N), take the underlying space of G to be the plane $\mathbf{R} \times \mathbf{R}$ with the y-axis deleted and the product on G to be $(a_1,b_1,)(a_2,b_2,) = (a_1a_2,\, a_1b_2 + b_1)$. Show that G is a topological group and that projection onto the nonzero real first-coordinate space is a morphism. Show also that if $H = \{(1,b): b \in \mathbf{R}\}$, then the coset space G/H is homeomorphic to $\mathbf{R}$, although this quotient map is not a morphism.

M Let (G,m) be a (algebraic) group, and let τ be a Hausdorff topology on the underlying set G. Show that the function on $G \times G$ sending (g_1,g_2) to $g_1g_2^{-1}$ in G is continuous on $\tau \times \tau$ to τ if and only if (G,m,τ) is a topological group. Thus you have proved that the two requirements for a topological group, continuity of both multiplication and inversion, may be replaced by the equivalent single requirement that the above function be a map.

N Show that every coset space is a homogeneous space. Do this by showing first that each element of G defines, by left multiplication, a homeomorphism of a coset space G/H of G.

P Prove that the evaluation morphism $e_t: C \rightarrow \mathbf{R}$ from the group of real-valued functions on $\mathbf{I}$ to $\mathbf{R}$ is homeomorphic to a direct-product projection for each $t \in \mathbf{I}$. That is, for each $t \in \mathbf{I}$, define a homeomorphism $h_t: C \rightarrow S_t \times \mathbf{R}$, where $S_t = Ker\,(e_t)$, so that if p_t is the projection of $S_t \times \mathbf{R}$ onto its second factor, then the following diagram is commutative:

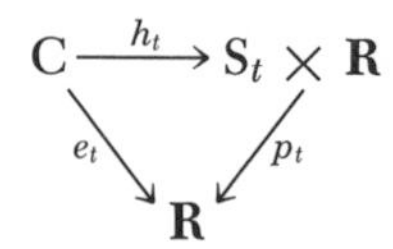

Q Show that an *algebraic* morphism $f: G \rightarrow G'$ of topological groups is open iff it is open at $e \in G$ (that is, carries neighborhoods of the identity in G to neighborhoods of the identity in G'). This important consequence of the continuity of the group operations means that we need only check openness at one point, which often simplifies arithmetic.

R An **embedding** of a space X in a space Y is a function $f: X \rightarrow Y$ which defines a homeomorphism of X with $Im\,(f)$ (by restriction of the range of f), where $Im\,(f)$ is given its relative topology in Y. Show that there is no embedding of S^1 in $\mathbf{R}$. Then argue that this gives a purely topological proof that the map $exp: \mathbf{R} \rightarrow S^1$ is not homeomorphic to a direct-product projection.

S Show that the torus $S^1 \times S^1$ is a 2-dimensional group manifold. That is, exhibit an embedding of the plane in the torus whose image is a nucleus of the torus.

T Prove that the property of being the underlying space for some topological group is a topological invariant. (This property, for a space X, is sometimes expressed by saying that X is a **group space.**) Exhibit a topological group structure for the interval $(0,1)$.

U Since $\mathbf{Z} \subset \mathbf{R}$, the gaussian integers $\mathbf{Z} \times \mathbf{Z}$ form a subgroup of the plane $\mathbf{R} \times \mathbf{R}$, the subgroup of all ordered pairs of integers. Exhibit an isomorphism of the torus with the quotient of the plane modulo the gaussian integers,

$$\frac{\mathbf{R}}{\mathbf{Z}} \times \frac{\mathbf{R}}{\mathbf{Z}} \cong \frac{\mathbf{R} \times \mathbf{R}}{\mathbf{Z} \times \mathbf{Z}}.$$

(Be sure to show that the inverse of your isomorphism is continuous.)

W Prove that the closure of a subgroup is always a subgroup, and that the closure of a normal subgroup is normal. Then give an example of a subgroup which is neither open nor closed.

The interior of a subgroup need not be a subgroup; for example, the interior of the singleton subgroup $\{0\}$ of $\mathbf{R}$ is empty. Is it true that the nonempty interior of a (normal) subgroup is also a (normal) subgroup?

PROBLEMS

AA Quaternions Consider $\mathbf{Q} = \mathbf{C} \times \mathbf{C}$, with the coordinatewise addition of complex pairs, $(c_1,c_2) + (d_1,d_2) = (c_1 + d_1, c_2 + d_2)$. Under this operation, $\mathbf{Q}$ is an abelian group; the identity is $(0,0)$, which we shall usually denote by "0." There is also a multiplication possible in $\mathbf{Q}$: a pair of complex pairs has product $(c_1,c_2)(d_1,d_2) = (c_1d_1 - \bar{c}_2d_2, c_2d_1 + \bar{c}_1d_2)$, where the operations on coordinates are those of the complex numbers, and the bar denotes complex conjugation. Elements of $\mathbf{Q}$ are called quaternions; the complex numbers (and therefore the reals) may be thought of as quaternions by the identification of $c \in \mathbf{C}$ with $(c,0) \in \mathbf{Q}$; on this subset of $\mathbf{Q}$ the sums and products defined in $\mathbf{Q}$ agree with those of $\mathbf{C}$ (or $\mathbf{R}$). The quaternion $(1,0)$ is a multiplicative identity for all of $\mathbf{Q}$; we shall usually denote this element by "1." But the multiplication in $\mathbf{Q}$ differs from that of $\mathbf{C}$ or $\mathbf{R}$ in that it is not commutative; $(0,1)(i,0) \neq (i,0)(0,1)$.

There is a conjugation in $\mathbf{Q}$, $\overline{(c_1,c_2)} = (\bar{c}_1, -c_2)$, and it has algebraic properties like those of complex conjugation; for all members q_1 and q_2 of $\mathbf{Q}$,

$$\bar{\bar{q}}_1 = q_1 \qquad \text{and} \qquad \overline{(q_1q_2)} = \bar{q}_2\bar{q}_1.$$

(Note, however, that conjugation in **Q** "preserves" products but *reverses* their order.) Furthermore, $q\bar{q}$ is always a nonnegative real! It is called the "norm" of q, $N(q)$, and $N(q) = 0$ iff $q = 0$. If $q \neq 0$, then $q([1/N(q)]\bar{q}) = (1,0)$, so there exist multiplicative inverses for the nonzero quaternions which form a group under this product.

Hence the quaternions, with these operations, have the following algebraic properties of the real and complex fields: $(\mathbf{Q},+)$ is an abelian group, $\mathbf{Q} - \{0\}$ is a group under the multiplication, and the distributive laws hold (but the multiplicative group is not abelian). A set having sums and products which satisfy these requirements is called a "division ring," or "skew field."

Now, $\mathbf{Q} = \mathbf{C} \times \mathbf{C}$ has a product topology, and with this topology it is homeomorphic to $\mathbf{R}^4$; the additive and multiplicative groups are both topological groups. The nonzero reals are a closed, normal, multiplicative subgroup; in fact, they constitute the center of **Q**. The nonzero complex numbers are a closed subgroup which is not normal. Conjugation is continuous, and the norm is an open epimorphism of the nonzero quaternions onto the multiplicative group of positive reals. The kernel of the norm morphism is homeomorphic to the 3-sphere S^3 of unit vectors in real 4-space, $S^3 = \{(x_1,x_2,x_3,x_4) \in \mathbf{R}^4\colon \sum_{i=1}^{4} x_i^2 = 1\}$. Hence S^3 possesses a topological group structure which is nonabelian. Summing this up, the group $\mathbf{Q} - \{0\}$ is (topologically) isomorphic to $S^3 \times \mathbf{R}$.

BB **Nuclei** A neighborhood of the identity in a topological group is called a nucleus. The family $\mathfrak{N}$ of nuclei of a topological group G has the following properties:

i $M, N \in \mathfrak{N}$ implies $M \cap N \in \mathfrak{N}$,
ii $M \subset N \subset G$ and $M \in \mathfrak{N}$ implies $N \in \mathfrak{N}$,
iii $N \in \mathfrak{N}$ implies that there exists $M \in \mathfrak{N}$ such that $MM^{-1} \subset N$,
iv $N \in \mathfrak{N}$ implies that for all $g \in G$, $g^{-1}Ng \in \mathfrak{N}$,
v $\bigcap \mathfrak{N} = \{e\}$.

Since a topological group is homogeneous, we might expect that a knowledge of its topology at one point would define it everywhere. This is true in the following strong sense: given a (algebraic) group G and a family $\mathfrak{N}$ of subsets of G having the above five properties, there exists a unique topology for G under which G is a topological group and $\mathfrak{N}$ is exactly the family of nuclei.

CC **Affine Groups** The (real) affine group (defined in Exercise L) has an obvious topology under which it is homeomorphic to $(\mathbf{R} - \{0\}) \times \mathbf{R}$. With this it becomes a topological group, and the quotient functions p and q become open maps (and p is a morphism). Discuss the corresponding groups obtained by using the complex or quaternionic division rings (see Prob. AA) in place of $\mathbf{R}$. Be sure to consider the new subgroups whose definitions follow from $\mathbf{R} \subset \mathbf{C} \subset \mathbf{Q}$.

DD **A Homeomorphism Group** The set G of homeomorphisms of $(0,1)$ onto itself is a group (of transformations) under the operation of composition. There is a metric $\mathfrak{d}$ on G, $\mathfrak{d}(g,h) = lub\ \{|g(x) - h(x)|: x \in (0,1)\}$, and G is a topological group with the metric topology. For each element x of $(0,1)$ there is an open map p_x of G onto $(0,1)$, $p_x(g) = g(x)$. The subgroup $p_x^{-1}(x) = H_x$ of homeomorphisms that "fix" x is called the "isotropy subgroup" of G at x; G/H_x is homeomorphic to $(0,1)$. In fact, G is homeomorphic to $H_x \times (0,1)$ for each $x \in (0,1)$. To show this you will need a "cross section" σ_x to the projection p_x; that is, a continuous map σ_x: $(0,1) \to G$ such that $p_x \circ \sigma_x$ is the identity on $(0,1)$.

The group G' of homeomorphisms of the closed unit interval $\mathbf{I}$ (when furnished with a similar metric) is (topologically) isomorphic to G.

EE **Isometries of S^n** An isometry g of a metric space M is a function g of M onto M which preserves distances; for all x, $y \in M$, $\mathfrak{d}(x,y) = \mathfrak{d}[g(x),g(y)]$, where $\mathfrak{d}$ is the metric on M. Every isometry is a homeomorphism. The set I of all isometries of M is clearly a group. Now let M be an n-sphere S^n (and so compact); a metric $\mathfrak{D}$ on I is defined by $\mathfrak{D}(g,h) = lub\ \{\mathfrak{d}[g(x),h(x)]: x \in M\}$, and I is a topological group under the metric topology, the group of "rigid motions" and "reflections" of S^n.

Think of S^n as the subset of unit vectors of $\mathbf{R}^{n+1}$, $(x_1, \ldots, x_n) \in S^n$ iff $\sum_{i=1}^{n+1} x_i^2 = 1$. An isometry of S^n extends to a unique isometry of $\mathbf{R}^{n+1}$, and this yields an algebraic isomorphism of I with the group O_{n+1} of all isometries of $\mathbf{R}^{n+1}$ which fix the origin. This isomorphism endows O_{n+1} with a topological group structure: it is called the "orthogonal group" on $\mathbf{R}^{n+1}$.

Compute (that is, completely describe) the group O_n for $n = 0$, 1, and 2.

We may regard each group O_n as a subgroup of O_{n+1} in the following way: if $g \in O_n$ and $x = (x_1, \ldots, x_n, x_{n+1})$, then $g(x)$ is to be the point of $\mathbf{R}^{n+1}$ whose first n coordinates are the n coordinates of

$g(x_1, \ldots, x_n)$ and whose last coordinate is x_{n+1}. Put more arithmetically, if $g(x_1, \ldots, x_n) = (y_1, y_2, \ldots, y_n)$, then $g(x_1, \ldots, x_{n+1}) = (y_1, y_2, \ldots, y_n, x_{n+1})$. The subgroup O_n is not normal in O_{n+1} if $n \geq 2$.

Let $N = (0, 0, \ldots, 1)$ be the "north pole" of S^n, and let the map $p_n: O_{n+1} \to S^n$ be the evaluation at N: $p_n(g) = g(N)$ for each $g \in O_{n+1}$. The equivalence relation on O_{n+1} induced by p_n is exactly the relation of belonging to the same coset of O_n in O_{n+1}. (In the illustration below, O_n acts on S^n by doing the "same thing" to each "horizontal" slice, a homeomorph of S^{n-1}, that it does to the "equatorial" S^{n-1}. Each element of O_n leaves both N and the "south pole" $-N$ fixed.) Use this to show that the coset space O_{n+1}/O_n is homeomorphic to S^n.

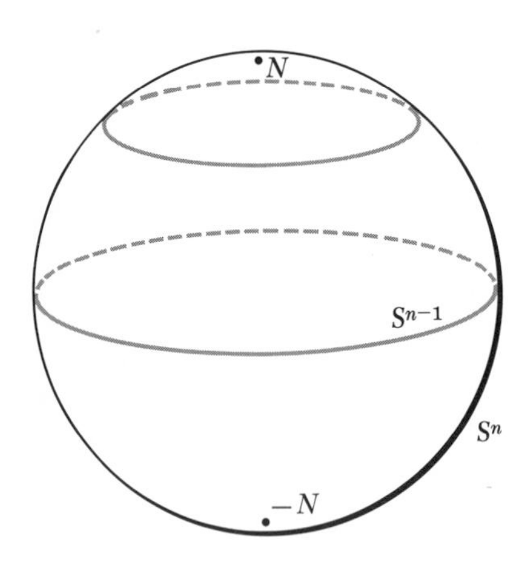

FF **The Lorentz Group** The "one-dimensional Lorentz group" L is the topological group of homeomorphisms of the plane $\{(x,t): x \text{ and } t \in \mathbf{R}\}$ which leave invariant the form $x^2 - t^2$ and do not interchange up (future) with down (past). That is, a homeomorphism h of the plane is in L iff $g(x,t) = [g_1(x,t), g_2(x,t)]$ implies $x^2 - t^2 = g_1(x,t)^2 - g_2(x,t)^2$ and $t < u$ implies $g_2(x,t) < g_2(x,u)$ for all real x,t, and u. (The members of L connect the various "normal" coordinate systems for the space-time of special relativity theory when the space involved is one-dimensional; x is the space variable, t is time, and the speed of light is 1.) The "restricted Lorentz group" is the subgroup L_0 of those members g of L for which $x < y$ implies $g_1(x,t) < g_1(y,t)$ for all real x, y, and t; hence right and left, as well as past and future, are invariant under members of L_0.

The members of L_0 are just the transformations which send (x,t) to

$$(x \cosh \theta + t \sinh \theta, x \sinh \theta + t \cosh \theta)$$

for the various real numbers θ. If we denote the above transformation by g_θ, then

$$g_\theta(x,t) = \frac{1}{(1 - \tanh^2 \theta)^{1/2}} (x + t \tanh \theta, x \tanh \theta + t).$$

The assignment of g_θ to each real θ defines an algebraic isomorphism of the additive group of real numbers onto L_0; clearly, this may be used to visit the structure of a topological group on L_0. To get L from L_0, adjoin the transform h: $h(x,t) = (-x,t)$.

(For a discussion of the three-dimensional Lorentz group, see pp. 146–149 of the reference to Weyl in this chapter.)

GG Mechanics: Hooke's Law The "state" of a set of n bodies, thought of as point masses which move about in an otherwise empty three-dimensional real world, is given by the momentary positions and velocities of the n points, all with respect to the fixed defining coordinate system of $\mathbf{R}^3$. Each position is an ordered triple of reals, and so is each velocity; hence a state of our n-body system is just a point of real $6n$-space. The set S of all states of the system is called state space; it is thus a subset of $\mathbf{R}^{6n}$. Of course, there are points of $\mathbf{R}^{6n}$ which do not correspond to allowable states of the system; for instance, no two bodies are allowed to be in the same place at a given moment. (Physical intuition suggests that S is an open subset of $\mathbf{R}^{6n}$.) If the laws of physics dictate the state $P_t(s)$ in which the system will be after the elapse of an amount of time t when it is in a state s at time $t = 0$, then P_t is a well-defined function from S to S for each $t \geq 0$. We further suppose that for each state s and each $t \geq 0$ there exists a unique state s' such that $P_t(s') = s$; that is, we assume P_t has an inverse, which we denote by P_{-t}, $P_{-t}(s) = s'$. Lastly, we assume that physical happenings are continuous in time and in state; that is, P_t is a homeomorphism of state space and $P_t(s)$ is a continuous function of argument pairs (t,s).

It is easy to see from the definition that $P_{t+u}(s) = P_t \circ P_u(s)$; accordingly, there is an algebraic morphism of the additive reals which carries t to P_t; its range is the group H of homeomorphisms of state space, and its image is called a "one-parameter subgroup of H," a "one-parameter group of transformations." If the quotient topology induced by this morphism from the reals is used in the one-parameter group, it becomes a topological group.

For an example of this situation, think of a body free to move in only one dimension, that is, along a real line, and let $x(t)$ denote the real coordinate of (the center of mass of) the body at time t. We adopt the notation $\dot{x}$ for the first derivative dx/dt of the function x, so $\dot{x}$ is the velocity of the body, and $\ddot{x}$ is then the acceleration d^2x/dt^2. For some positive integer k, Newton's k-th law claims that $F = m\ddot{x}$, where F is (the measurement of) a force on the body and M is (the measurement of) the mass of the body, $M > 0$. If one end of a relaxed spring is attached to our body at $x = 0$ and the other end is fixed, Hooke's law says that when the spring is distended by movement of the body to a point x, the force the spring exerts on the body is $F = -Kx$, where $K > 0$ is a constant depending on the particular spring used.

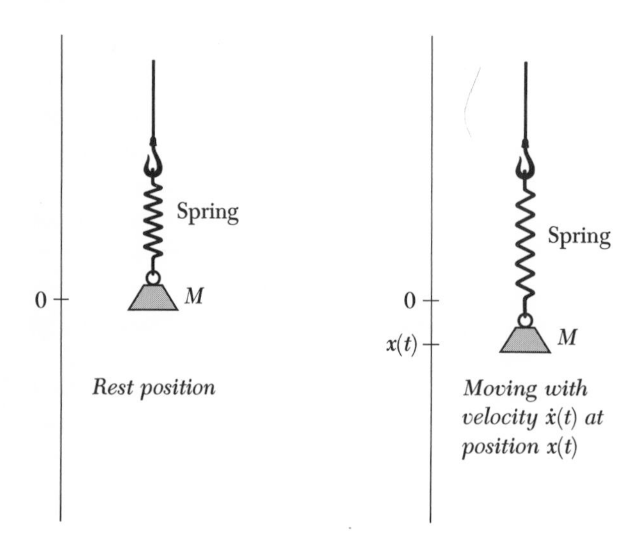

Putting this all together, we find that a body in a state (a,b) at time zero will be in the state $(x,\dot{x})$ at time t, where if $J = \sqrt{K/M}$, then

$$x = a \cos Jt + \frac{b}{J} \sin Jt,$$

$$\dot{x} = -aJ \sin Jt + b \cos Jt.$$

The homeomorphism P_t is given by $P_t(a,b) = (x,\dot{x})$. It is easy to verify that $P_{t+u} = P_t \circ P_u$; accordingly, we have defined a one-parameter group of transformations of the state space $\mathbf{R}^2$; this group is isomorphic to S^1.

The "orbit" or "trajectory" of a state s is the set $\{P_t(s): t \in \mathbf{R}\}$, an equivalence class of states under the action of the group. State space is partitioned by its orbits; what are they like for Hooke's law? How would you describe the quotient space of state space which this equivalence defines?

HH **The Second Isomorphism Theorem** State and prove an analog for topological groups of Prob. II.HH; make the assumption that every subgroup in sight is closed.

Compactness and Connectedness

CHAPTER VI

In our study of metric spaces (Chap. III) the two important notions of compactness and path-connectedness were defined. These were properties of spaces that were preserved by continuous functions; if a subset S of the domain of a map f was compact or path connected, then so was $f(S)$. But these definitions were too narrow for our present purposes; compactness, for instance, was defined only for subsets of $\mathbf{R}^n$, when the latter was given the usual metric. If an equivalent bounded metric were given $\mathbf{R}^n$ (and that is always possible), then every subset would be bounded, and since the topologies are identical for equivalent metrics, every closed set would be closed and bounded. In particular, $\mathbf{R}^n$ itself would be closed and bounded; clearly the Heine-Borel theorem would no longer be true if compact subsets of an arbitrary metric space were defined to be the closed bounded subsets. Hence the definition we borrowed from the calculus for compactness will not generalize even to metric spaces. We shall remedy this in the present chapter.

On the other hand, the definition of path-connectedness can be applied without change to any topological space; X is **path connected** iff whenever x and y are points of X there exists a map $f\colon \mathbf{I} \to X$ with $f(0) = x$ and $f(1) = y$. But there is another important concept, that of "connectedness," which also corresponds roughly to the everyday idea of being in one piece. Each path-connected space is connected, so the newer concept is the more general; nevertheless connectedness, like path-connectedness, is still preserved by every map.

CONNECTEDNESS

Recall that a partition of a set X is a family $\mathbb{S}$ of subsets of X which is mutually disjoint and whose union is X. This may be translated to say that each element x of X is in exactly one member of $\mathbb{S}$. An **open partition** of a space X is, then, a partition of X whose members are open subsets of X. In these terms, a space X is defined to be **connected** iff every open partition of X contains X as a member. This means that the only open partitions of X are the trivial ones, $\{X\}$ and $\{X, \varnothing\}$. Suppose X is not connected, so that there is a partition $\mathbb{S}$ with $X \notin \mathbb{S}$; clearly, $\mathbb{S}$ contains a nonempty subset A of X. Then $B = \bigcup(\mathbb{S} - \{A\})$ is open and $\{A,B\}$ is also an open partition of X; hence *a space X is connected iff it cannot be expressed as the union of two disjoint, proper, open subsets.* ∎ Furthermore, A and B are complements of one another, so they are both closed in X; we could have said that X is connected iff it cannot be expressed as the union of two disjoint, proper, closed subsets. A space which is not connected is called **disconnected,** and a two-member open partition of a space X is a **separation** of X. Hence X is disconnected iff there exists a nontrivial separation of X. Still another characterization is that *X is connected iff the only subsets of X which are both open and closed are $\varnothing$ and X.* ∎

An example of a disconnected subspace of the plane is the complement X of the two axes; the four quadrants of X form a nontrivial open partition of X. We shall now prove that the unit interval $\mathbf{I} = [0,1]$ is connected. Let $\mathbb{S}$ be a separation of $\mathbf{I}$, and let A be the member of $\mathbb{S}$ with $0 \in A$. Form the set $J = \{x \in \mathbf{I}\text{: } [0,x] \subset A\}$; the goal is to show $J = \mathbf{I}$. Let $b = lub\, J$, so that if $y < b$, then $[0,y] \subset A$. Clearly, $b \in A$, since A is closed and every neighborhood of b contains members of A. But if $b < 1$, then, since A is open, there is a $\delta > 0$ with $(b - \delta,\ b + \delta) \subset A$, and this is contradictory, since $[0,\ b - \delta] \subset A$ implies $[0,\ b + \delta/2] \subset A$, or $b + \delta/2 \in J$. Hence $b = 1$, since $b > 1$ is impossible. But if $b = 1$, then $A = \mathbf{I}$, because $A \supset J \supset [0,1)$ and the complement of A in $\mathbf{I}$ must be an open subset of $\{1\}$. This guarantees that each separation of $\mathbf{I}$ is trivial, so $\mathbf{I}$ is connected. We shall say that $\mathbf{I}$ is a connected subset of $\mathbf{R}$; more generally, a *subset* S of a space X is called **connected** if S as a *subspace* is a connected space.

Our next result is suggested by the fact that two boards are connected if they are nailed together somewhere. The union of two overlapping connected sets is connected. More precisely, if S and T are connected (as subsets of some universal space U), and $S \cap T \neq \varnothing$, then $S \cup T$ is connected. To see this, assume A is both open and closed in $S \cup T$; if $A \cap S = S$, then $A \cap T \supset S \cap T \neq \varnothing$, so $A \supset T$ and $A = S \cup T$. On the other hand, if $A \cap S \neq S$, then $A \cap S = \varnothing$, since S is connected; a similar argument shows $A \cap T = \varnothing$, and A must be empty. Hence $S \cup T$ is connected. An immediate corollary is the more general statement that *if a family* $\mathcal{S}$ *of connected sets has the property that every pair S,T of its members overlap, then* $\bigcup \mathcal{S}$ *is connected.* For, if A is open and closed in $\bigcup \mathcal{S}$ and $A \neq \varnothing$, then there is some connected member S of $\mathcal{S}$ lying entirely in A (why?). Thus, if $T \in \mathcal{S}$, then $T \subset A$, and $A = \bigcup \mathcal{S}$ is connected. ■

The relation between points of a space T of belonging together to a connected subset, xRy iff there is a connected $S \subset T$ with $\{x,y\} \subset S$, is an equivalence relation. A member C of the associated partition $\mathcal{C}(T)$ of T is called a **component** of T. A component C of T is clearly a maximal connected subset of T. If T is connected, then $\mathcal{C}(T) = \{T\}$, or T is disconnected iff $\mathcal{C}(T)$ has more than one member. If $t \in T$, then the **component of** t **in** T is the member of $\mathcal{C}(T)$ in which t lies. Let S be a connected subset of the space T, and let A be both open and closed in the closure S^- of S. Now, $A \cap S$ is either $\varnothing$ or S; assume $A \cap S = S$ (or apply this argument to $S^- - A$). A is closed in S^- and $A \supset S$, so $A \supset S^-$, and therefore $A = S^-$. Thus S^- is connected if S is; *the closure of a connected set is connected. Components are always closed,* since they are maximal connected sets. ■

The continuous image of a connected set is connected, because the inverse image of an open partition of the image set is an open partition of the domain set. ■ In fact, we may see here a guiding motivation for the definition of connectedness: a space is connected iff every map from it to a discrete space is constant. Put another way, the two-element set $\{0,1\}$ with discrete topology is a kind of minimal disconnected set, and if f is a map from T to $\{0,1\}$, then $f^{-1}(0)$ and $f^{-1}(1)$ are both open and closed in T; if neither of these sets is empty, then T is disconnected.

Exercise A

COMPONENTS OF GROUPS

Let G be a topological group, and for each $g \in G$ let C_g be the component of g in G. We assert that C_g is the translate $L_g(C_e)$ of the component of e by L_g; $L_g(C_e)$ is a continuous image of C_e, so it is connected, and $g \in L_g(C_e)$, so $C_g \supset L_g(C_e)$. Similarly, $L_{g^{-1}}(C_g)$ is a connected set containing

e, so $C_e \supset L_{g^{-1}}(C_g)$; the left translation L_g preserves set inclusion, so $L_g(C_e) \supset L_g \circ L_{g^{-1}}(C_g) = C_g$; thus $L_g(C_e) = C_g$.

We now show that C_e is a closed normal subgroup. It is clearly closed since all components are closed, but C_e^{-1} is connected and contains e, so $C_e^{-1} \subset C_e$. Since inversion preserves set inclusion, we have $(C_e^{-1})^{-1} \subset C_e^{-1}$, so $C_e^{-1} = C_e$. Also, if g and h are both in C_e, then gC_e is a connected set containing g and gh, so $gh \in C_e$. This shows that C_e is a subgroup. $C_e g = C_g$ is connected, so $g^{-1}(C_e g)$ is connected and contains e; C_e is normal. The quotient group G/C_e is called the **group of components** of G, denoted by $\mathcal{C}(G)$. We shall regard it as a group without topology, though G does bestow one on it (see Prob. AA).

A continuous function must carry components into components, so a morphism $f: G \to G'$ of topological groups must define a morphism $\mathcal{C}(f)$ of the group of components of G to the group of components of G'; $\mathcal{C}(f)(gC_e)$ is the component of $f(g)$ in G'. Further, if f' is another morphism and the composite $f' \circ f$ is defined, then $\mathcal{C}(f' \circ f) = \mathcal{C}(f') \circ \mathcal{C}(f)$, and surely $\mathcal{C}(1_G) = 1_{\mathcal{C}(G)}$. Thus the "function" $\mathcal{C}$ may be described as assigning to each topological group a group, and to each morphism a morphism, such that composites are preserved and identities go to identities (or "identities are preserved"):

$$\begin{array}{ccccc} G & \xrightarrow{f} & G' & \xrightarrow{f'} & G'' \\ {\scriptstyle q}\downarrow & & {\scriptstyle q'}\downarrow & & {\scriptstyle q''}\downarrow \\ \mathcal{C}(G) & \xrightarrow{\mathcal{C}(f)} & \mathcal{C}(G') & \xrightarrow{\mathcal{C}(f')} & \mathcal{C}(G'') \end{array}$$

This sounds very much like a morphism of some structure; and indeed it is, with the proper language—that of "categories." We shall not formalize this idea, but several other "functions" like $\mathcal{C}$ will be seen as we go along.

Exercises B and P

PATH COMPONENTS

If T is a space, a **path** in T is a map $a: \mathbf{I} \to T$ from $[0,1]$ into T. The path a is said to **begin** at $a(0)$ and **end** at $a(1)$, or to be a path from $a(0)$ to $a(1)$, or to be a path connecting $a(0)$ and $a(1)$. If a is a path from s to t in T, then there is an **inverse path** b from t to s in T: $b(u) = a(1 - u)$. Hence the relation on T of being connected by a path, sRt iff there is a path beginning at s and ending at t, is symmetric. The constant path at s,

$a(\mathbf{I}) = \{s\}$, shows sRs. This relation is also transitive. Let a be a path from r to s, and b a path from s to t, in T; then there is a **product path** $a \cdot b$ from r to t with $a \cdot b(u) = a(2u)$ if $0 \leq u \leq ½$ and $a \cdot b(u) = b(2u - 1)$ if $½ \leq u \leq 1$. This construction depends on the simple homeomorphism $h: [0,1] \to [0,2]: h(u) = 2u$, the homeomorphism of $[0,1]$ with $[1,2]$ which adds 1 to everything, and the gluing together of continuous functions which agree on the overlap of their closed domains. Draw a picture of it.

Hence R is an equivalence; for each $t \in T$ the class of elements path connected to t is called the **path component** of t in T, denoted by $[t]$. Since $a(\mathbf{I})$ is connected for each path a, $[t]$ is the union of a family of connected sets which overlap at t; $[t]$ is connected. The path components of T may not be the same as the components (see Exercise D); probably the path components correspond more closely to the intuitive notion of "the pieces of T." Obviously, in many cases they are the same.

For each space T let $\pi_0(T)$ denote the *set* of path components of T. If $f: T \to U$ is continuous, then $f([t])$ is path connected for each $t \in T$. This is so because if a is a path from t to s, then $f \circ a: \mathbf{I} \to U$ is a path from $f(t)$ to $f(s)$. Hence f induces a function $\pi_0(f): \pi_0(T) \to \pi_0(U): \pi_0(f)([t]) = [f(t)]$, a path component of U (the vertical arrows in the diagram represent the respective quotient functions):

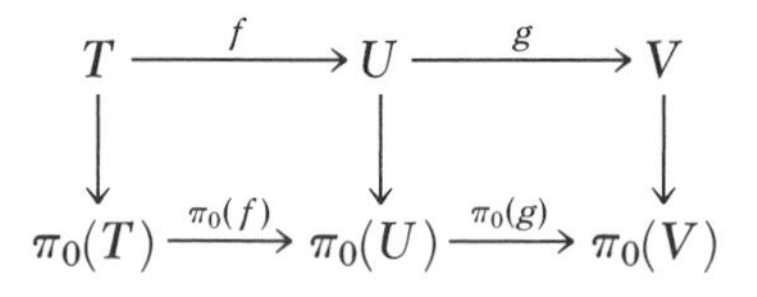

$$\begin{array}{ccccc} T & \xrightarrow{f} & U & \xrightarrow{g} & V \\ \downarrow & & \downarrow & & \downarrow \\ \pi_0(T) & \xrightarrow{\pi_0(f)} & \pi_0(U) & \xrightarrow{\pi_0(g)} & \pi_0(V) \end{array}$$

Just as does the component "function" $\mathcal{C}$, the path-component "function" π_0 preserves both compositions of maps and identities. Clearly, if f is a homeomorphism of T with U, then $\pi_0(f)\pi_0(f^{-1}) = \pi_0(f \circ f^{-1}) = \pi_0(1_U)$, etc., so $\pi_0(f): \pi_0(T) \cong \pi_0(U)$. Hence the isomorphism class of $\pi_0(T)$ is a topological invariant of T, and with this *algebraic invariant* of topological spaces, we may hope to distinguish between nonhomeomorphic spaces. Explicitly, T is not homeomorphic to U if one has fewer pieces than the other. These comments may, of course, be made for $\mathcal{C}(T)$ as well.

If G is a topological group, then the path component G_0 of e in G is a normal subgroup: If a and b are paths beginning at e, then $a^{-1}: \mathbf{I} \to G: a^{-1}(u) = a(u)^{-1}$ ends at $a(1)^{-1}$ and the path ab, $ab(u) = a(u)b(u)$, ends at $a(1)b(1)$; the path $g^{-1}ag$, which has values $g^{-1}a(u)g$ for each $u \in \mathbf{I}$, ends at $g^{-1}a(1)g$. ■ It is an exercise to check that $\pi_0(G) = G/G_0$; it has a "quotient" group structure and is called the **group of path components** of G. Just as before, the "function" π_0 sends morphisms of topological groups to morphisms of the (algebraic) groups of components, and identities go to identities.

Exercises C, D, and E

COMPACTNESS

A closed bounded subset of $\mathbf{R}^n$ is called compact. The Heine-Borel theorem (see Chap. III; see Prob. III.FF for the general case) states that the compactness of a subset T of $\mathbf{R}^n$ is equivalent to the property that each cover of T by open sets contains a finite subcover. This latter property is stated in purely topological terms; it does not mention boundedness, which is a metric concept. Hence we may use this theorem to generalize the notion of compactness to apply to an arbitrary topological space.

If (T,τ) is a space, then $\mathcal{S}$ is an **open cover** of T iff $\mathcal{S} \subset \tau$ and $\bigcup \mathcal{S} = T$. A **finite open cover** $\mathcal{R}$ of T is an open cover with only a finite number of members. (*Warning:* The *members* of $\mathcal{R}$ may well be infinite sets!) The space (T,τ) is **compact** (or "bicompact") iff for each open cover $\mathcal{S}$ of T there exists a finite cover $\mathcal{R}$ of T with $\mathcal{R} \subset \mathcal{S}$; then $\mathcal{R}$ is called a **subcover** of T in $\mathcal{S}$. A subset A of T is **compact** if A is compact (as a space) with the relative topology. The Heine-Borel theorem now assures us that if $A \subset \mathbf{R}^n$ this broader definition of compactness agrees with our former one. Of course, there are immediately many examples of compact and noncompact subsets of $\mathbf{R}^n$. For another example, consider any infinite set, say $\mathbf{R}$, with its discrete topology (every subset is open); the family of all singleton subsets is an open cover in which there is no finite subcover. Conversely, every finite space is compact, as is every indiscrete space.

The continuous image of a compact space is compact. The proof is quite the same as for the special case of subspaces of $\mathbf{R}^n$ (see Chap. III). That is, an open cover of the image defines an open cover of the domain, and a finite open subcover of that domain cover yields a finite subcover in the cover of the image. ∎

Exercises H and Q

ONE-POINT COMPACTIFICATION

The real line $\mathbf{R}$ is not compact (with the usual topology). However, we may manufacture a compact space $\mathbf{R}^*$ which contains $\mathbf{R}$ as a (proper) subspace by "adjoining the point at infinity." Specifically, let ∞ stand for an object which is not in $\mathbf{R}$, and let the set $\mathbf{R}^* = \mathbf{R} \cup \{\infty\}$ include all the real numbers and ∞. Give $\mathbf{R}^*$ the topology μ^* which contains all the open sets of the usual topology μ for $\mathbf{R}$, plus all subsets M of $\mathbf{R}^*$ which are complements (in $\mathbf{R}^*$) of compact subsets of $\mathbf{R}$: $\mu^* = \mu \cup \{M: \mathbf{R}^* - M \text{ is compact in } \mathbf{R}\}$.

Rather than checking directly that μ^* is a topology for $\mathbf{R}^*$ and that $\mathbf{R}^*$ is compact ($\mathbf{R}$ is clearly a subspace of $\mathbf{R}^*$), we resort to a picture which gives an intuitive feeling for this construction. Let S^1 be the unit circle in the plane, with the usual relative topology. Define the map $m: S^1 \to \mathbf{R}^*$ as $m(0,1) = \infty$ and if $x \neq (0,1)$, $m(x)$ is the unique point at the intersection of the real axis $\mathbf{R} \subset \mathbf{C}$ and the line through x and $(0,1)$. That m is a homeo-

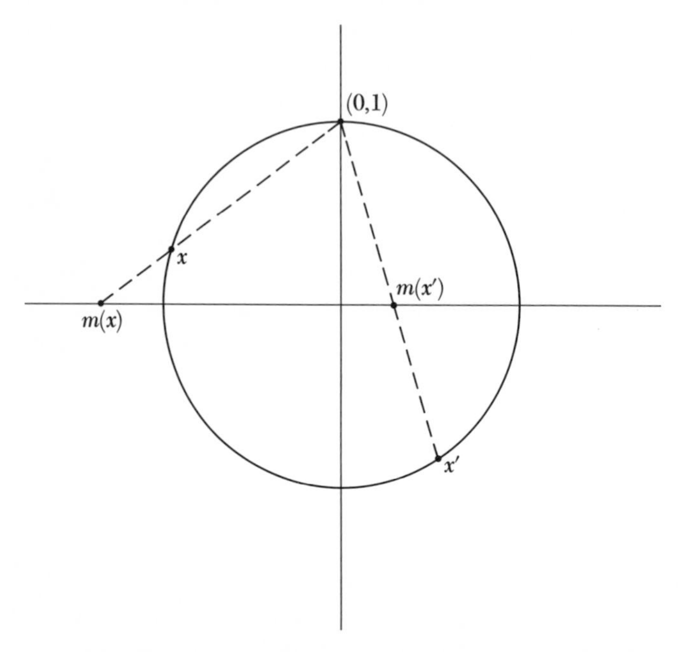

morphism when restricted to $S^1 - \{(0,1)\}$ is geometrically clear. But m makes a set whose complement in $\mathbf{R}^*$ is compact correspond to an open set at $(0,1)$, and conversely. Thus m is a homeomorphism of $\mathbf{R}^*$ with S^1.

This construction is an instance of a general method of building from a space T a compact space T^* which has T as a subspace. The **one-point compactification** (T^*,τ^*) of a topological space (T,τ) is the set $T^* = T \cup \{\infty\}$, where ∞ denotes an object not in T, with the topology defined as follows: all members of τ are in τ^* and each set $U \subset T^*$ is in τ^* if $U' = T^* - U$ is a closed compact subset of T.

Let us first check that this does give a topology τ^* for T^*. Clearly, $\varnothing$ and (since $\varnothing$ is trivially compact) T^* are in τ^*. If $U, V \in \tau^*$ and both U and V are in τ, then $U \cap V \in \tau \subset \tau^*$. Also, if $U \in \tau$ and V' is closed and compact, then $V \cap T$ is open in T, so $U \cap V \in \tau \subset \tau^*$. But if both U' and V' are closed and compact in T, then so is $U' \cup V' = (U \cap V)'$, since *the union of two compact sets is always compact* (why?). ■ Hence τ^* is closed under finite intersections. Now we examine the union of a family $\mathcal{V} \subset \tau^*$. If $\infty \notin \bigcup \mathcal{V}$, so that $\mathcal{V} \subset \tau$, then $\bigcup \mathcal{V} \in \tau \subset \tau^*$. But if $\infty \in U \in \mathcal{V}$, then $(\bigcup \mathcal{V})'$ is a closed subset of the compact set $U' \subset T$, and *every closed subset of a compact set is compact.* This is true because an open cover $\mathcal{S}_0$ of $(\bigcup \mathcal{V})'$ can be augmented to an open cover $\mathcal{S}_1 = \mathcal{S}_0 \cup [U' - (\bigcup \mathcal{V})']$, and a finite subcover in $\mathcal{S}_1$ surely yields a finite subfamily of $\mathcal{S}_0$ which covers $(\bigcup \mathcal{V})'$. ■ Thus $\bigcup \mathcal{V}$ has a complement which is closed and compact in T, $\bigcup \mathcal{V} \in \tau^*$, and τ^* is a topology for T^*.

Obviously, T is a subspace of T^*, and T^* is compact. To see this, let $\mathcal{S}$ be an open cover of T^*; $\infty \in \bigcup \mathcal{S}$, so there is an $S_0 \in \mathcal{S}$ with $\infty \in S_0$, and S_0' is compact. But $\mathcal{S}$ is an open cover of S_0', so there are members $S_1, S_2, \ldots, S_n$ of $\mathcal{S}$ with $S_0' \subset S_1 \cup S_2 \cup \cdots \cup S_n$; hence $T^* \subset S_0 \cup S_1 \cup \cdots \cup S_n$, and $\mathcal{S}$ contains a finite subcover. Now let U be a compact Hausdorff space with $u \in U$ and $f\colon T \to U - \{u\}$ a homeomorphism. Then f extends to a map $F\colon T^* \to U$ (why?). By Exercise Q, F is closed and thus is a homeomorphism. The example given above could have been

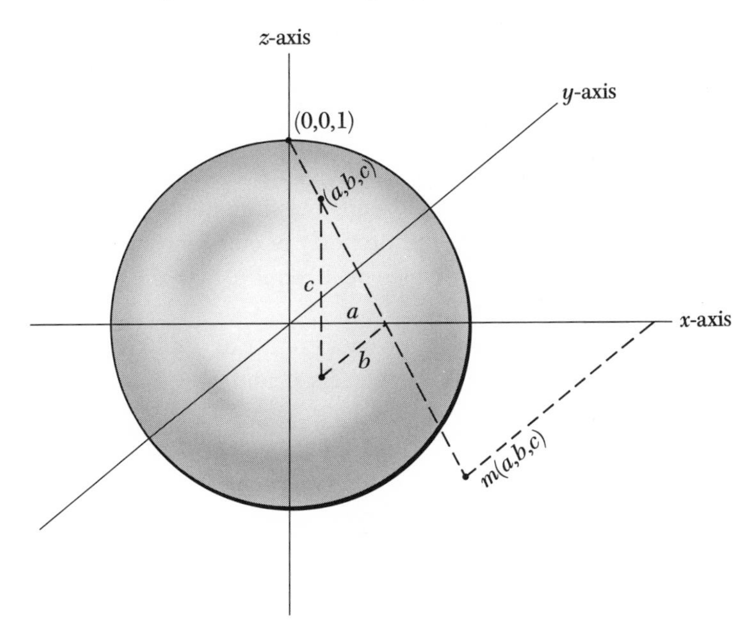

abbreviated, then, by observing that $S^1 - \{(0,1)\}$ is a homeomorph of **R**. A very similar construction shows that the "Riemann sphere" $S^2 \subset \mathbf{R}^3$ is (a homeomorph of) the one-point compactification $\mathbf{C}^*$ of the complex plane $\mathbf{C} = \mathbf{R}^2$. The added point ∞ of $\mathbf{C}^*$ corresponds to the "north pole" (0,0,1) of S^2, and the projection m is defined analogously to the lower-dimensional case of S^1. Conversely, a 2-sphere "becomes" a 2-plane when a single point is deleted. You may visualize the process by starting with a rubber plane, molding its edge (which it does not have) upward to form a cup, and then, finally, $S^2 - \{(0,0,1)\}$; the "point at infinity" now will just fill the hole. A surprise in this example is that both S^1 and S^2 are homogeneous. This is not always the case; examine in your mind, for instance, the one-point compactification of a subspace of the plane which is the union of two parallel lines.

Exercise J

REGULARITY AND T_3-SPACES

Suppose K is a compact subset of a Hausdorff space T, and let $t \notin K$. The compact set K is necessarily closed in T. We now argue that disjoint open sets U and V may be found in T with $t \in U$ and $K \subset V$; thus the defining property of T_2-spaces, the ability to separate points with open sets, still is valid when one point is replaced by a compact set. For each $k \in K$ choose open sets U_k and V_k with $t \in U_k$, $k \in V_k$, and $U_k \cap V_k = \varnothing$; then $\{V_k: k \in K\}$ is an open cover of K. Choose a finite subcover $V_{k_1}, V_{k_2}, \ldots, V_{k_n}$ and define $V = \bigcup\{V_{k_i}: i = 1, 2, \ldots, n\}$; V is open and $K \subset V$. Similarly, if $U = \bigcap\{U_{k_i}: i = 1, 2, \ldots, n\}$, then U is open and $t \in U$. But if $x \in U$, then x can be a member of no V_{k_i}, since it is in every U_{k_i}; hence $x \notin V$. This implies $U \cap V = \varnothing$ and ends our argument.

Now we wish to turn this property of compact subsets of T_2-spaces into a new separation axiom, which will require that the property be held by every closed subset of a space.

A space T is **regular** iff, given a point t of T and a closed subset S of T with $t \notin S$, there are disjoint open sets U and V of T with $t \in U$ and $S \subset V$.

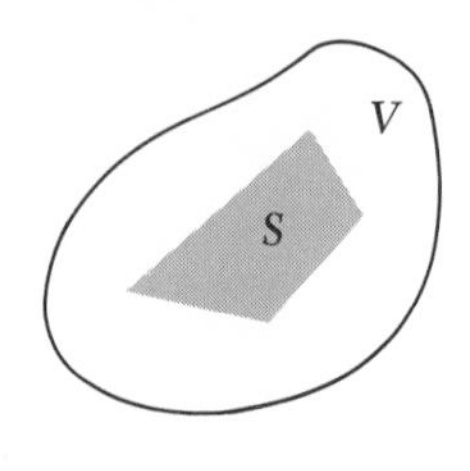

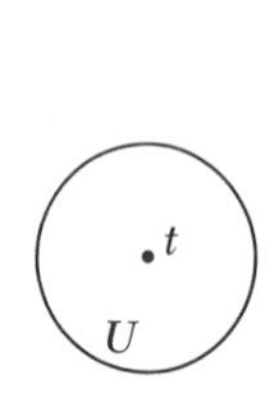

This definition may be rephrased: a space T is regular iff for each $t \in T$ and neighborhood M of t there is a closed neighborhood N of t with $N \subset M$. The equivalence of these definitions is immediate if we let S' correspond to M, V' to N, and U to N° (the interior of N); do the proof in your head.

Since singletons are closed sets in a T_1-space, every regular T_1-space is Hausdorff (or T_2), but a discrete space having two members is a regular space which is not T_0 (or T_1 or T_2). However, we can now add a new separation axiom to our list: a T_3-**space** is one which is both T_1 and regular. As observed above, each T_3-space is T_2 (and T_1 and T_0).

A topological property is called **hereditary** if it holds for every subspace of a space X when it holds for X. An example is T_1-ness: if each singleton is closed in X, then each singleton of a subset A of X is the intersection of a closed subset of X with A, and so is closed. *Regularity and T_3-ness are hereditary properties;* since T_1-ness is hereditary, we need only prove this statement for regularity. If X_0 is a subspace of a regular space X_1, and S_0 is closed in X_0 with $x \in X_0 - S_0$, then there is a closed subset S_1 of X_1 with $S_1 \cap X_0 = S_0$. Because x cannot be a member of S_1, we may find open disjoint sets U_1 and V_1 in X_1 with $x \in U_1$ and $S_1 \subset V_1$. Thus there are open disjoint subsets $U_0 = U_1 \cap X_0$ and $V_0 = V_1 \cap X_0$ in X_0 with $x \in U_0$ and $S_0 \subset V_0$. Hence X_0 is regular when X_1 is regular. ■

Examples of regular and T_3-spaces abound. *Every metric space is T_3.* One has only to find a ball $\mathfrak{b}(\varepsilon,x)$ disjoint from a closed set S; then the closed ball of radius $\varepsilon/2$ at x is a closed neighborhood of x which is disjoint from S. ■ Also, *every compact Hausdorff space is a T_3-space.* This is easy to see; a closed subset of such a space is always compact, and we showed above that compact subsets of Hausdorff spaces could be separated from points with disjoint open sets. ■ Hence our list of examples of T_3-spaces includes all the metric spaces and all subspaces of compact Hausdorff spaces.

The above fact is worth inspection. It says that, for Hausdorff spaces, compactness implies regularity. But the conclusion, regularity, is a "local" property in the sense that a space T is regular iff, for each point $t \in T$ and neighborhood M of t, t has a closed neighborhood N lying in M. The same closed neighborhood N would do for all larger neighborhoods $M' \supset M$, so we can decide if a space T is regular or not by examining its character in some neighborhood (perhaps quite small) of each of its points. Compactness, on the other hand, cannot be established by merely looking at some region about each point; it is a property of the whole space, viewed all at once. Such a property as compactness is called "global." We shall not give a formal definition of the terms "local" and "global," but they will be useful notions, even if they are imprecise.

Now, our fact at hand has a global hypothesis, but a local conclusion.

You may suspect, then, that the statement can be generalized or strengthened. To do that, we need a new definition: a space T is called **locally compact** iff there is at least one compact neighborhood of each point of T. Clearly, if T is compact, then T is locally compact. However, $\mathbf{R}^n$ is locally compact yet not compact. Every *closed* subspace (that is, closed subset with relative topology) of a locally compact space is also locally compact, because the intersection of a closed set with a compact set must be compact.

An example of a subspace T of the plane which is not locally compact is the union of the right half-plane $\{(x,y): x > 0\}$ with the origin $\{0\}$. Every neighborhood of the origin in this space must fail to be compact by being too close to points of the y-axis. More specifically, if N is a neighborhood of 0 in T, then there is an $\varepsilon > 0$ with $\mathcal{b}(\varepsilon,0) \cap T \subset N$. Let the sequence $S_1, S_2, \ldots$ of open subsets of T be defined by the rule $S_i = [\mathcal{b}(\varepsilon/2,0) \cap T] \cup \{(x,y): x > 1/i\}$; obviously, $\bigcup\{S_1, S_2, \ldots\} = T$. However, the union of any finite subfamily $\{S_{i_1}, S_{i_2}, \ldots, S_{i_n}\}$ of this open cover of N is simply the largest member of the subfamily; since for no index i is $N \subset S_i$, there exists no finite subcover for N. This shows that there is no compact neighborhood of 0 in T.

There is the following connection between the notions of compactness and local compactness: *a Hausdorff space T is locally compact iff its one-point compactification T^* is Hausdorff.* If T^* is Hausdorff and $x \in T$, let U and V be disjoint open sets containing x and ∞, respectively; the complement of V in T^* is a closed, and so compact, neighborhood of x which lies inside T. Conversely, if T is locally compact and Hausdorff, then to show T^* to be Hausdorff we need only find disjoint open sets U and V separating the added point ∞ from an arbitrary point x of T. But if K is a compact neighborhood of x in T, then K is closed (since T is T_2), so $U = K^\circ$ and $V = K'$ will do very well. ■

But look: *a locally compact Hausdorff space T is regular,* since it is a subspace of the compact Hausdorff space T^*. ■ This is the strengthened statement we wished; its hypothesis is now a local requirement.

There is another remark possible here: *at each point of a locally compact Hausdorff space the closed compact neighborhoods form a local base.* That is, for each point t of a locally compact Hausdorff space T, and each neighborhood U of t, there is a closed compact neighborhood V of t, $V \subset U$. The proof is trivial. Let C be a compact neighborhood of t, and choose a neighborhood V of t with $V^- \subset U \cap C$ (we can always find V, since T is regular). The closed subset V^- of the compact set C must be compact. ■

Exercise M

TWO APPLICATIONS TO TOPOLOGICAL GROUPS

As a final example of regularity, we show that *every group space is* T_3. Since left multiplications are homeomorphisms, it need only be shown that if U is a nucleus (a neighborhood of e) in a topological group G, then there is a closed nucleus V^- lying in U. But recall that there is a nucleus V inside U, with $VV^{-1} \subset U$. We assert $V^- \subset U$, for if $g \in V^-$, then every neighborhood of g intersects V, and so $gV \cap V \neq \varnothing$. Choose a point $gv_1 = v_2$ in this intersection. Clearly, $g = v_2 v_1^{-1} \in VV^{-1} \subset U$, so $V^- \subset U$. ∎

We have observed (in Chap. V) that every quotient map of a group onto its coset space is an open map. The proof of this amounts to the fact that if A and B are subsets of a group and A is open, then AB is open; hence, when H is a subgroup and A is open, $q(A)$ is open in G/H, since $q^{-1}[q(A)] = AH$ is open in G. But such quotient maps are not necessarily closed. Consider $\mathbf{C}/\mathbf{R}$, the quotient of the plane obtained by projection onto the real factor. The closed subset $A = \{(x,1/x): x \in \mathbf{R}\}$ is carried to the positive reals; this is equivalent to the fact that the product of A with the kernel of q, the imaginary axis, in the additive group $\mathbf{C}$ is the open right half-plane $\{(x,y): x > 0\}$ (can you see why?), and this half-plane is not closed in $\mathbf{C}$. This is thus an example of two closed subsets of $\mathbf{C}$ whose product in $\mathbf{C}$ is not closed. We can assert, however, that *the product of a closed set with a compact set is closed* (but not necessarily compact) in any topological group. We shall need some notation for a fairly complex proof. Let A be closed and B compact, and let $y \notin AB$, so $e \notin y^{-1}AB$. For each $b \in B$, then, the complement of the closed set $y^{-1}Ab$ is a nucleus, and we may find an open nucleus V_b such that $V_b V_b^{-1} \subset (y^{-1}Ab)'$. Obviously, the family $\{bV_b: b \in B\}$ is an open cover of B. Choose a finite subcover $\{b_i V_{b_i}: i = 1, \ldots, n\}$, and let $V = \bigcap\{V_{b_i}: i = 1, \ldots, n\}$. We claim yV is a neighborhood of y which is disjoint from AB. Otherwise suppose there are elements $v \in V$, $a \in A$, and $b \in B$ with $yv = ab$, and let $b = b_i v_i \in b_i V_{b_i}$, an element of the finite cover for B. Then $yv = ab_i v_i$, $yvv_i^{-1} = ab_i$, and $vv_i^{-1} = y^{-1}ab_i$. This is contradictory, since $vv_i^{-1} \in V_{b_i}(V_{b_i})^{-1}$, which is disjoint from $y^{-1}Ab_i$. Therefore each element y of the complement of AB has a neighborhood yV lying in that complement, so AB is closed. ∎

Now we can claim that $q: G \to G/H$ *is a closed map whenever H is compact,* for $q(A)$ is closed for a closed set A iff $q^{-1}[q(A)] = AH$ is closed, which is the case when H is compact. ∎ Note that this result is valid even when H is not normal and G/H is merely a coset space.

PRODUCTS

The cartesian product of two sets, $A_1 \times A_2$, was defined (in Chap. I) to be a set of ordered pairs $A_1 \times A_2 = \{(a_1,a_2): a_1 \in A_1 \text{ and } a_2 \in A_2\}$. This is set isomorphic to the set of all functions f from $\{1,2\}$ into $A_1 \cup A_2$ with $f(1) \in A_1$ and $f(2)$ in A_2. There is an isomorphism which assigns to each such function f the pair $[f(1),f(2)]$; the inverse isomorphism defines a function g for each ordered pair (a_1,a_2) in $A_1 \times A_2$, where g has the values $g(1) = a_1$ and $g(2) = a_2$. This suggests that we may regard each complex number as a function from $\{1,2\}$ into **R**, for example. To continue, if A_1, A_2, and A_3 are sets, and S is the set of all functions f from $\{1,2,3\}$ into $A_1 \cup A_2 \cup A_3$ for which $f(i) \in A_i$, $i = 1$, 2, and 3, then there are set isomorphisms $(A_1 \times A_2) \times A_3 \cong S \cong A_1 \times (A_2 \times A_3)$ given by the correspondences

$$[(a_1,a_2),a_3] \leftrightarrow f \leftrightarrow [a_1,(a_2,a_3)]$$

where $f(i) = a_i$ for $i = 1$, 2, and 3. This shows the "associativity up to isomorphism" of our definition of products of sets. It also suggests the following generalization of the product of a family of sets; here the family may be infinite.

An **indexed family of sets** is a function A whose values are sets. The domain L of A is called the **indexing set,** or **set of indices;** if $\lambda \in L$, then the value $A(\lambda)$ is usually denoted by A_λ. (This agrees with the standard notation for a sequence of sets, $A_1, A_2, A_3, \ldots$; here L is the positive integers.) If A is an indexed family of sets with indexing set L, the **direct** (or "cartesian") **product** of A is $\times A = \{f: f$ is a function from L into $\bigcup\{A_\lambda: \lambda \in L\}$ with $f(\lambda) \in A_\lambda$ for each $\lambda \in L\}$; that is, a member of the direct product is a function f which chooses one member $f(\lambda)$ from each indexed set A_λ. If, for instance, $L = \{1,2\}$, then the direct product of A is just the set of functions described above whose domain is $\{1,2\}$ and whose range is $A_1 \cup A_2$; $f \in \times A$ implies $f(1) \in A_1$ and $f(2) \in A_2$. As we argued above, our new definition of the direct product of A yields, in case $L = \{1,2\}$, a set isomorphic to $A_1 \times A_2$. In fact, this new definition agrees (up to isomorphism) with the previously defined product of a finite number of sets $A_1, A_2, \ldots, A_n$, where any association whatsoever may be used. Associativity is "built into" the new definition. Henceforth, *all* direct products will be understood in this new sense. We shall, however, continue to use such notation as $A_1 \times A_2$, and we shall often substitute "$\times\{A_\lambda: \lambda \in L\}$" for "$\times A$" in order to display the function A by giving its values A_λ. Each set A_λ is called the λ-th **factor,** or **coordinate set,** of the direct product, and the values $f(\lambda)$ of an element f of the product will frequently be written as f_λ and called the λ-th **coordinate** of f.

The space $\mathbf{R}^n$, for example, though it was defined as a set of ordered n-tuples of reals, may now be thought of as the set of all functions from $\{1, 2, \ldots, n\}$ to $\mathbf{R}$; an n-tuple $(x_1, x_2, \ldots, x_n) = x$ corresponds to the function f with $f(i) = x_i$ for $i = 1, 2, \ldots, n$. Thus $f(i)$ is the i-th coordinate of the n-tuple x, which we henceforth confuse with the function f. Put another way, an ordered n-tuple is neither more nor less than a finite sequence of length n, and of course, sequences are functions.

An example with an infinite index set is $\times\{\mathbf{R}_i: i = 1, 2, 3, \ldots\}$, the set of all *infinite* sequences of real numbers; each $\mathbf{R}_i$ is the same set $\mathbf{R}$. A direct product each of whose factors is the same set A, taken over the index set L, is often written A^L, with exponential notation. Is it clear that if both A and L are finite sets of cardinality $\bar{A}$ and $\bar{L}$, then the number of elements in the direct product is $\bar{A}^{\bar{L}}$? Thus the set of all real sequences is called $\mathbf{R}^N$, where N denotes the **natural numbers** $1, 2, 3, \ldots$.

We remark that if each factor of a direct product over a nonempty indexing set is a nonempty set, then the product is nonempty. This is simply a restatement of the axiom of choice (see Chap. I). For each index $\lambda \in \mathrm{L}$ there is a **projection,** or **evaluation,** p_λ of the direct product $\times A$, when $A = \{A_\lambda: \lambda \in L\}$, into its factor A_λ, $p_\lambda: \times A \to A_\lambda: p_\lambda(f) = f(\lambda)$. That each of these projection functions is onto is clear, at least when $\times A \neq \varnothing$.

PRODUCTS OF GROUPS

If each factor A_λ of a direct product $\times A$ over L is a group, then $\times A$ is a group with the pointwise multiplication of functions, $(fg)(\lambda) = f(\lambda)g(\lambda)$, the latter multiplication taking place in A_λ. The identity of $\times A$ is the function e with $e(\lambda) = e_\lambda$, the identity of A_λ; the inverse *in the group* $\times A$ of f is f^{-1}, $f^{-1}(\lambda) = f(\lambda)^{-1}$, where the latter inversion is in A_λ. (Notice here the conflict of this notation with that for the relational inverse to the function f as well as that for a functional inverse of f.) The associativity of the pointwise multiplication is trivial. Each projection $p_\lambda: \times A \to A_\lambda$ is an epimorphism.

A familiar example of this is the set F of all real-valued functions of a real variable. Clearly, F is just the direct product $\times\{\mathbf{R}_r: r \in \mathbf{R}\}$, the set of all functions on the indexing set $\mathbf{R}$ whose value at each index r lies in the

r-th factor $\mathbf{R}_r = \mathbf{R}$. Since each factor is the group of additive reals, F is an abelian group with the familiar pointwise addition of functions which you first saw defined in the calculus.

PRODUCTS OF SPACES

Now consider the problem of building an appropriate topology for $\times\{A_\lambda: \lambda \in L\}$, given a topology τ_λ for each A_λ. A natural expectation is that the projections p_λ all be continuous. Hence, if τ is a topology for the product such that each p_λ is continuous, then the set $\{p_\lambda^{-1}(S): S \in \tau_\lambda\}$ must be contained in τ. The **product topology** ρ for the direct product is generated by the subbase $\{p_\lambda^{-1}(S): \lambda \in L \text{ and } S \in \tau_\lambda\}$ consisting of all the inverse images (under projection) of the open sets of the factors. It is, by construction, the smallest topology for the product on which each projection is continuous. Recall that a base for ρ is offered by the family of finite intersections of members of the subbase. Each subbase member is, for particular λ_0 and some open set $S_{\lambda_0} \in A_{\lambda_0}$, of the form $\{f: f(\lambda_0) \in S_{\lambda_0}\}$, a subset of the direct product. Therefore a finite intersection of subbase elements is a product $\times\{S_\lambda: \lambda \in L\}$ of open sets $S_\lambda \subset A_\lambda$, one for each λ, where $S_\lambda = A_\lambda$ except for a finite subset of indices λ. This makes it clear that *the projections p_λ are open maps*, with the product topology, because the projection p_μ, evaluated on a base element $\times\{S_\lambda: \lambda \in L\}$, gives S_μ, which is open in A_μ (by the definition of the base). But if a function carries each member of a base (for the domain topology) to an open set of the range, then it is an open function (why?). ∎

Again the set F of real-valued functions of a real variable provides us with an example. A subbasic element of the product topology ρ for F is depicted in the diagram; it corresponds to an open set W of $\mathbf{R}_r = \mathbf{R}$ for some index r and consists of all $f \in F$ for which $f(r) \in W$. We denote this member of the subbase by (r, W); thus (r, W) is the set of all those functions in F whose graphs go through the "slot" $r \times W$ of vertical width W which lies above the point r of the domain axis. The intersection of a finite collection $\{(r_i, W_i): i = 1, 2, \ldots, n\}$ of these subbasic sets is, of course, the set of functions whose graphs go through each of the slots (r_i, W_i); this is the general picture of an element of the base for the product topology on F. Notice that in the diagrams the horizontal axis is the indexing set, and the vertical line through an index r is the r-th coordinate space $\mathbf{R}_r$. Do

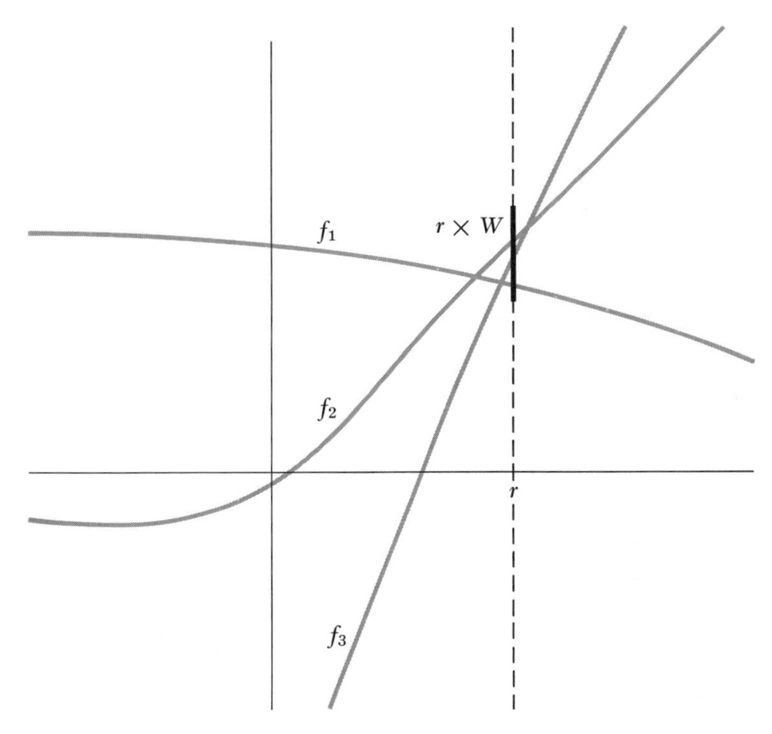

Three members of (r, W)

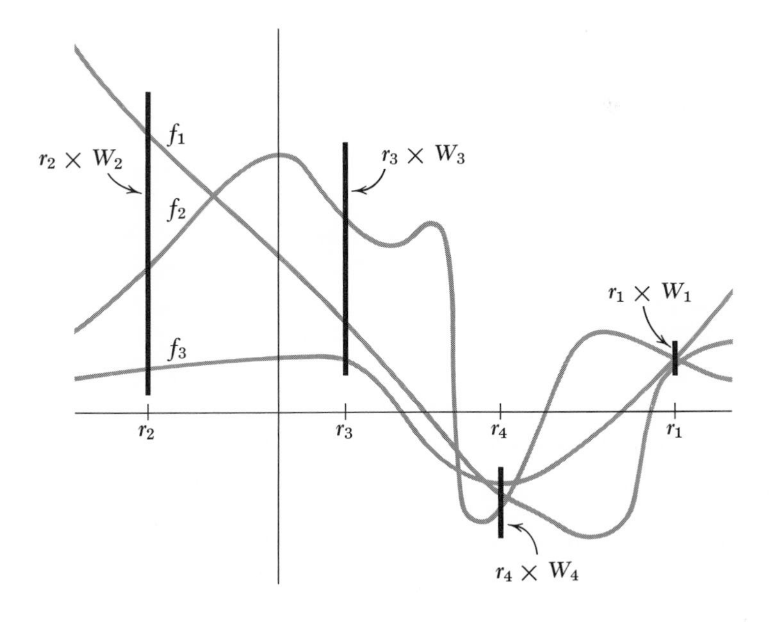

not be confused by these diagrams; the plane is not the direct product in question, and the pictures are not of F, merely of a few elements of that huge set.

Exercises K, L, and N

CROSS SECTIONS

If $p: X \to Y$ is a map, a **cross section** s to p is a map $s: Y \to X$ which is a right inverse for p, $p \circ s = 1_Y$. Clearly, if s exists, then p is onto (Theorem I.8). The cross section s is said to be **at** a point $x \in X$ if x is a value of s; then $x = s \circ p(x)$, since $x = s(y)$ implies $p(x) = p \circ s(y) = y$. Since restrictions of continuous functions are always continuous, the restriction of a map p to the image of its cross section s must be continuous, and therefore a homeomorphism (why?); $s(Y)$ is a homeomorph of Y. If, for example, p is the absolute-value function on the plane, $p(z) = |z|$, then one cross section s, defined on the nonnegative real numbers, assigns $(r,0) \in \mathbf{R}^2$ to each $r \geqslant 0$.

Now consider the projection p_μ of a direct-product space $\times A$ onto a factor space A_μ, and let f be an arbitrary member of the product. We shall construct a cross section s_μ for p_μ at f, with the image B of s, a homeomorph of A_μ, lying in $\times A$. Consider the subset $B = \{g: g(\lambda) = f(\lambda) \text{ if } \lambda \neq \mu\}$ of the direct product. Clearly, $f \in B$; and if g and h are members of B with $p_\mu(g) = g(\mu) = h(\mu) = p_\mu(h)$, then $g = h$. Hence p_μ, restricted to B, is 1-1. Further, if $a \in A_\mu$, there is a member g of B with $g(\mu) = a$ [and $g(\lambda) = f(\lambda)$ for all $\lambda \neq \mu$], so $p_\mu(B) = A_\mu$. Hence there is a two-sided inverse function s for the restriction $p_\mu|_B$. Now, s is continuous if the inverse image $s^{-1}(S)$ is open for every element S of a base for B. But it is obvious that the intersections with B of elements of a base for the product form a base for B. Hence, let $\times\{S_\lambda: \lambda \in L\}$ be a base element for $\times A$, so that each S_λ is open in A_λ. A quick calculation shows that $s^{-1}(\times\{S_\lambda: \lambda \in L\}) = p_\mu(B \cap \times\{S_\lambda: \lambda \in L\}) = S_\mu$ (or $\varnothing$), an open set of A_μ. This construction of a cross section to the projection of a direct product onto one of its factors is of general usefulness and should be well understood. An intuitive grasp of it is offered by a picture drawn in the product $A_1 \times A_2$ of two real intervals. The process described yields, for each point f of the product, a homeomorph B_i of A_i, with $f \in B_i$, $i = 1$ and 2.

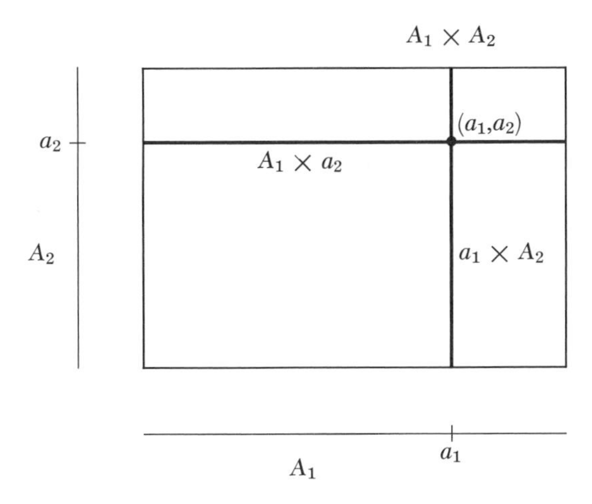

PRODUCTIVE PROPERTIES

A property of topological spaces is called **productive** if whenever each factor of a direct product has the property, so does the product. Examples of productive properties are provided by the separation axioms T_0, T_1, and T_2. This is easy to see. Suppose f and g are distinct elements of the product, so that there exists an index λ_0 with $f(\lambda_0) \neq g(\lambda_0)$. If the factor A_{λ_0} is, for instance, T_1, then there is a neighborhood N of $f(\lambda_0)$ with $g(\lambda_0) \notin N$; hence $p_{\lambda_0}{}^{-1}(N)$ is a neighborhood of f which does not contain g. The change of wording necessary for a proof in case A_{λ_0} is T_0 or T_2 is obvious: *if each factor A_λ is T_i, then the product must also be T_i, for $i = 0$, 1, or 2.* ∎

An example of a topological property which is not productive is the property of finiteness. If, for instance, each factor of a product is $A_i = \{0,1\}$, a two-member set, and the indexing set is the set N of natural numbers (or positive integers), then the direct product $\{0,1\}^N$ is the set of sequences each of whose terms is 0 or 1. These sequences are, in an obvious way, in 1-1 correspondence with the set of all infinite binary decimal expressions for numbers between 0 and 1, an infinite set.

Exercise Y

CONNECTED PRODUCTS

Connectedness is a productive property of spaces; in fact, the direct-product space $\times A$ of an indexed family of nonempty spaces is connected iff each member of the family is a connected space. It is easy to see that if A_μ is disconnected, then so is $\times A$. Let S_μ be a nonempty, proper, open and closed subset of A_μ. Since p_μ is continuous, $p_\mu^{-1}(S_\mu)$ is a nonempty, proper, open and closed subset of $\times A$, which is thus disconnected (supply the arguments in your mind).

Conversely, let each factor space be connected, and suppose S is a nonempty open and closed subset of $\times A$. We shall show that $S = \times A$. Choose a point $f \in S$. Since S is open, there must be a member T of the base for the product topology with $f \in T \subset S$. Hence $T = \times\{T_\lambda : \lambda \in L\}$, where each T_λ is open in the factor A_λ, and for all but a finite number of indices λ, $T_\lambda = A_\lambda$; by renaming indices if necessary, we assume that only for $\lambda = 1, 2, \ldots, n$ is $T_\lambda \neq A_\lambda$. Now define $U = \times\{U_\lambda : \lambda \in L\}$, where $U_\lambda = T_\lambda$ if $\lambda \neq n$, and $U_n = A_n$; thus $T \subset U$.

We shall show that $U \subset S$, and hence, by induction (that is, after n such steps), $\times A \subset S$, which means that $\times A$ is connected because S was an arbitrary nonempty open and closed subset of $\times A$. Let $g \in U$, so $g(\lambda) \in T_\lambda$ if $\lambda \neq n$. Define $\bar{g}$ by $\bar{g}(\lambda) = g(\lambda)$ if $\lambda \neq n$, and $\bar{g}(n) = f(n)$. Clearly, $\bar{g} \in T$, so $\bar{g} \in S$. Let B_n be the image of a cross section to p_n at $\bar{g}$; B_n is a homeomorph of A_n and thus is connected (in the relative topology). Since $g \in B_n \cap S$ and $B_n \cap S$ is both open and closed, $B_n \subset S$; therefore $g \in S$. But g was an arbitrary point of U, so $U \subset S$. ∎

We recapitulate this proof: it was shown that if g differed from some member of S by only one coordinate (that is, one value), then g was in S; hence all points of the product differing by a finite number of coordinates were in S. But every element of the product differed by at most a finite number of coordinates from some member of S. Draw a picture of this proof in $\mathbf{R}^3$!

TYCHONOFF FOR TWO

Suppose that the direct product $\times\{A_\lambda : \lambda \in L\}$ has one factor A_μ which is not compact. Then there is an open cover $\mathcal{S}$ of A_μ which contains no finite subcover. Therefore the family $\{p_\mu^{-1}(S) : S \in \mathcal{S}\}$ of subbasic open sets covers the direct product, and it certainly has no finite subcover. This shows that when a single factor of a direct product is not compact the product also fails to be compact.

We shall now prove that *the product of a finite number of compact spaces is itself a compact space.* It will suffice to prove this for every product $A \times B$ of two compact spaces, since then the direct product $A_1 \times A_2 \times \cdots \times A_n$ of a finite collection of compact spaces would be homeomorphic to a direct product $(A_1 \times A_2 \times \cdots \times A_{n-1}) \times A_n$ of just two compact spaces. (This is an inductive assertion; could you prove it?) Suppose we are given an open cover $\mathcal{S}$ of $A \times B$, and that both factor spaces are compact. For each $a \in A$ the homeomorph $a \times B$ of B in $A \times B$ is compact, and for each $b \in B$ there is a member $S_{(a,b)}$ of $\mathcal{S}$ with $(a,b) \in S_{(a,b)}$. Since $S_{(a,b)}$ is open, it contains a rectangular element

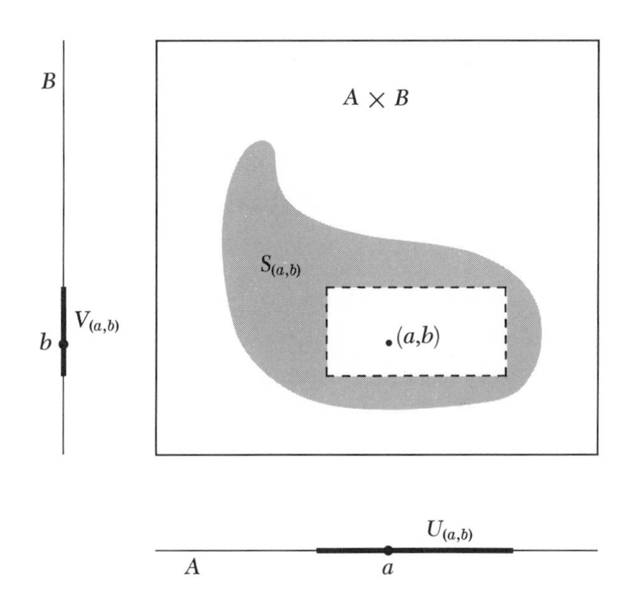

$U_{(a,b)} \times V_{(a,b)}$ of the base for the product topology such that $a \in U_{(a,b)}$ and $b \in V_{(a,b)}$. Fixing a for the moment, choose a finite subcover $\{V_{(a,b_j)}: j = 1, 2, \ldots, n_a\}$ of the open cover $\{V_{(a,b)}: b \in B\}$ of the compact set $a \times B$, and let $U_a = \bigcap \{U_{(a,b_j)}: j = 1, 2, \ldots, n_a\}$. When this has been done for each $a \in A$, choose a finite subcover $\{U_{a_1}, U_{a_2}, \ldots, U_{a_m}\}$ of the open cover $\{U_a: a \in A\}$ of A. We now claim that $\{S_{(a_i,b_j)}: 1 \leq i \leq m \text{ and } 1 \leq j \leq n_{a_i}\}$ is a finite subcover of $\mathcal{S}$ for $A \times B$. Finite it is, and a subset of $\mathcal{S}$ as well; we need only show that each point (a,b) of the direct product is in some set $S_{(a_i,b_j)}$. But if $a \in A$, then $a \in U_{a_k}$ for some k, $1 \leq k \leq m$, and then

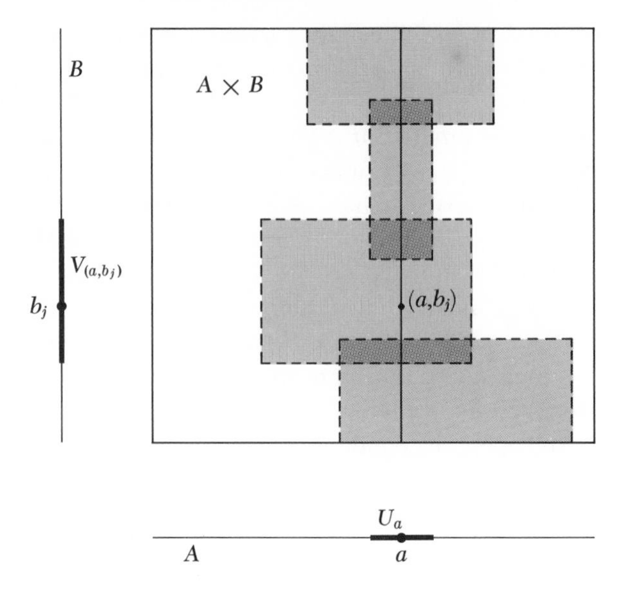

(a_k,b) is a member of $U_{(a_k,b_l)} \times V_{(a_k,b_l)}$ for an integer l between 1 and n_{a_k}. This says $b \in V_{(a_k,b_l)}$, so

$$(a,b) \in U_{a_k} \times V_{(a_k,b_l)} \subset U_{(a_k,b_l)} \times V_{(a_k,b_l)} \subset S_{(a_k,b_l)};$$

sets of the latter type therefore cover $A \times B$. ■

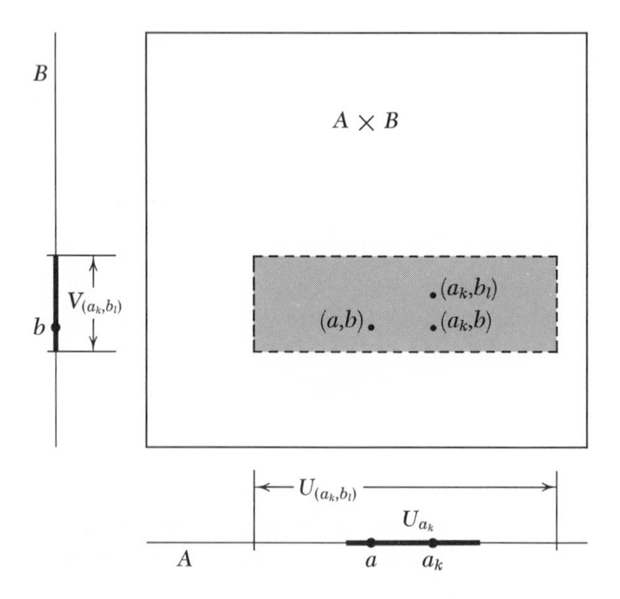

The plan of the proof is this: an open neighborhood U_a is constructed for each $a \in A$ with the property that a finite subset of $\mathfrak{S}$ covers $U_a \times B$, and then a finite number of the sets $\{U_a\}$ are found to cover A; the conclusion follows immediately.

This theorem is called the Tychonoff theorem; it is valid for arbitrary direct products, but the proof in the general case is more difficult. We shall use it only in applications where there are a finite number of factors.

As an example of its use, we belatedly show that the Heine-Borel theorem is valid in $\mathbf{R}^n$, that a closed and bounded (in the usual metric) subset of $\mathbf{R}^n$ is compact in our newer, topological sense (this was proved in Chap. III for the case $n = 1$; see also Prob. III.FF). Let K be a closed subset of the ball of radius ε at the origin of $\mathbf{R}^n$. Clearly, K is a closed subset of the cube C, centered at the origin and with sides of length 2ε; since closed subsets of compact spaces are compact, it will suffice to prove C compact. But C is (homeomorphic to) the direct product of n copies of the closed interval $[-\varepsilon,\varepsilon] \subset \mathbf{R}$, and that interval has been shown (in Chap. III) to be compact.

Exercise Z

REFERENCES AND FURTHER TOPICS

The material of this chapter is discussed in any introductory text for topology, and the references for Chap. IV apply here as well.

Our most glaring omission is a proof of the Tychonoff theorem for arbitrary indexing sets, that compactness is a productive property. This has been called the most important theorem of point-set topology and has many uses in analysis as well as in topology. Two differing proofs of it are offered by

J. L. Kelley, *General Topology*, chap. 5 (Princeton, N.J.: Van Nostrand, 1955).

Every proof of the Tychonoff theorem must use the axiom of choice, since the theorem implies that axiom. A new proof of the Tychonoff theorem which makes this use explicit, and does not require the axiom at all in some special cases (such as a countable product of real compact sets), is given by

P. A. Loeb, *A new proof of the Tychonoff theorem*, Amer. Math. Monthly 72 (1965), 711–717.

Incidentally, this reference is the first in this book to a journal article. It is given in the form usually used by mathematicians: author, *Title,* abbreviated name of journal followed by the volume number (date of volume), first–last page numbers.

For a delightful discussion of compactness, which argues that this notion generalizes that of finiteness, see

E. Hewitt, *The rôle of compactness in analysis,* Amer. Math. Monthly 67 (1960), 499–516.

Another important omission from this chapter is the topic of sequential compactness: a space is "sequentially compact" if every sequence in it has a convergent subsequence. Reminiscent of this, the Bolzano-Weierstrass theorem states that every infinite subset of a compact set of real numbers has a limit point in the compact set. In general, a space is said to have the "Bolzano-Weierstrass property" if every infinite subset has a limit point. It is a theorem that for metric spaces the three notions of compactness, sequential compactness, and the Bolzano-Weierstrass property are equivalent. An excellent exposition of these matters is

G. Simmons, *Introduction to Topology and Modern Analysis,* pp. 120–128 (New York: McGraw-Hill, 1963).

Any compact space in which a given space is dense is called a "compactification" of the given space; the one-point compactification is one such, and it is minimal in some sense of the word. But there are many other useful constructions of compactifications; the texts of Kelley and Simmons, cited above, both discuss the "Stone-Čech compactification."

It is not difficult to see that if a space has the property that for each point x and each neighborhood N of that point there is a map from the space to the unit interval which is zero at x and identically 1 outside N, then that space is regular. This suggests the definition of a new separation axiom; the space above is called "completely regular." Still another and stronger separation axiom, that of "normality," demands that for each pair of disjoint closed subsets of a space there exist a pair of disjoint open sets, each containing one of the closed sets. Urysohn's lemma implies that normal T_1-spaces are completely regular; they are obviously regular. It is a theorem that every Hausdorff topological group is completely regular, yet there are examples of nonnormal groups. You can consult the references offered in Chap. IV for these new topological ideas. For the complete regularity of groups, see

Montgomery and Zippin, *Topological Transformation Groups,* pp. 29–30 (New York: Interscience, 1955).

EXERCISES

A Show that every indiscrete space is connected. The components of a discrete space are points (that is, singleton subsets); what are the components of the space of rational numbers (with the usual relative topology)? Is this space discrete?

B Verify in detail that a morphism of topological groups induces a well-defined algebraic morphism of their component groups.

C Prove that the path component of a point x in a space X is contained wholly within the component of x. Do this by showing that a path-connected set must be connected.

D Show that the subset of the plane consisting of the graph of $sin\ (1/x)$ for $x \in (0,1]$ together with the part of the y-axis from $(0,-1)$ to $(0,1)$ is connected. (It is not path connected, by Exercise III.K.)

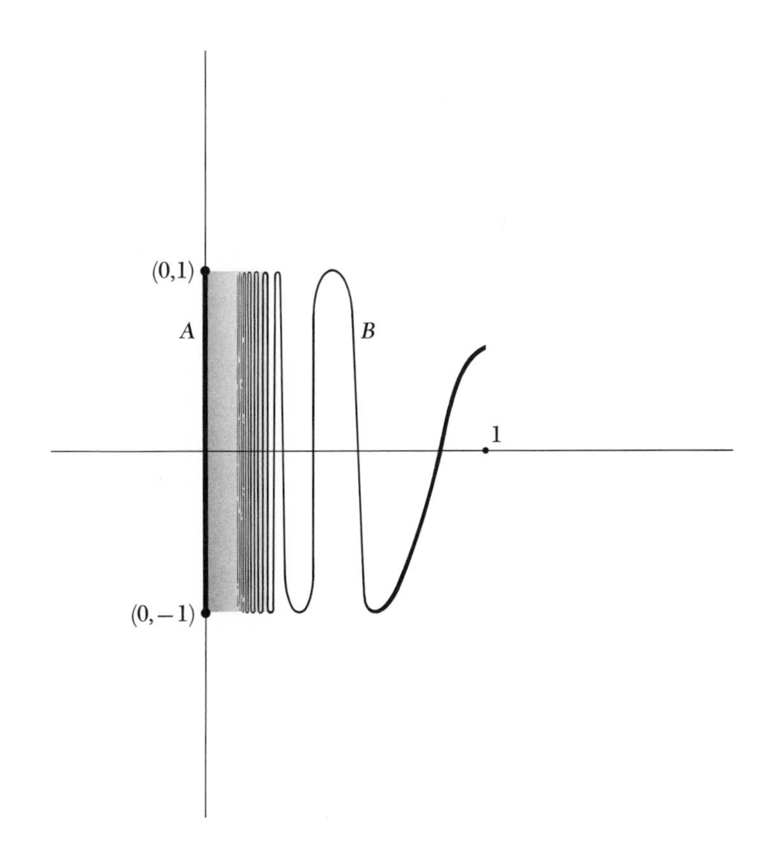

E Verify in detail that $\pi_0(f)$ is well defined and that π_0 preserves compositions and identities.

F If G is a topological group, show that $\pi_0(G) = G/G_0$.

G Show that the product of two path-connected spaces is path connected. Then show, in general, that for any spaces A and B, $\pi_0(A \times B)$ is set isomorphic to $\pi_0(A) \times \pi_0(B)$. (The "function" π_0 may thus be said to preserve direct products.)

H Invent an infinite open cover $\mathcal{S}$ for the open unit disc in the plane such that $\mathcal{S}$ contains no finite subcover. Then do the same for the closed upper half-plane.

J Show that $\{\infty\}$ is a component of T^*, the one-point compactification of T, if T is compact. If T is not compact but is connected, is T^* necessarily connected?

K Prove that the following theorem, which is valid for finite direct products, is true in general: *a function f whose range is the direct product $\times\{A_\lambda: \lambda \in L\}$ is continuous if and only if the composite $p_\lambda \circ f$ is continuous for each $\lambda \in L$*, where p_λ is the projection of the product onto its λ-th factor. ■

L The set $F = \times\{\mathbf{R}_r: r \in \mathbf{R}\}$ can be metrized as follows: let each factor space $\mathbf{R}_r$ be given the same bounded metric $\mathfrak{d}_r$ (as in Exercise III.A), and define the metric $\mathfrak{d}$ on F by

$$\mathfrak{d}(f,g) = lub\ \{\mathfrak{d}_r[f(r),g(r)]: r \in \mathbf{R}\}.$$

You may presume that $\mathfrak{d}$ is a metric; draw a picture (like those of this chapter) to illustrate the relation of an element of the base for the product topology ρ on F with an element of the base for the metric topology μ. Then prove that $\rho \subset \mu$ and $\rho \neq \mu$.

M Show that the set R of rational numbers, with its usual topology, is not locally compact. Since R is T_2, R^* must not be T_2. Prove directly that R^* is not T_2; is it T_0 or T_1?

N Let $A = \{A_\lambda: \lambda \in L\}$ be an indexed family of spaces, and for each $\lambda \in L$ let σ_λ be a subbase (or base) for the topology of A_λ. Show that sets of the type $\times\{S_\lambda: \lambda \in L\}$, where $S_\lambda = A_\lambda$ for all $\lambda \neq \mu$ and $S_\mu \in \sigma_\mu$, form a subbase for the product topology on $\times A$.

P Show that every nucleus of a connected group generates the whole group (see Exercise II.K for a definition of "generates").

Q Prove that each compact subset of a Hausdorff space is closed. Then construct an example of a space (non-Hausdorff, of course) with a compact subset which is not closed.

Now let $f: X \to Y$ be a map from a compact space X to a Hausdorff space Y; show that f is necessarily closed. Combine this result with Corollary IV.7 to show that $Im\ (f)$ must have the quotient topology. Do you see how this result could have saved you labor, for instance, in your proof that $\mathbf{I}/\{0,1\}$ is homeomorphic to S^1?

R Use Exercise Q to give a proof of the theorem that *a continuous one-to-one correspondence from a compact space to a Hausdorff space is a homeomorphism.* ■ Then show by example that this statement becomes false if the domain is only assumed to be locally compact or the range only assumed to be T_1.

S A "Lindelöf space" is one on which every open cover contains a countable subcover. Evidently, every compact space is Lindelöf; show that every second-countable space (see Exercise L) is Lindelöf.

T A sequence $s_1, s_2, \ldots$ in a space X is said to **converge** to $x \in X$ if for every neighborhood W of x there exists an integer N such that $i \geq N$ implies $s_i \in W$. A space is said to satisfy the "first axiom of countability," or to be "first countable," if at each point of the space there is a countable (finite or denumerably infinite) local base. That is, X is first countable if for each $x \in X$ there exists a sequence $W_1, W_2, \ldots$ of neighborhoods of x such that if N is an arbitrary neighborhood of x, there is some integer i with $W_i \subset W$. Metric spaces are first countable.

Show that if X is a first-countable space and x is a point in the closure A^- of a subset A of X, then there is a sequence in A which converges to x.

U A **subsequence** of a sequence $s_1, s_2, \ldots$ is a subset $s_{n_1}, s_{n_2}, \ldots$ of that sequence, arranged in increasing order; $i < j$ implies $n_i < n_j$. Show that every sequence in a compact first-countable space (see Exercise T) has a convergent subsequence. Hence such a space is "sequentially compact." (*Hint:* Assume that a sequence in a space has no convergent subsequence and prove that the space is not compact. The open cover you construct should contain, for each term of the sequence, an open set which includes only finitely many terms; the complement of the set of all terms of the sequence also ought to belong to the cover, but is it open?)

W Show that the metric topology μ (defined in Exercise L) on F makes it into a topological group.

Y Prove that regularity and T_3-ness are productive properties.

Z Prove that the product of a finite collection of spaces is locally compact if and only if each factor space is locally compact.

PROBLEMS

AA Spaces of Components Prove that if T is a space and $\mathcal{C}$ is the collection of components of T, then $\mathcal{C}$ is totally disconnected in the quotient topology; that is, the components of $\mathcal{C}$ are just its points (or rather, singleton subsets) (note Exercise A). This seemingly natural statement is tricky to prove. Do not make the false assumption that if x and y lie in different components of a space, then there exists a both open and closed subset of the space with x in that subset and y in its complement. What condition on T will force $\mathcal{C}$ to be discrete?

BB Relations of Planes with Spheres Generalize the textual examples to show in detail that for all n, $(\mathbf{R}^n)^*$ is homeomorphic to S^n (and therefore "S^n minus a point is $\mathbf{R}^n$"). Complete this picture by observing that by deleting a point from $\mathbf{R}^n$ you get a space homeomorphic to $S^{n-1} \times \mathbf{R}$.

CC Products of Topological Groups Let $A = \{A_\lambda: \lambda \in L\}$ and $B = \{B_\lambda: \lambda \in L\}$ be two indexed families of spaces, indexed by the same set L. Use Exercise N to show that the correspondence of (f,g) with $f \times g$ defines a natural homeomorphism between $\times A \times \times B$ and $\times \{A_\lambda \times B_\lambda: \lambda \in L\}$.

Now employ Exercise M to show that if G_λ is a topological group for each $\lambda \in L$, then $\times \{G_\lambda: \lambda \in L\}$ is a topological group (with the multiplication and the topology of a direct product). This, together with Exercise W, exhibits two distinct and nontrivial topologies with each of which the group F is a topological group.

Show that ρ is not a locally compact topology for F. Since it is a group, this provides us with an example to show that the T_3 property does not imply local compactness.

DD Normality A space is "normal" if for each two disjoint closed subsets A and B there exist disjoint open sets U and V with $A \subset U$ and

$B \subset V$. Show that every compact Hausdorff space is normal. (This is another instance of compact sets' enjoying a property of finite sets.)

EE Compact Groups Show that G is compact if both H and G/H are compact (H need not be normal).

FF T_3-spaces Make a direct argument, without the use of compactifications, that a locally compact T_2-space is T_3.

GG Pointwise Convergence Let Y^X be the set of all functions from X to Y; show that a sequence $f_1, f_2, f_3, \ldots$ of elements of Y^X converges to a function f in the product topology iff for each $x \in X$ the sequence $f_1(x), f_2(x), f_3(x), \ldots$ of points of Y converges to $f(x)$. The convergence of the values of a sequence of functions for each point of their common domain is called the "pointwise convergence" of the sequence of functions. It is the familiar definition of convergence of functions from the calculus. Thus you have shown that a sequence of functions of Y^X converges in the product topology iff it converges pointwise. Because of this fact, the product topology is often called the "topology of pointwise (or coordinatewise) convergence."

HH Completeness A sequence $s_1, s_2, s_3, \ldots$ of points in a metric space $(M,\mathfrak{d})$ is a "Cauchy sequence" iff for each positive real ε there exists an integer N such that $\mathfrak{d}(s_i,s_j) < \varepsilon$ whenever both i and j are greater than N. The metric space M is called "complete" iff every Cauchy sequence in M converges to a limit point in M. For examples, **I**, **Z**, and **R** are complete, while the rationals and the open interval $(0,1)$ are not.

Every compact metric space is complete, and a subset of a complete metric space is complete (as a subspace) iff it is closed.

JJ Path Components of Products Show that the set $\pi_0(\times\{A_\lambda: \lambda \in L\})$ of path components of a direct product of spaces is set isomorphic to the direct product $\times\{\pi_0(A_\lambda): \lambda \in L\}$ of the sets of path components of the factors.

Is a similar theorem true for components?

KK Product Functions We denote by X^L a direct product over the indexing set L, each factor of which is X. Let L, X, and Y be sets and $f: X \to Y$ a function. The function $f^L: X^L \to Y^L$ is defined by its values: for each element h of X^L (so $h: L \to X$), $f^L(h) = f \circ h$. Show that if f is continuous, then f^L is continuous when the product sets are given the product topologies. Furthermore, if $X \xrightarrow{f} Y \xrightarrow{g} Z$, then

$(g \circ f)^L = (g^L) \circ (f^L)$; also $(1_X)^L$ is the identity function on X^L. Thus a set L assigns to each set a product set and to each function a "product" function, so that compositions and identities are preserved. If one is dealing with spaces instead of sets, and continuous functions between them, L assigns spaces and continuous functions to these. What can be said if X and Y are groups and f is a morphism?

Function Spaces

CHAPTER VII

You have seen several situations already in which a set of functions was the object under scrutiny. One instance was the group of all permutations of some set, and another was the metric space C of all real-valued maps on $\mathbf{I}$. Yet a third was our definition of a direct product as a family of functions on an indexing set. Further examples abound: the set of all linear transformations from one vector space to another, or morphisms from one group to another, is a mathematical example; physicists are often interested in the group of all "symmetries" of space-time which leave invariant a particular physical quantity or expression.

We now focus our attention on a quite general sort of family of functions, a family of functions whose common domain is one topological space and whose common range is another space. Our interest lies in the construction of topologies for such a set of functions, topologies which relate in natural ways with the topologies of the common domain and range spaces. Although some of our definitions and statements will apply in this generality, we shall have little interest in discontinuous functions; accordingly, we restrict our entire discussion to families of continuous functions.

We wish to examine a set F of maps from a fixed domain space X to a fixed range space Y. Clearly, F is a subset of the direct product $\times\{Y_x: x \in X\}$, where each factor $Y_x = Y$; F inherits its **pointwise** or **product topology** ρ as a subspace of this direct product. However, the definition of ρ does not depend on the topology of X, but only on that of Y. The defining subbase for ρ consists of sets of the form

$$(x,S) = \{f \in F: f(x) \in S\},$$

where $x \in X$ and S is an open subset of Y, and the definition of neither this subbasic set nor the class of all such sets (the subbase) involves the topology on the domain X of the members of F. Such a topology for the set F of functions is bound to have ridiculous properties when a familiar topology on X forces on us a strong intuition of "nearness" in F.

THE DEFINITION

Now, every singleton subset $\{x\}$ of X is compact, as is every finite subset $\{x_1, x_2, \ldots, x_n\} \subset X$ (why?). And Edwin Hewitt has suggested that, in many ways, compact sets are a natural generalization of finite sets (see the references for Chap. VI). If, in place of the singleton subsets used in the defining subbase for the product topology, we use compact sets, a new sort of topology for F arises. The **compact-open topology** (abbreviated **c-o topology**) on F is the unique topology generated by the subbase of all sets of the form $(K,S) = \{f: f \in F \text{ and } f(K) \subset S\}$, where K is compact in X and S is open in Y. Since the compact sets of X are determined by their possible open covers, the topology on X helps determine the family of subbase elements, and hence the c-o topology, on F. If we abbreviate a subbase element as $(K,S) = \{f \in F: f(K) \subset S\}$, then a basic set for the c-o topology is of the form $\bigcap \{(K_i,S_i): i = 1, 2, \ldots, n$; each K_i is compact in X and each S_i is open in $Y\}$, a finite intersection of elements of the defining subbase. Is it clear that this topology for F is exactly the relative topology which F inherits from the c-o-topologized space of all continuous functions from X to Y?

The c-o topology has, among the members of its defining subbase, every element of the defining subbase for the pointwise, or product, topology on F; hence *the product topology is contained in the c-o topology* (ρ is sometimes called the "point-open topology"). ∎ Usually the product topology is strictly smaller. As an example, we examine the set F of continuous functions from $\mathbf{R}$ to $\mathbf{R}$, furnished with three topologies: the pointwise topology ρ, the c-o topology κ, and also the metric topology μ which is defined as follows. If f and g are members of F, and there is an $x \in \mathbf{R}$ with $|f(x) - g(x)| \geq 1$, let $\partial(f,g) = 1$; otherwise let

$$\partial(f,g) = lub\,\{|f(x) - g(x)|: x \in \mathbf{R}\}.$$

The function ∂ is well defined; that it is a metric on F is perhaps intuitively clear (it was described in Exercise VI.L as the metric induced on F by the bounded metric defined in Exercise III.A for $\mathbf{R}$).

As was observed above, $\rho \subset \kappa$; we shall show that $\rho \subset \kappa \subset \mu$, and that each inclusion is strict (that is, these topologies are unequal). First, $\rho \not\supset \kappa$. An illustration of the c-o subbasic element

$$\{f \in F: f([1,2]) \subset (3,4)\} = C$$

shows that $f \in C$ iff the part of the graph of f which lies above $[1,2]$ is in

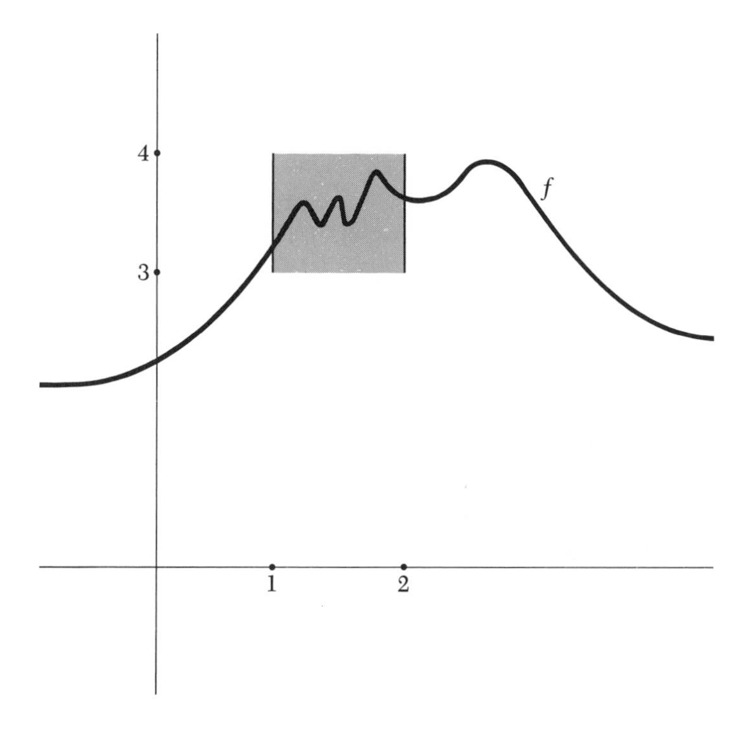

the open strip between $y = 3$ and $y = 4$. But no base element $P = \bigcap\{(x_i,S_i): i = 1, 2, \ldots, n\}$ of ρ could lie inside C; for if x_i and x_{i+1} are adjacent among the x's, then it is easy to construct a function in P which has values greater than 4 between x_i and x_{i+1}. (See illustration on the next page and also those on page 151.) Such a function is not in C, so $P \not\subset C$ if P is any base element of ρ whatsoever, and hence $\rho \not\supset \kappa$.

Now, to show $\kappa \subset \mu$, let (K,S) be an element of the subbase for κ, $f \in (K,S)$, and define a function $g: K \to \mathbf{R}$: let $g(k)$ equal the distance from $f(k)$ to the (closed) complement S' of S (in $\mathbf{R}$). It is easy (see Exercise B) to show that g is a well-defined continuous function from K into $\mathbf{R}$,

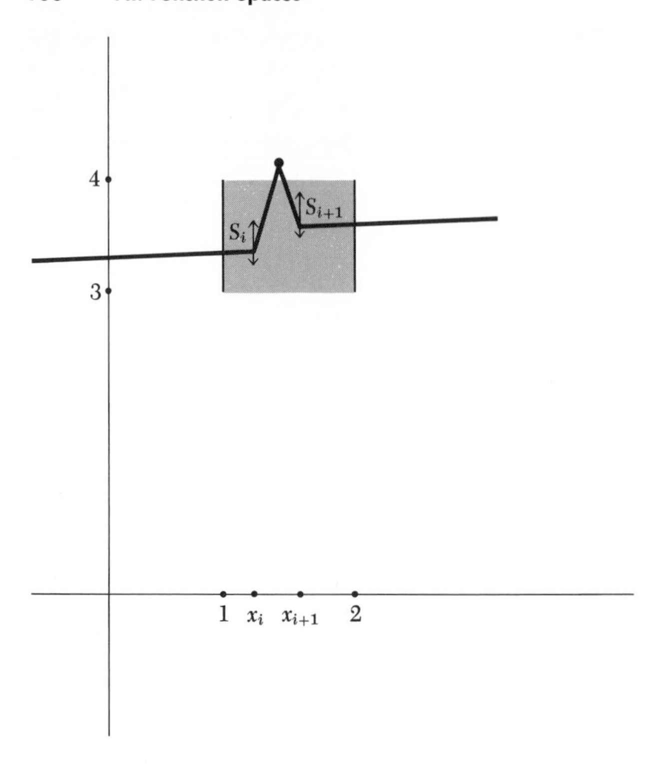

4
3
S_i
S_{i+1}
1
x_i
x_{i+1}
2

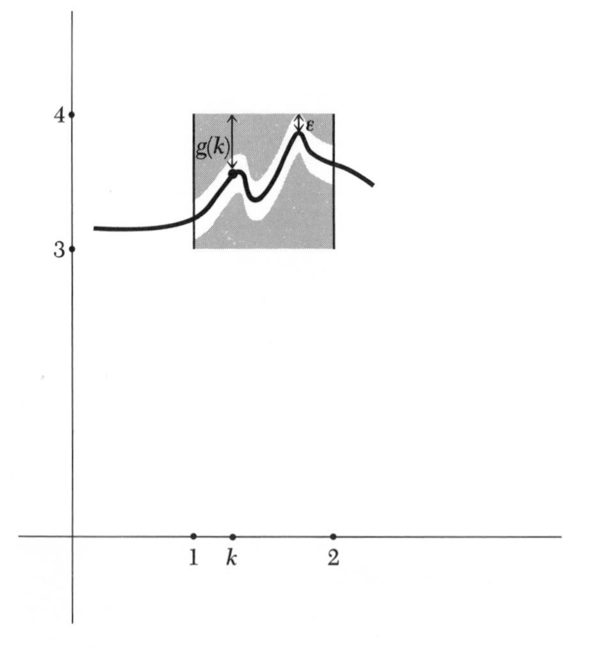

4
3
$g(k)$
ε
1
k
2

and g is always nonnegative. Since K is compact, g takes on a minimal value ε; and $\varepsilon > 0$, because if $g(k) = 0$, then $f(k) \notin S$ (remember that S is open). The number ε is called the **distance** from $f(K)$ to S'. But then for all $k \in K$ the distance from $f(k)$ to S' is at least ε. Therefore the ball $\mathfrak{b}(\varepsilon,f)$ lies inside (K,S), since all functions whose values are within ε of the corresponding values of f must map K into S. This shows that (K,S) is a μ-neighborhood of each of its elements, so $(K,S) \in \mu$. Clearly, then, $\kappa \subset \mu$. To show that $\kappa \not\supset \mu$ is Exercise C.

Each of these topologies for F is constructed in a more or less "natural" way. Each will be useful when it expresses a particular situation at hand better than the others; but κ, even though its definition may seem unmotivated and complicated, is often the most useful of the three. For instance, for each real number r the multiplication function M_r determined by r, $M_r(t) = rt$, is continuous. We might expect that as r gets close to s the function M_r should "get close to" M_s, but in the metric topology μ, the ball $\mathfrak{b}(1,M_s)$ contains no function M_r with $r \neq s$. That is, if $r \neq s$ then the distance from M_r to M_s is always 1. This will be clear if, given $r \neq s$, there is always some $t \in \mathbf{R}$ for which $|rt - st| \geq 1$, or $|t| \geq 1/|r - s|$; $t = 1/(r - s)$ will do nicely. Thus μ is too large a topology to adequately describe the "nearness" of M_r to M_s, when r is near s. However, given a c-o base element C which contains M_s, there is a $\delta > 0$ such that $|r - s| < \delta$ implies $M_r \in C$. We shall assume that C is in the defining subbase for κ; for a finite intersection of these we can take the smallest of the corresponding numbers δ. Thus let K be compact and S open in $\mathbf{R}$, with $sK \subset S$, or $M_s \in (K,S)$. If $b > 0$ is an upper bound for the set $\{|k|: k \in K\}$, and ε is the distance from sK to S', then $|r - s| < \delta = \varepsilon/b$ means that $|rk - sk| \leq |r - s|b < \varepsilon$ for all $k \in K$. But, if for every k, $|rk - sk| < \varepsilon$, then $rK \subset S$, and so $M_r \in (K,S)$. This shows that κ is a small enough topology that every neighborhood of M_s contains all the functions M_r for which r is sufficiently close to s.

On the other hand, the pointwise topology is perhaps too small. For instance, inside every ρ-neighborhood of the identity function are functions which are zero except on a finite family of intervals of total length 1 (or total length as small as you like). For instance, let $f(x) = x$ for all x, and let $f \in \bigcap\{(x_i,S_i): i = 1, 2, 3\}$, so that $x_i \in S_i$ for each i. The function g depicted below is zero except on three intervals, centered at the points x_i, each interval of length ⅓. On these intervals the graph of g rises so that $g(x_i) = x_i$, then falls sharply back to zero, as one goes from left to right (see Chap. VI for the interpretation of this diagram). The c-o topology has no such failing. For instance, let $K = [100,200]$ and $S = (99,201)$ be inter-

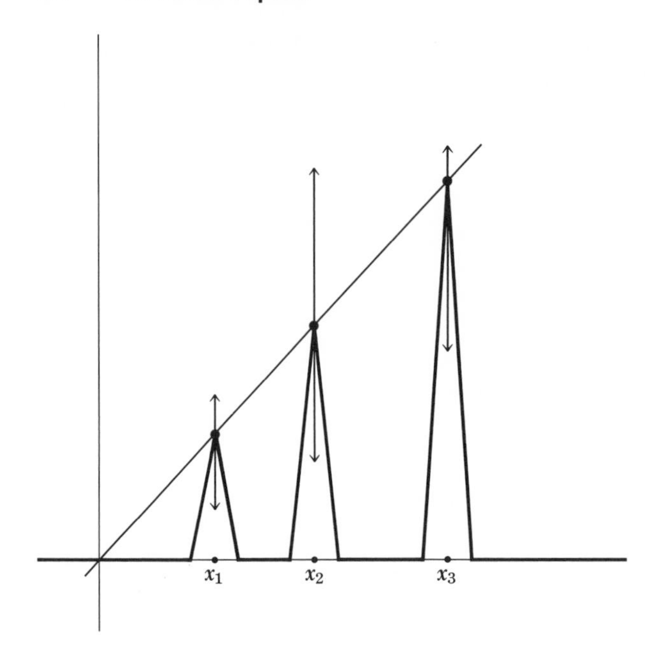

vals defining the subbasic element (K,S) of κ. Then $f \in (K,S)$, and every function g in (K,S) must be nonzero on K, an interval of length 100. Hence the c-o topology is able to "separate" the identity function from the set W of all continuous functions which are nonzero only on a set of total length 1. To put it more precisely, the closure of W in κ does not contain f, but the closure of W in ρ does, since every ρ-neighborhood of f intersects W.

The sets of functions we shall be dealing with henceforth are all sets of continuous functions, and we have just argued that the c-o topology is usually the most natural for such a set. Accordingly, we adopt the notation Y^X for the set of all *continuous* functions from a space X into a space Y, furnished with the *c-o topology.* Such a space Y^X is often called a **mapping space.** This "exponential" notation has been previously introduced (in Chap. VI) to describe a direct product, $Y^X = \times\{Y_x: x \in X\}$, where each factor space is $Y_x = Y$, and X is merely a set, without topology. But this is really a special application of our new definition, since the set X may always be given the discrete topology. This causes every function from X to Y (that is, every element of the direct product) to be continuous; furthermore, the c-o topology is here the same as the product topology, because X is discrete and its only compact sets are the finite sets.

Exercises A, B, and C

ADMISSIBLE TOPOLOGIES

Let F be a set of continuous functions from X to Y. The **evaluation function** e: $F \times X \to Y$ carries each pair (f,x) to $e(f,x) = f(x) \in Y$. Since the topologies for X and Y are thought of as fixed for this discussion, the continuity of the evaluation function depends solely on the topology chosen for the set F. A given topology for F is called **admissible** iff e is continuous. In the language of the calculus, this means the joint continuity of e, simultaneously in both its variables, f and x. To fix this idea, we consider the weaker requirement that e be separately continuous in each of its variables; of course, if the argument f is held fast at a point f_0 of F, then the continuity of $e(f_0,x) = f_0(x)$ in the variable x is equivalent to the continuity of f_0 itself, which we have assumed. But if the second argument of e is held fast at a point x_0 of X, the continuity of $e(f,x_0)$ as a function of its first variable f is established if each inverse image of an open set of Y is open in F. If S is open in Y, this inverse image is the set $\{f: e(f,x_0) \in S\} = \{f: f(x_0) \in S\}$, which is exactly the subbase element (x_0,S) of the product topology on F. Hence e is separately continuous iff (x_0,S) is open in F for each $x_0 \in X$ and each open set S of Y; that is, iff the topology of F contains the product topology. We might even have defined the product topology for F as the smallest topology for which e is separately continuous; it exists and is unique.

The more stringent requirement that e be jointly continuous in both its arguments, that F be admissibly topologized, is not so easily characterized in general. But when X has a plentiful supply of compact sets, so that there are enough c-o subbasic sets, we get a concise answer: if X is locally compact and Hausdorff then the c-o topology is the smallest admissible topology.

We first show that *the c-o topology is always smaller than any admissible topology for F*, whether X is locally compact Hausdorff or not (remember that "smaller than" here means "contained in"; this does not imply inequality). Suppose τ is an admissible topology for F, and let κ be the c-o topology on F, with (K,S) a member of the defining subbase for κ. If $f \in (K,S)$ and $k \in K$, then $(f,k) \in e^{-1}(S)$; hence there is a τ-open set U_k and an open set V_k of X with $(f,k) \in U_k \times V_k \subset e^{-1}(S)$. The family $\{V_k: k \in K\}$ is an open cover of the compact set K; let $V_1, V_2, \ldots, V_n$ be a finite subcover. Define $U = U_1 \cap U_2 \cap \cdots \cap U_n$; U is τ-open in F, and we claim $U \subset (K,S)$. That is nearly obvious. Let $g \in U$ and $k \in K$, say $k \in V_i$; then $g \in U_i$, so $g(k) \subset e(U_i \times V_i) \subset S$. Thus $f \in U \subset (K,S)$, (K,S) is a τ-neighborhood of each of its points, $(K,S) \in \tau$, and finally, $\kappa \subset \tau$. ∎

Of course, this does not show that there need exist any admissible topology for F; it shows only that if there is one, then it contains κ (but note Exercise G). However, we now prove that if X is a locally compact Hausdorff space then κ is admissible. Let S be open in Y and let $(f,x) \in e^{-1}(S)$, so that $f(x) \in S$. Since f is continuous, $f^{-1}(S)$ is a neighborhood of x; furthermore, whenever X is locally compact and Hausdorff, there is a closed compact neighborhood K of x with $K \subset f^{-1}(S)$. Suppose $g \in (K,S)$, or $g(K) \subset S$; clearly, $g(x) \in S$, so $(g,x) \in e^{-1}(S)$. Let K° be the interior of K; $(f,x) \in (K,S) \times K^\circ \subset e^{-1}(S)$, and $e^{-1}(S)$ must be a neighborhood of each of its points. Thus $e^{-1}(S)$ is open and e is continuous. We have just shown that *if X is a locally compact Hausdorff space and F is a set of continuous functions from X to a space Y, there exists a unique smallest admissible topology κ for F; this is exactly the relative topology for F in Y^X.* ∎

Exercises D, E, and G

GROUPS OF MATRICES

A real $n \times n$ matrix M is, for some positive integer n, an arrangement of n^2 real numbers in a square array:

$$M = (m_{ij}) = \begin{bmatrix} m_{11} & m_{12} & \cdots & m_{1n} \\ m_{21} & m_{22} & \cdots & m_{2n} \\ \cdots & \cdots & \cdots & \cdots \\ m_{n1} & m_{n2} & \cdots & m_{nn} \end{bmatrix}$$

This may be regarded, in an obvious way, as an element of real n^2-space. Thus the set $\mathfrak{M}_n$ of all $n \times n$ matrices has a metric topology μ under which it is homeomorphic to real n^2-space.

Each $n \times n$ matrix M defines a linear transformation, a function on $\mathbf{R}^n$ to $\mathbf{R}^n$, in the following fashion. If $x = (x_1, x_2, \ldots, x_n)$ is a real n-vector, then the value of M at x is the real n-vector whose i-th coordinate is

$$[M(x)]_i = \sum_{j=1}^{n} m_{ij}x_j.$$

This function M: $\mathbf{R}^n \to \mathbf{R}^n$ is continuous iff each projection of it into a factor space $\mathbf{R}$ of its range is continuous; that is, iff $[M(x)]_i$ is continuous in x, which it clearly is (more details are given below). Therefore $\mathfrak{M}_n$, as a set of continuous functions with common domain and range, may be given the product topology ρ or the c-o topology κ. One warning here: The metric topology μ is also a product topology, that of the product of n^2 copies of **R**. But that "product" topology μ for $\mathcal{M}_n$ regards each matrix as a function from a finite set of n^2 elements into **R**, whereas the product topology ρ for $\mathcal{M}_n$ regards matrices as functions from $\mathbf{R}^n$ to $\mathbf{R}^n$.

As always, $\rho \subset \kappa$. And if μ is admissible, then $\kappa \subset \mu$, since κ is smaller than any admissible topology for $\mathcal{M}_n$. Now, μ is admissible iff e is continuous; that is, iff $M(x)$ is continuous simultaneously in M and in x, when the domain of e has the product metric topology of real $(n^2 + n)$-space. We may check this separately for each coordinate $[M(x)]_i$ of the values $M(x)$. But the i-th coordinate of the value of e is

$$[M(x)]_i = \sum_{j=1}^{n} m_{ij}x_j,$$

and this is a continuous function simultaneously of the real numbers m_{ij} and x_j, since real multiplication and addition are continuous, and each m_{ij} or x_j is a (value of a) continuous function on $(n^2 + n)$-space, namely, the projection onto one of the coordinate spaces of the domain of e. We are claiming, then, that, given $\varepsilon > 0$, we can find a $\delta > 0$ such that if each coordinate m_{ij}' of M' is within δ of m_{ij}, and each $|x_j' - x_j| < \delta$, then

$$\left| \sum_{j=1}^{n} m_{ij}'x_j' - \sum_{j=1}^{n} m_{ij}x_j \right| < \varepsilon.$$

Of course, a complete description of δ as a function of $1 + n^2 + n$ variables (ε, each m_{ij}, and each x_j) would be tedious. We have argued, rather, that e is a composite of continuous functions.

There is a surprise in this example: all three of these topologies are the same! We prove this by showing that $\mu \subset \rho$. Recall first that if $x^i = (0, 0, \ldots, 0, 1, 0, \ldots, 0)$ denotes the vector of $\mathbf{R}^n$ whose coordinates are zero except for the i-th coordinate, which is 1, $x_j^i = \delta_j^i$,† then $M(x^i)$ is the vector whose coordinates are just the entries of the i-th column of M. Thus let S_ε^i denote the neighborhood of the i-th column vector of M which consists of all vectors whose entries are within $\varepsilon > 0$ of those of that column of M. Then $\bigcap\{(x^i, S_\varepsilon^i): i = 1, 2, \ldots, n\}$ is the member of the defining subbase for ρ which contains all the matrices of M_n whose entries are within ε of their corresponding entries in M. It is obvious now that there is a ρ-neighborhood of M inside any given μ-neighborhood of M, or $\mu \subset \rho$. The

† The symbol δ_j^i denotes the **Kronecker delta function** of two variables i and j whose value is 1 if $i = j$ and 0 if $i \neq j$.

trick of this proof is the use of the fact that, when the image of a (finite!) basis is known for a linear transformation, the matrix of that transform is determined.

A glance at the definition of the product of two $n \times n$ matrices—$(m_{ij})(m_{ij})$ has as its (i,j)-th entry $\Sigma_{k=1}^{n} m_{ik}m_{kj}$—convinces us that the multiplication of matrices defines a continuous function from $\mathcal{M}_n \times \mathcal{M}_n$ to $\mathcal{M}_n$. Of course, some matrices do not have inverses, but the ones with nonzero determinant are exactly the invertible ones. Again, the recipe for the determinant of a matrix is a sum of products of real numbers, each of which real numbers is the projection of the matrix (regarded as a real n^2-vector) on one factor of the product space $\mathbf{R}^{n^2}$. Since projections of product spaces are continuous functions, as are real addition and multiplication, the determinant function is a map from $\mathcal{M}_n$ to $\mathbf{R}$. Now, $\mathbf{R} - \{0\}$ is an open subset of $\mathbf{R}$, and the set $GL(n,\mathbf{R}) = \{M \in \mathcal{M}_n: |M| \neq 0\}$ of invertible $n \times n$ matrices is thus an open subset of $\mathcal{M}_n$. A rephrasing of this argument is that if $|M| \neq 0$, then there exists an $\varepsilon > 0$ such that, if the entries of N are within ε of those of M, then $|N| \neq 0$. The set $GL(n,\mathbf{R})$ is called the **full** (or **general**) **linear group** on $\mathbf{R}^n$; it is clearly a group, and its multiplication is continuous. Furthermore, an argument similar to that for matrix multiplication shows that the inversion function is continuous from $GL(n,\mathbf{R})$ to itself (see Exercise H). Since every metric space is Hausdorff (see Exercise V.G), the full linear group is a topological group; in fact, it is a group manifold.

Since the determinant function preserves products, $|MN| = |M|\,|N|$, and is an open map (see Exercise J), it is a morphism of $GL(n,\mathbf{R})$ onto $\mathbf{R}$. Its kernel, $SL(n,\mathbf{R}) = \{M \in \mathcal{M}_n: |M| = 1\}$, is called the **special linear group;** it is a closed normal subgroup of the full linear group. By the Quotient Theorem for Topological Groups, the quotient group $GL(n,\mathbf{R})/SL(n,\mathbf{R})$ is topologically isomorphic to the multiplicative group of nonzero reals.

Exercises H and K

TOPOLOGICAL TRANSFORMATION GROUPS

In Chap. II a group whose elements were functions on some set S and whose group operation was composition was called a permutation group. In a sense, such groups were less abstract, although Cayley's theorem (see Prob. II.EE) asserted that every group was isomorphic to some permutation group. By analogy, we define a **topological transformation group** (or **group**

of transformations), abbreviated **ttg**, to be an admissible group G of functions on a fixed Hausdorf space X with composition as the group operation. It is clear that the members of G must all be homeomorphisms of X onto itself, since the group contains the products and inverses of its elements. Notice that the definition of a ttg involves G, X, and the evaluation function. A more formal definition is that a ttg is a pair (G,X), where G is a topological group whose elements are permutations of X, X is a Hausdorff space,

- **i** For all f and g in G, and for all $x \in X$, $(fg)(x) = f[g(x)]$,
- **ii** Each $f \in G$ is a homeomorphism of X onto X, and
- **iii** The evaluation function is continuous on $G \times X$ to X. (See also Prob. CC.)

The group G is said to **act on** X; the evaluation is called the **action of** G **on** X. Notice that every ttg G on X must contain the identity 1_X on X, which is the identity element of G.

We have already studied examples of groups of transformations; the action of $GL(n,\mathbf{R})$ on $\mathbf{R}^n$ is admissible, and the groups $SL(n,\mathbf{R})$, together with all other subgroups of $GL(n,\mathbf{R})$, provide further examples. One subgroup of $GL(n,\mathbf{R})$ of considerable interest is the orthogonal group $O_n = \{M \in \mathcal{M}_n \colon M^T = M^{-1}\}$, the set of $n \times n$ matrices for which the transposed matrix is the inverse matrix. [The transposed matrix M^T of M has as its (i,j)-th entry the number m_{ji}. Using the equations $(MN)^T = N^TM^T$ and $(MN)^{-1} = N^{-1}M^{-1}$, we may quickly verify that O_n is indeed a subgroup of the full linear group.] It is a familiar fact of linear algebra that the orthogonal matrices are just those corresponding to linear transformations which preserve lengths of vectors, just the isometries of $\mathbf{R}^n$ which fix the origin. The determinant of each orthogonal matrix is ± 1 (although not all matrices having determinant 1 are orthogonal). The **special orthogonal group** $SO_n = O_n \cap SL(n,\mathbf{R})$ is the subgroup of O_n of those matrices of determinant $+1$. It is sometimes called the group of "rotations of $\mathbf{R}^n$ about the origin," or the group of "rigid motions of $\mathbf{R}^n$ which fix the origin."

A ttg G on X is called **transitive** if for each two points x and y of X there is an element $g \in G$ which carries x to $g(x) = y$. Since every linear transformation in $\mathcal{M}_n$ leaves the origin fixed, no group of matrices is transitive on $\mathbf{R}^n$. However, each member of $GL(n,\mathbf{R})$ is invertible, and hence is 1-1; if we regard $GL(n,\mathbf{R})$ as a ttg on $\mathbf{R}^n - \{0\}$, then it is indeed transitive. A proof of this begins with the observation that if y is a nonzero real multiple of x, $y = \lambda x$, then the matrix $(m_{ij}) = \lambda\delta_j{}^i$ will do. If no such λ exists, then the two nonzero vectors x and y may be *interchanged* by some member of the group. We need the theorem that, if $\{a_1, a_2, \ldots, a_n\}$ and $\{b_1, b_2, \ldots, b_n\}$ are two bases for the vector space $\mathbf{R}^n$, then there exists a unique matrix M (corresponding to a unique linear transformation) such that $M(a_i) = b_i$ for $i = 1, 2, \ldots, n$. Furthermore, the independent set $\{x = a_1, y = a_2\}$ may be extended to a basis $\{x, y, a_3, \ldots, a_n\}$, which can be carried to the basis $\{y, x, a_3, \ldots, a_n\}$ by some member of $GL(n,\mathbf{R})$. Thus the full linear group is transitive on $\mathbf{R}^n - \{0\}$.

Certainly O_n is not transitive on $\mathbf{R}^n - \{0\}$, for an orthogonal matrix cannot carry x to y if these vectors have differing norms, $\|x\| \neq \|y\|$; here $\|x\| = (\Sigma_{i=1}^n x_i^2)^{1/2}$ is the norm, or "length," of the vector x. However, O_n and SO_n are both transitive on the unit sphere $S^{n-1} \subset \mathbf{R}^n - \{0\}$; a proof of this may proceed quite as the above treatment of the full linear group. Geometrically, you may visualize an appropriate orthogonal transform as a rotation of $\mathbf{R}^n$ about an axis perpendicular to x and to y.

Let G be a transitive ttg on X, and let x_0 be a point of X. There is a **projection** p: $G \to X$ of G into X defined by $p(g) = g(x_0)$; p is clearly continuous, since the topology of G is admissible; and p is onto, since G is transitive. The closed subset $p^{-1}(x_0) = H$ is a subgroup of G called the **isotropy group of** G **at** x_0; it is the subgroup of elements of G which fix x_0. Each inverse image $p^{-1}(y)$ of a point of X is a coset of H, because $g(x_0) = y = h(x_0)$ implies $g \circ h^{-1}(x_0) = x_0$. This subgroup is not usually normal; nevertheless, the map p always factors through the quotient map q of G onto its coset space G/H:

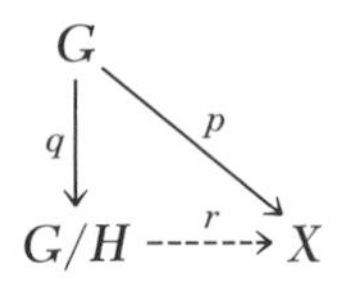

That is, there exists a map r: $G/H \to X$ with $r \circ q = p$; r is always 1-1 and onto. The map r will be open, and will thus be a homeomorphism, if p is open, since then $r(\mathfrak{U}) = p[q^{-1}(\mathfrak{U})]$ is open for each open set $\mathfrak{U}$ of G/H. Another hypothesis guaranteeing that r is a homeomorphism is that G be compact, for then G/H is also compact, and we know that a 1-1 map of a compact space onto a Hausdorff space is a homeomorphism. A direct proof that r is a closed map when G is compact merely observes that when $\mathfrak{U}$ is closed in G/H, then $q^{-1}(\mathfrak{U})$ is closed in G, and so is compact; therefore $p[q^{-1}(\mathfrak{U})] = r(\mathfrak{U})$ is compact in X, and thus is closed (since X is Hausdorff). We collect these facts in a formal pronouncement.

QUOTIENT THEOREM FOR TRANSFORMATION GROUPS *Let G be a ttg on X, $x_0 \in X$, and let $H = \{g \in G: g(x_0) = x_0\}$. Then H is a closed subgroup* *of G, and the function r: $G/H \to X$ which assigns $r(gH) = g(x_0)$ to each coset gH of H in G is a well-defined 1-1 onto map. The spaces X and G/H are homeomorphic if G is compact or if the projection p: $G \to X$ is open, where $p(g) = g(x_0)$.* ∎

As an example, we examine the ttg O_n on the space $S^{n-1} \subset \mathbf{R}^n$. We choose the special point $N \in S^{n-1}$, $N = (0, 0, \ldots, 0, 1)$. It is easy to compute that a matrix M of O_n which fixes N must have a 1 in its lower right-hand corner, $M_{nn} = 1$. Some equally easy matrix arithmetic shows that the balance of the right-hand column and of the bottom row of M must be all zeroes, $M_{in} = M_{ni} = 0$ if $i \neq n$. This in turn implies that the $(n - 1) \times (n - 1)$ matrix $\overline{M}$ resulting from the deletion of the bottom row and right-hand column from M is a member of O_{n-1}; this may be converted into a proof that the isotropy subgroup $\{M \in O_n: M(N) = N\}$ of O_n at $N \in S^{n-1}$ is topologically isomorphic to O_{n-1}, and we identify O_{n-1} with this subgroup of O_n, $O_{n-1} \subset O_n$. Hence we have a 1-1 onto map $r: O_n/O_{n-1} \to S^{n-1}$. Now let the map $f: GL(n,\mathbf{R}) \to \mathfrak{M}_n$ be defined by $f(M) = M^T - M^{-1}$; $O_n = f^{-1}(0)$ (0 denotes the matrix with entries all zeroes) is a closed subset of the open subspace $GL(n,\mathbf{R})$ of $\mathfrak{M}_n$, so that O_n is closed in $\mathfrak{M}_n$. Furthermore, if $x^i \in \mathbf{R}^n$ is the vector with entries $x_j^i = \delta_j^i$, so that x^i has a 1 as its i-th coordinate and zeroes elsewhere, then $\|x^i\| = 1$; the value $M(x^i)$ of a matrix is just the i-th column of M. If $M \in O_n$, so that M preserves norms, then $\|M(x^i)\| = \|x^i\| = 1$ for $i = 1, 2, \ldots, n$; hence no entry of M can be greater than 1, and O_n is a closed bounded subset of $\mathfrak{M}_n$. Since $\mathfrak{M}_n$ is a homeomorph of real n^2-space, O_n is compact, and we infer that O_n/O_{n-1} "is" an $(n - 1)$-sphere.

Exercises L and P

THE EXPONENTIAL LAW $(Z^Y)^X \cong Z^{X \times Y}$

A real-valued function $f: \mathbf{R}^2 \to \mathbf{R}$ of two real variables defines, for each fixed real number x_0, a function $g_0: \mathbf{R} \to \mathbf{R}$ of one real variable by $g_0(y) = f(x_0,y)$. This natural construction is often used; for example, in the calculus, the derivative of g defines a partial derivative of f. Of course, choice of a different value $x_1 \neq x_0$ for the first variable of f yields, in general, a different function g_1, $g_1(y) = f(x_1,y)$. Hence the function g constructed from f by fixing its first variable is itself a function of the "fixed" value of that first variable. More specifically, the construction may best be regarded as a function Φ which assigns to each function f: $\mathbf{R}^2 \to \mathbf{R}$ a new function $\Phi(f)$: $\mathbf{R} \to \mathbf{R}^{\mathbf{R}}$: $\Phi(f)(r)$ is the real-valued function on $\mathbf{R}$ which sends $s \in \mathbf{R}$ to $f(r,s) = \Phi(f)(r)(s)$.

This may seem a needless complication of a really simple situation, but we are searching for an appropriate generality. Let X, Y, and Z be topological spaces, and let Z^Y and $Z^{X \times Y}$ be the spaces of maps from Y and $X \times Y$ into Z, each with the c-o topology. Now build the space $(Z^Y)^X$ of continuous functions from X into Z^Y, again with the c-o topology. Using the above outline, construct the function Φ: $Z^{X \times Y} \to (Z^Y)^X$; Φ is called an **association,** and if f is a function from $X \times Y$ to Z, then $\Phi(f)$ is the **associated function** of f; we shall denote $\Phi(f)$ by $\bar{f}$ for simplicity. Thus $\bar{f}(x)$ is a function from Y to Z, $\bar{f}(x)(y) = f(x,y)$. It is not obvious that f is a member of $(Z^Y)^X$ at all; that is, is $\bar{f}(x)$ continuous for each x, and is $\bar{f}$ continuous?

THE EXPONENTIAL LAW FOR MAPPING SPACES *If X, Y, and Z are spaces, and Y is locally compact and Hausdorff, then the association Φ: $Z^{X \times Y} \to (Z^Y)^X$ is a set isomorphism (that is, a 1-1 correspondence).*

Proof Since the proof is rather long, we shall break it into four pieces.

i $\bar{f}(x)$ is continuous The function $\bar{f}(x)$ may be thought of as the restriction of the continuous function f to $x \times Y$, $\bar{f}(x)(y) = f(x,y)$; since $x \times Y$ is a homeomorph of Y, $\bar{f}(x)$ is continuous.

ii $\bar{f}$ is continuous It will suffice to show that if (K,S) is a member of the defining subbase for the c-o topology on Z^Y, then $\bar{f}^{-1}(K,S)$ is open in X. By the definition of $\bar{f}$, $\bar{f}^{-1}(K,S) = \{x \in X: \bar{f}(x)(K) \subset S\} = \{x \in X: f(x \times K) \subset S\}$. Suppose $x_0 \in \bar{f}^{-1}(K,S)$; since f is continuous, $f^{-1}(S)$ is open and $f^{-1}(S) \supset x_0 \times K$. Thus for each $k \in K$ there is an open set $U_k \subset X$ and an open set $V_k \subset Y$ with $(x_0,k) \in U_k \times V_k \subset f^{-1}(S)$. The family $\{V_k: k \in K\}$ covers the compact set K. Let $V_1, V_2, \ldots, V_n$ be a finite subcover. Define $U = U_1 \cap U_2 \cap \cdots \cap U_n$ and $V = V_1 \cup V_2 \cup \cdots \cup V_n$. (This step looks familiar; can you close the book now and finish the proof that $\bar{f}$ is continuous?) It is easy to check that $(x_0,K) \subset U \times V \subset f^{-1}(S)$, and thus that $x_0 \in U \subset \bar{f}^{-1}(K,S)$. Hence $\bar{f}^{-1}(K,S)$ is a neighborhood of each of its elements x_0, and so it is open; $\bar{f}$ is continuous.

iii Φ is 1-1 Trivially, if f and g are distinct members of $Z^{X \times Y}$, so that there is a point (x_0,y_0) with $f(x_0,y_0) \neq g(x_0,y_0)$, then $\bar{f}(x_0)(y_0) \neq \bar{g}(x_0)(y_0)$; hence $\bar{f}(x_0) \neq \bar{g}(x_0)$ and $\bar{f} \neq \bar{g}$.

iv Φ is onto Suppose $\varphi \in (Z^Y)^X$. Define a function f from $X \times Y$ to Z by $f(x,y) = \varphi(x)(y)$. We must show f to be continuous. But f is the composite function $e \circ \tilde{\varphi}$, where $\tilde{\varphi}$: $X \times Y \to Z^Y \times Y$ is defined by

$\varphi(x,y) = [\varphi(x),y]$ and e: $Z^Y \times Y \to Z$ is the evaluation, $e[\varphi(x),y] = \varphi(x)(y)$. Since $\tilde{\varphi}$ is clearly continuous in each coordinate, it is continuous into the product $Z^Y \times Y$; and because Y is locally compact and Hausdorff, the c-o topology is admissible for Z^Y. Hence e is continuous, and so is f. Obviously, $\bar{f} = \varphi$ is a value of Φ, and Φ is therefore onto. ∎

Our motivating example for the exponential law is now clear: a real-valued function f of two real variables is an element $f \in \mathbf{R}^{\mathbf{R}\times\mathbf{R}}$, and we now know that, for each x, $\bar{f}(x)$ is continuous in y; and, furthermore, $\bar{f}$ is continuous from $\mathbf{R}$ to $\mathbf{R}^{\mathbf{R}}$. To check the continuity of any function $\bar{f}$: $\mathbf{R} \to \mathbf{R}^{\mathbf{R}}$, we need only check the function f: $\mathbf{R} \times \mathbf{R} \to \mathbf{R}$: $f(x,y) = \bar{f}(x)(y)$; $\bar{f}$ is continuous iff f is continuous. We postpone further examples until the next chapter, where they will be abundant.

One corollary statement to the exponential law should be made now, though. The natural homeomorphism of $X \times Y$ with $Y \times X$ defines the central set isomorphism in the chain

$$(Z^Y)^X \cong Z^{X\times Y} \cong Z^{Y\times X} \cong (Z^X)^Y,$$

which is valid when both X and Y are locally compact Hausdorff spaces.

REFERENCES AND FURTHER TOPICS

Our exponential law claimed only a set isomorphism of $(Z^Y)^X$ with $Z^{X\times Y}$, under the condition that Y be Hausdorff and locally compact. But if X is also Hausdorff these assumptions suffice to conclude that these two mapping spaces are homeomorphic. We shall not need this stronger statement in our work, and the proof that the association Φ is continuous and open is long. Perhaps the most accessible proof is offered in

S. -T. Hu, *Elements of General Topology*, chap. V (San Francisco: Holden-Day, 1964).

Three other sources for a general development of the theory of mapping spaces, each more difficult than the one above, are

J. Dugundji, *Topology*, chap. XII (Boston: Allyn and Bacon, 1966).

S. -T. Hu, *Homotopy Theory*, pp. 73–77 (New York: Academic Press, 1959).

J. L. Kelley, *General Topology*, chap. 7 (Princeton, N.J.: Van Nostrand, 1955).

For excellent and thorough discussions of groups of matrices and ttg's, see

C. Chevalley, *Theory of Lie Groups* (Princeton, N.J.: Princeton University Press, 1946).

Montgomery and Zippin, *Topological Transformation Groups* (New York: Interscience, 1955).

L. Pontrjagin, *Topological Groups* (Princeton, N.J.: Princeton University Press, 1939).

(Both the Princeton University Press books have been published in paperback editions, as well as cloth.)

If X is a compact Hausdorff space, the set of all continuous complex-valued functions on X becomes, with the c-o topology, an additive topological group. Moreover, it has a ring multiplication and the scalar multiplication of a complex vector space (both pointwise), and these multiplications are continuous simultaneously in their factors; such a space is called a topological algebra. In fact, this algebra contains complete information about X, and X may even be recovered from it. To give details of this would carry us far afield; a very digestible account is given in

G. F. Simmons, *Topology and Modern Analysis,* part three (New York: McGraw-Hill, 1963).

EXERCISES

A Prove that the c-o topology on the set C of continuous functions from $\mathbf{I} = [0,1]$ into $\mathbf{R}$ is the same as the metric topology for C described in detail in an example of Chap. III. The metric on C is given by

$$\mathfrak{d}(f,g) = lub\ \{|f(x) - g(x)| : x \in \mathbf{I}\}.$$

(*Hint:* Show that a subbase or base for each topology is contained in the other topology.)

B Show in detail that the function g defined in the first example of this chapter (on page 167) is well defined, continuous, and positive.

C Show that the c-o topology κ on the set F of continuous functions from $\mathbf{R}$ to $\mathbf{R}$ does not contain the metric topology μ for F. Do this by showing first that the ball $\mathfrak{b}(\frac{1}{2}, 1_{\mathbf{R}}) \in \mu$ contains no member of the defining *base* for κ.

D Cayley's theorem (see Prob. II.EE) asserts that if G is a group and L is the set of left multiplications of G, $L = \{L_g: g \in G\}$, then L is a group of functions with composition as the group operation, and $L \cong G$. Prove that the set of left-multiplication functions of the circle group S^1, furnished with the c-o topology, is homeomorphic to the *space* S^1; this, together with Cayley's theorem, shows that L is a topological group and that it is (topologically) isomorphic to S^1.

E Prove that the product topology ρ is not admissible on the set C of all maps from $\mathbf{I}$ into $\mathbf{R}$. (*Hint:* Show first that under the evaluation function the inverse image in $C \times \mathbf{I}$ of an ε-ball in $\mathbf{R}$ could contain no set of the form $[\bigcap\{(x_i,S_i): i = 1, 2, \ldots, n\}] \times B$, where $x_i \in \mathbf{I}$ and S_i is open in $\mathbf{R}$ for each index i, and B is an open ball in $\mathbf{I}$. Then observe that every subset of $C \times \mathbf{I}$ which is open in its product topology must contain a subset of the above form.)

F Prove that if F is a set of maps from a compact space X into a metric space $(Y,\mathfrak{d}_0)$, then the c-o topology for F is the same as the metric topology defined by

$$\mathfrak{d}_1(f,g) = lub\ \{\mathfrak{d}_0[f(x),g(x)]: x \in X\}.$$

(A model for this proof is given by Exercise A.)

G Show that the discrete topology on a set F of maps from X to Y is always admissible.

H Give a complete argument that multiplication of matrices defines a continuous function on $\mathfrak{M}_n \times \mathfrak{M}_n$ to $\mathfrak{M}_n$. Then sketch a proof that inversion of matrices is continuous from $GL(n,\mathbf{R})$ into itself.

J Prove that the determinant function $det: \mathfrak{M}_n \to \mathbf{R}$ is open.

K Define a right-inverse map $s: \mathbf{R} \to \mathfrak{M}_n$ for the determinant function such that s, when restricted to the positive reals $\mathbf{R}_+$, is a monomorphism from the multiplicative group of positive reals onto a normal subgroup of the group $G = \{M \in GL(n,\mathbf{R}): det\ (M) > 0\}$. Conclude from your work that G is topologically isomorphic to $\mathbf{R}_+ \times SL(n,\mathbf{R})$.

L Show that whenever G is a ttg on X and H is the set of all homeomorphisms of X, the inclusion of G in H is continuous if H is given the c-o topology.

M Let G be a topological group, and let H be the set of all homeomorphisms of the space G, furnished with the c-o topology. Use Exercise L to show that the left multiplications of G define a map $s\colon G \to H$ of G into its own set of homeomorphisms.

N Now prove that a group G is topologically isomorphic to a ttg (on G) which has the c-o topology. This is a generalization of Cayley's theorem (see Exercise D and Prob. II.EE) to topological groups. (In the same sense, the topology of G is the pointwise topology, too.)

P Argue in detail that the coset space SO_n/SO_{n-1} is homeomorphic to the sphere S^{n-1}.

Q Let G be a transitive ttg on X and $x_0 \in X$, and let $p\colon G \to X$ be the evaluation at x_0, $p(g) = g(x_0)$. A "local cross section" to p at x_0 is a map $s\colon S \to G$ whose domain is a neighborhood S of x_0 in X and whose composite $p \circ s$ with p is the identity on S, $p \circ s(x) = x$ for every $x \in S$. Show that if a local cross section to p exists, then G/H is homeomorphic to X, where H is the isotropy subgroup at x_0.

R Establish that the group G of all homeomorphisms of the open interval $(0,1)$ is a ttg on $(0,1)$ when it is given the metric topology defined by

$$\mathfrak{d}(f,g) = lub\ \{|f(x) - g(x)| : 0 < x < 1\}.$$

(See Prob. V.DD.)

S If G is the ttg of Exercise R and p is the evaluation at $\frac{1}{2} \in (0,1)$, there exists a "global" cross section $s\colon (0,1) \to G$ to p. Hence show that G is homeomorphic to $H \times (0,1)$, where $H = p^{-1}(\frac{1}{2})$ (but note that G is not *isomorphic* to this direct product).

T Exhibit isomorphisms of the transformation groups $GL(1,\mathbf{R})$ and $GL(2,\mathbf{R})$ with more familiar topological groups. (Be sure your isomorphisms are both continuous and open.)

PROBLEMS

AA A Product on X^X Show that if X is a locally compact Hausdorff space, K is compact, and W is open in X, with $K \subset W$, then there exists a compact subset L of W such that K lies in the interior of L (in other words, L is a compact neighborhood of K). Use this result to show that composition defines a continuous product on X^X.

Now suppose that X is a locally compact group, and let $Y = \{f \in X^X: f(e) = e\}$. Show that the function $h: X \times Y \to X^X$ defined by $h(x,f) = L_x \circ f$ is a homeomorphism (it does not, however, preserve the group operation).

BB Metrics for O_n Show that the metric topology on O_n, considered as a subset of real n^2-space, is the same as the metric topology built for O_n in Problem V.EE. Are the metrics themselves the same?

CC Noneffective Transformation Groups A somewhat broader definition is often given for a ttg G on X. It differs by allowing the elements of G to be homeomorphisms of a larger space $Y \supset X$, provided that the restriction of $g \in G$ to X is a homeomorphism of X onto X. This difference amounts to allowing a nontrivial subgroup S of elements which fix each point of X; that is, there may be $g \in G$ with $g(x) = x$ for all $x \in X$, yet $g \neq e$. When this definition is used, a group satisfying our earlier definition is called an "effective" ttg. The connection of the two definitions will be clear if S is shown to be a closed normal subgroup whenever G is a ttg in the broader sense; then G/S is an effective ttg on X. But if $S_x = \{g \in G: g(x) = x\}$, then $S = \bigcap \{S_x: x \in X\}$, and each S_x is a subgroup, so S is a subgroup. If x and y are distinct points of X, then there is a neighborhood U of y with $x \notin U$; the set $\{g \in G: g(x) \in U\}$ is open, since evaluation at x is continuous. This shows $G - S_x$ to be a neighborhood of each of its points (why?); hence S_x is closed and S is closed. That S is normal is obvious. Show that G/S is an effective ttg on X.

DD Unitary Geometry Our discussion for $\mathfrak{M}_n$ may be applied to the set $\mathfrak{M}_n(\mathbf{C})$ of all $n \times n$ matrices with complex entries. The metric, product, and c-o topologies are identical for $\mathfrak{M}_n(\mathbf{C})$, which is homeomorphic to the direct product of n^2 copies of $\mathbf{C}$. [The "length" or "norm" of $x = (x_1, \ldots, x_k) \in \mathbf{C}^k$ is $\|x\| = (\Sigma_{i=1}^k |x_i|^2)^{1/2}$, and the distance from x to y in $\mathbf{C}^k$ is $\|x - y\|$.] The group $GL(n,\mathbf{C})$ of invertible members of $\mathfrak{M}_n(\mathbf{C})$ is a topological group. A new operation, conjugation, is possible on $\mathfrak{M}_n(\mathbf{C})$: if M is a matrix with complex entries, $M = (m_{ij})$, then $\overline{M} = (\overline{m}_{ij})$. Conjugation yields an automorphism of $GL(n,\mathbf{C})$ with itself; since $\mathbf{R} \subset \mathbf{C}$, we have $GL(n,\mathbf{R}) \subset GL(n,\mathbf{C})$, and a complex matrix M is a member of $GL(n,\mathbf{R})$ iff $M = \overline{M}$. A matrix M in $GL(n,\mathbf{C})$ is called "complex orthogonal" if $M^{-1} = M^T$, and the subgroup of all such matrices is the "complex orthogonal group" $O(n,\mathbf{C})$. Clearly, $O_n = GL(n,\mathbf{R}) \cap O(n,\mathbf{C})$. The kernel of the determinant morphism on $GL(n,\mathbf{C})$ is the "special linear group" $SL(n,\mathbf{C})$.

A "unitary matrix" M is one for which $\overline{M}^T = M^{-1}$; the "unitary group" of all such matrices is denoted by U_n; $U_n \cap O(n,\mathbf{C}) = O_n$. It is an easy exercise of complex linear algebra to show that each unitary matrix defines an isometry of $\mathbf{C}^n$; members of U_n correspond exactly to the "unitary transformations," those which preserve the norm of $\mathbf{C}^n$. Hence the members of U_n all send the unit sphere in $\mathbf{C}^n$ onto itself. Since $\mathbf{C}^n$ may be regarded as a $2n$-dimensional vector space over the reals, that unit sphere is of dimension $2n - 1$. The "special unitary group" is $SU_n = U_n \cap SL(n,\mathbf{C})$, and $SO_n = SU_n \cap O(n,\mathbf{C})$. Each of the subgroups we have defined in $GL(n,\mathbf{C})$ is closed there. Furthermore, U_n is bounded in $\mathfrak{m}_n(\mathbf{C})$, and so is compact; hence SU_n is also compact.

The coset spaces U_n/U_{n-1} and SU_n/SU_{n-1} are homeomorphic to the unit sphere S^{2n-1} in $\mathbf{C}^n$, a space homeomorphic to $\mathbf{R}^{2n}$.

EE Connected Groups Prove that if the subgroup H of G and the coset space G/H are both connected, then G must be connected. Then show that the groups SO_n, SU_n, and U_n are all connected and that O_n has exactly two components, $n \geq 1$ (SU_n and U_n are discussed in Prob. DD above).

FF Operator Norms If $M \in \mathfrak{m}_n$, let the "norm" of M be

$$\|M\| = lub\ \{\|M(x)\|: x \in S^{n-1}\},$$

and define $\mathfrak{d}(M,M') = \|M - M'\|$. This defines a metric on $\mathfrak{m}_n$; is it the same as, or equivalent to, the metric $\mathfrak{m}_n$ inherits from real n^2-space? (Compare Prob. BB.)

GG Character Groups Let G be an abelian topological group and let G^* be the set of all (continuous) morphisms of G into S^1. Then the pointwise product in G^*, $(f + f')(g) = f(g) + f'(g)$, is an abelian group operation on G^*. Furthermore, G^* is a topological group when it is given the c-o topology; it is called the "character group" of G.

If G is compact, then G^* is discrete; in fact, an appropriate subbasic set can be found to contain just the identity of G^*.

A surprising duality lies here: if G is discrete, then G^* is compact; this is true since G^* may be regarded as a closed subset of a direct product of circles, one for each element of G; the Tychonoff theorem implies G^* is compact.

Calculate the character groups of the integers, the reals, and the circle.

HH Homeomorphisms of the Circle Let G be the group of all homeomorphisms of the circle, J the interval $(0,1)$, and H the group of all homeomorphisms of J which fix the point $\frac{1}{2} \in J$ (note Exercise S). When each group of homeomorphisms is given the c-o topology, G is homeomorphic (but not isomorphic) to $S^1 \times J \times H$.

JJ A Subbase for the c-o Topology If X is Hausdorff and σ is a subbase for the topology of Y then $\{(K,S): K \text{ is compact in } X \text{ and } S \in \sigma\}$ is a subbase for the c-o topology on Y^X.

The Fundamental Group

CHAPTER VIII

There is a fundamental difference between an inner tube and a balloon. Both hold air, yet there is a hole in the air in an inner tube; one could never chain a balloon to a tree. As idealized topological subspaces of real 3-space, these two surfaces are not homeomorphic; no amount of stretching and reforming, without tearing, will deform a balloon into an inner tube. Yet you will find this highly intuitive fact equally difficult to prove. Many methods we have used thus far to show two spaces to be topologically distinct have involved only local properties, such as T_1-ness, regularity, or local compactness; but the torus and the sphere are both 2-manifolds, and they are locally alike. Global properties, such as compactness, connectedness, or the group-space property, might be useful; in fact, the torus is a group space, while the sphere is not, but this too is very difficult to prove.

You may now be impatiently thinking, "But the hole! Why not define 'hole' and then count the holes in the balloon and the inner tube?" Of course, each surface has a hole through which it can be filled with air, but these are supposed to be sealed off. And also, each surface has an "inside," a "hole" in it which holds the air, but this is not the difference you had in mind. On the face of it, the remark that we could chain an inner tube to a tree is not pertinent either, for it involves not the torus alone, but rather how it fits into real 3-space; the tree and the chain ought not be used to define an *intrinsic* property of the surface of the torus itself.

If a rubber band is laid on the surface of the torus so that it "goes around the hole once," it is intuitive that after the band is rearranged by sliding it around arbitrarily on the surface, with stretching, shrinking, kinking, and doubling it back on itself allowed (but not breaking the band), in its new position it will still go around the hole exactly once. This is the physical picture that guides our next project. We shall first make precise mathematics of the notion of a loop in a space and a continuous deformation of that loop. Then loops which cannot be deformed into one another will be regarded as essentially different. These different loops will be seen to form a group; hence we shall be able to assign to each topological space a group which "counts" the number of essentially distinct ways in which loops can "encircle holes" in the space, and this abstract group will be a topological invariant of the space, called its "fundamental group." For instance, to the 2-sphere we shall assign the singleton group; each two loops on the sphere can be deformed into one another. To the torus we assign the group $\mathbf{Z} \times \mathbf{Z}$ of gaussian integers; the pair (a,b) of integers corresponds to a loop which goes around the rim a times and then through the spokes b times. This will solve our problem; since their fundamental groups differ, the torus and the 2-sphere are not homeomorphic.

THE LOOP SPACE Ω

Let X be a space and x_0 an arbitrary but *fixed* point of X. A **loop at x_0 in X** is a map $a\colon \mathbf{I} \to X$ such that $a(0) = a(1) = x_0$; that is, a loop at x_0 is a path which begins and ends at x_0. The set Ω of all loops at x_0 in X is a subset of the space $X^{\mathbf{I}}$ of all paths in X, and Ω inherits the relative c-o topology from $X^{\mathbf{I}}$. With this topology, Ω is the **loop space of X at x_0**.

In our discussion of path components (in Chap. VI) we introduced a product between paths: if a and b are paths in X and $a(1) = b(0)$, then the product path $a \cdot b$ of a and b is the map $(a \cdot b)\colon \mathbf{I} \to X$ defined as a combined function (see Chap. I),

$$(a \cdot b)(t) = \begin{cases} a(2t) & \text{if } 0 \leq t \leq 1/2, \\ b(2t-1) & \text{if } 1/2 \leq t \leq 1. \end{cases}$$

When both a and b are loops at x_0, $a \cdot b$ is clearly also a loop at x_0. A physical picture is in order. If images of loops are thought of as wires and t represents time, then the loop a describes the position at various times of a bead which travels along the wire. At time t the bead is at the point $a(t)$ on the wire (in the space X). Then the product path $a \cdot b$ describes the position at various times of a bead which first goes around the wire

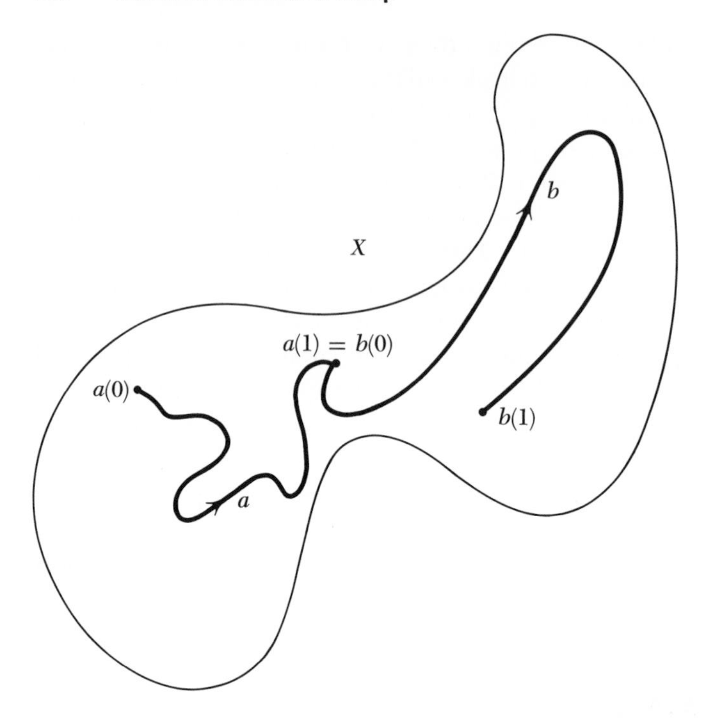

corresponding to a at twice the speed appropriate to the loop a; the $(a \cdot b)$-bead thus completes its circuit of the a-wire by time $t = \frac{1}{2}$. Then it travels along the b-wire at double the speed of the b-bead, until at time $t = 1$ it has completed its entire trip and is back at x_0.

If T is any space and $m: T \times T \to T$ is a map, then m is called a **multiplication on** T. The above formation of product paths yields a function $m: \Omega \times \Omega \to \Omega$; we shall show that this function is continuous. Suppose (K, W) is a subbasic element of the c-o topology on Ω; that is, K is a compact subset of $\mathbf{I}$, W is open in X, and (K, W) consists of all loops a such that $a(K) \subset W$. Define $L = 2(K \cap [0,\frac{1}{2}])$ and $M = 2(K \cap [\frac{1}{2},1]) - 1$; these sets are compact subsets of $\mathbf{I}$ since K is. You can easily check that $m^{-1}(K, W) = (L, W) \times (M, W)$, that is, $(a \cdot b)(K) \subset W$ iff both $a(L) \subset W$ and $b(M) \subset W$. Hence m^{-1} carries subbasic sets in Ω to basic open sets in $\Omega \times \Omega$, and m is therefore continuous (and open, too); it is a multiplication on Ω. This multiplication is certainly not a group operation on Ω. There is no identity, and therefore there are no inverse elements, and the associative law does not hold. A check of these facts, though, leaves us with the feeling that m is "nearly" a group product. For example, the difference between $(a \cdot b) \cdot c$ and $a \cdot (b \cdot c)$ is entirely one of differing speeds along the

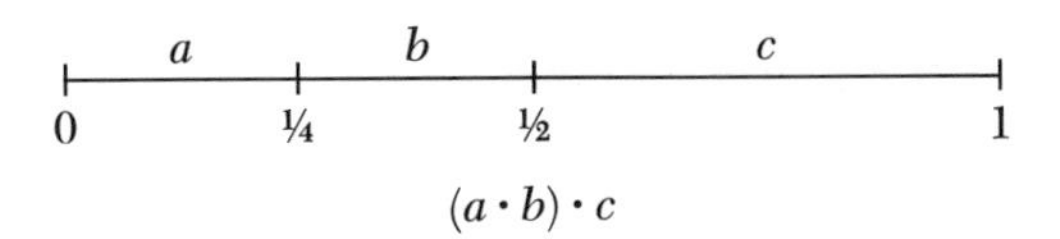

$(a \cdot b) \cdot c$

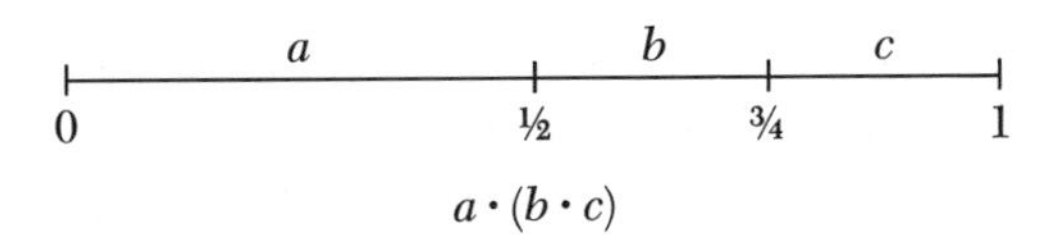

$a \cdot (b \cdot c)$

same route. In $(a \cdot b) \cdot$ c, the loop a is gone around at quadruple speed; in $a \cdot (b \cdot c)$ it is gone around at double speed. To make this analogy explicit, we use the definition of the multiplication on Ω to write out a description of $(a \cdot b) \cdot c$. By definition, this combined function is

$$[(a \cdot b) \cdot c](t) = \begin{cases} (a \cdot b)(2t) & \text{if } 0 \leq t \leq \frac{1}{2}, \\ c(2t - 1) & \text{if } \frac{1}{2} \leq t \leq 1; \end{cases}$$

and

$$(a \cdot b)(s) = \begin{cases} a(2s) & \text{if } 0 \leq s \leq \frac{1}{2}, \\ b(2s - 1) & \text{if } \frac{1}{2} \leq s \leq 1. \end{cases}$$

Substituting $s = 2t$, we get

$$[(a \cdot b) \cdot c](t) = \begin{cases} a(4t) & \text{if } 0 \leq 2t \leq \frac{1}{2}, \\ b(4t - 1) & \text{if } \frac{1}{2} \leq 2t \leq 1, \\ c(2t - 1) & \text{if } \frac{1}{2} \leq t \leq 1. \end{cases}$$

This may be restated as

$$[(a \cdot b) \cdot c](t) = \begin{cases} a(4t) & \text{if } 0 \leq t \leq \frac{1}{4}, \\ b(4t - 1) & \text{if } \frac{1}{4} \leq t \leq \frac{1}{2}, \\ c(2t - 1) & \text{if } \frac{1}{2} \leq t \leq 1. \end{cases}$$

You are asked in Exercise A to write out a similar description of the loop $a \cdot (b \cdot c)$.

Exercises A and B

THE GROUP $\pi_0(\Omega)$

The space Ω of loops at a point x_0 in a space X is a topological space in its own right; thus it now makes sense to speak of a path $p: \mathbf{I} \to \Omega$ in Ω. It helps to think of p as a parameterized family p_u of loops at x_0 in X, where the loop $p_u = p(u)$ in X changes continuously as u varies in $\mathbf{I}$. The path p represents a "continuous deformation" of the loop p_0 into the loop p_1; hence a path component of Ω is a set of loops in X, each of which may be continuously deformed into each other in that set.

We now examine the set $\pi_0(\Omega)$ of path components of Ω. The map m: $\Omega \times \Omega \to \Omega$ induces a function $\pi_0(m)$: $\pi_0(\Omega \times \Omega) \to \pi_0(\Omega)$. Since the continuous image of a path-connected space is path connected, each path component of $\Omega \times \Omega$ is carried by m into some *single* element of $\pi_0(\Omega)$. If $[(a,b)]$ is the path component of $(a,b) \in \Omega \times \Omega$, then $\pi_0(m)[(a,b)] = [m(a,b)] = [a \cdot b]$ is the path component of $a \cdot$ b in Ω. Since the "function" π_0 preserves direct products (see Exercise VI.G), there is a set isomorphism φ of $\pi_0(\Omega) \times \pi_0(\Omega)$ with $\pi_0(\Omega \times \Omega)$; it is defined by $\varphi([a],[b]) = [(a,b)]$. Therefore the composite function $m_* = \varphi \circ \pi_0(m)$ is a product on $\pi_0(\Omega)$:

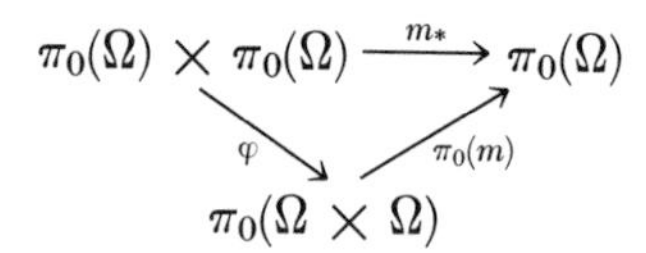

We now claim that m_* is a group operation on the set $\pi_0(\Omega)$! The identity for m_* is the component $[e]$ of the constant map $e: \mathbf{I} \to X$: $e(t) = x_0$ for all t. To show that this works, we shall define a path in Ω beginning at $a \cdot e$ and ending at a; thus $m_*([a],[e]) = [a \cdot e] = [a]$. Let the path $p: \mathbf{I} \to \Omega$ be defined by letting the loop $p(u)$, for each $u \in \mathbf{I}$, have values $p(u)(t) = a[2t/(1 + u)]$ if $0 \leq t \leq (1 + u)/2$ and $p(u)(t) = x_0$ if $(1 + u)/2 \leq t \leq 1$. Clearly, for each u, $p(u)$ goes along a at speed $2/(1 + u)$ until time $(1 + u)/2$ and then stands still until time 1, each path $p(u)$ is a loop at x_0, $p(0) = a \cdot e$ and $p(1) = a$. But is p continuous (in its argument u)? By the exponential law (Chap. VII), p is a continuous function from $\mathbf{I}$ into $X^{\mathbf{I}}$ iff it is the associated function $p = \bar{f}$ of a map $f \in X^{\mathbf{I} \times \mathbf{I}}$.

In that case, of course, $f(u,t) = p(u)(t)$; the continuity of p is equivalent to that of f. The combined function f is continuous iff it is continuous on each of the two closed subsets of its domain described by $t \leq (1 + u)/2$ and

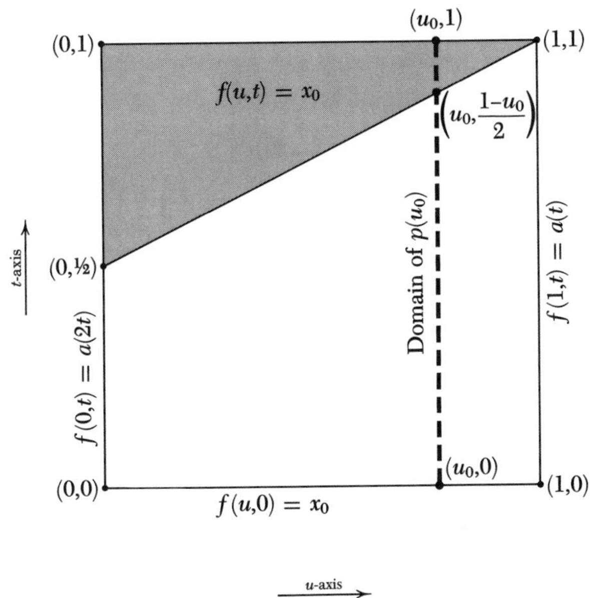

Domain square of f

$t \geq (1 + u)/2$. On the latter set, f is constant, and so is continuous. On the former, f is a composite function $f(u,t) = a[2t/(1 + u)]$; the argument of the map a is the quotient of continuous real-valued functions (with denominator never zero). It was an exercise (see Exercise V.D) to show that such a function is continuous from $\mathbf{I}^2 = \mathbf{I} \times \mathbf{I}$ to $\mathbf{I}$. Thus f is continuous, $\bar{f} = p$ is continuous, and p is a path in Ω from $e \cdot a$ to a. A similar construction yields a path from $a \cdot e$ to a, hence $[e]$ is an identity for the product m_* on $\pi_0(\Omega)$.

The trick used to prove p continuous is of general efficacy. A map f: $\mathbf{I}^2 \to X$ which defines a path p from one loop a at x_0 to another loop b is called a **homotopy** of a with b, and a and b are said to be **homotopic loops.** Two loops a and b are homotopic iff they are connected by a path in Ω, by the exponential law. The relation between loops of being homotopic is thus the same *equivalence* relation as that of their belonging to the same path component of Ω.

The associativity of m_* will thus be demonstrated if, given three loops a, b, and c at x_0, a homotopy g: $\mathbf{I}^2 \to X$ is constructed such that $g(0,t) = [(a \cdot b) \cdot c](t)$ and $g(1,t) = [a \cdot (b \cdot c)](t)$ for each $t \in \mathbf{I}$. Then the associated function $\bar{g} \in \Omega^{\mathbf{I}}$ will be a path in Ω from $(a \cdot b) \cdot c$ to $a \cdot (b \cdot c)$.

Formally, we now define g by

$$g(u,t) = \begin{cases} a\left(\dfrac{4t}{1+u}\right) & \text{if } 0 \leq t \leq \dfrac{1+u}{4} \\ b(4t - u - 1) & \text{if } \dfrac{1+u}{4} \leq t \leq \dfrac{2+u}{4} \\ c\left(\dfrac{4t-u-2}{2-u}\right) & \text{if } \dfrac{2+u}{4} \leq t \leq 1. \end{cases}$$

However, the definition of g may be understood better from the diagram of its domain square. The continuity of g on each of the three salient closed subspaces of its domain (pictured in the diagram) is clear. And a simple

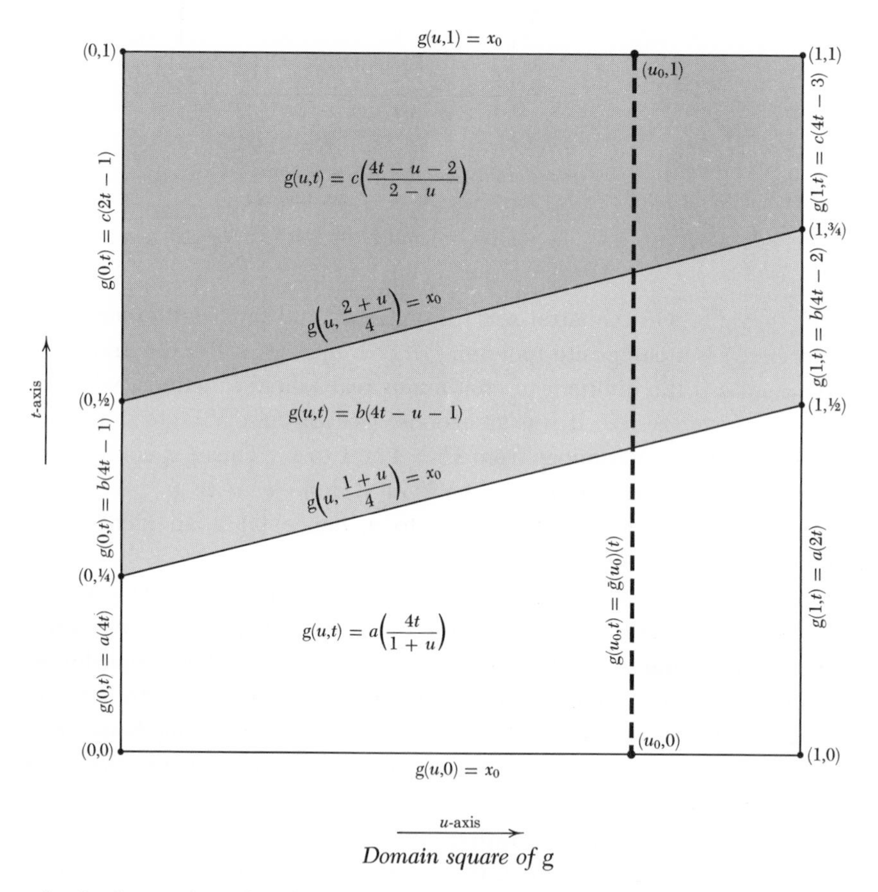

Domain square of g

check shows that the three seemingly different definitions for g do agree where they overlap; therefore g is continuous, and thus is a homotopy of $(a \cdot b) \cdot c$ with $a \cdot (b \cdot c)$.

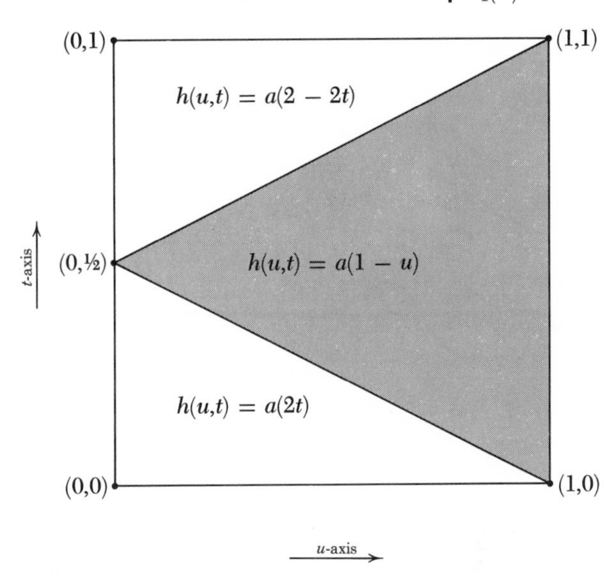

We merely sketch the diagram for the domain square of a homotopy h which, for a given loop a, exhibits an inverse $[\bar{a}]$ for $[a] \in \pi_0(\Omega)$. The loop $\bar{a}$ "goes around a backwards"; $\bar{a}(t) = a(1 - t)$ for each $t \in \mathbf{I}$. Can you complete the diagram and provide the discussion which proves the existence of $[a]^{-1}$ for each $a \in \Omega$?

Exercises C, D, and E

THE FUNDAMENTAL GROUP $\pi_1(X)$

We have shown that m_* is a group operation for $\pi_0(\Omega)$. This group is defined when a space X and a **base point** (or **distinguished point**) x_0 of X are given; it is denoted $\pi_1(X,x_0)$. Clearly, if $[x_0]$ is the path component of x_0 in X, then $\pi_1(X,x_0) = \pi_1([x_0],x_0)$, since every loop at x_0 in X must have its image in $[x_0]$. It is less obvious, but equally true, that if $x_1 \in [x_0]$, then $\pi_1(X,x_0)$ is isomorphic to $\pi_1(X,x_1)$. We shall prove that an isomorphism θ is obtained from an arbitrary path p from x_1 to x_0 as follows: if $[a] \in \pi_1(X,x_0)$, so that a is a loop at x_0 in X, let $\theta[a] = [p \cdot (a \cdot \bar{p})]$, where $\bar{p}(t) = p(1 - t)$ gives the path backwards along p from x_0 to x_1.

The class $\theta[a]$ of loops at x_1 is thus the component of the loop space of X at x_1 which contains $p \cdot (a \cdot \bar{p})$, that loop which first goes from x_1 to x_0 along p (at double speed), then around the loop a at x_0, then back along $\bar{p}$

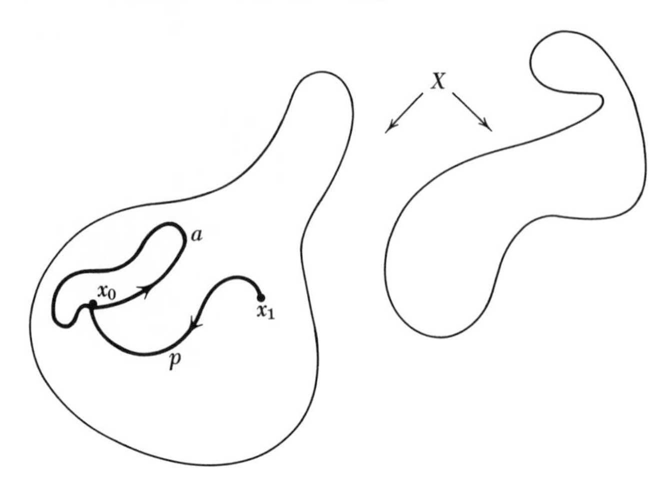

to x_1. The loop $p \cdot (a \cdot \bar{p})$ is called the **translate** of a along p. To verify that θ is well defined we must show that if a loop $b \in [a]$, then $p \cdot (a \cdot \bar{p})$ is homotopic to $p \cdot (b \cdot \bar{p})$. We sketch the values of a homotopy H of $p \cdot (a \cdot \bar{p})$ with $p \cdot (b \cdot \bar{p})$, which is defined by a given homotopy h of a with b. As you

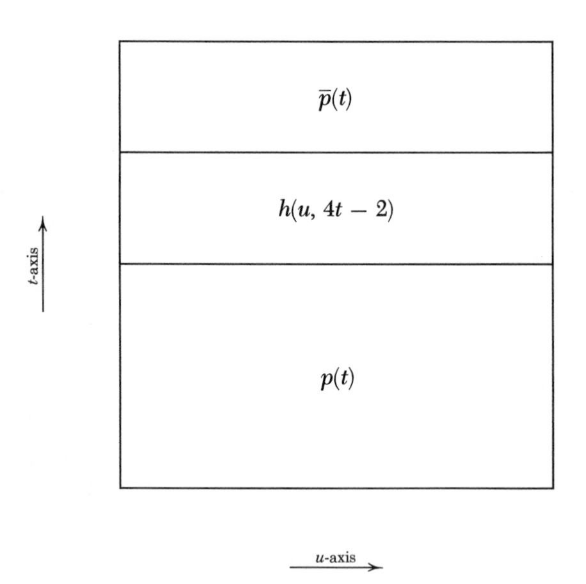

provide the details for this diagram, keep in mind that H is a homotopy "in $\pi_1(X,x_1)$," that is, a homotopy between loops at x_1, *not* at x_0.

The next diagram is a beginning of a description of a homotopy which shows that θ is a morphism; that is, for loops a and b at x_0, it shows a homot-

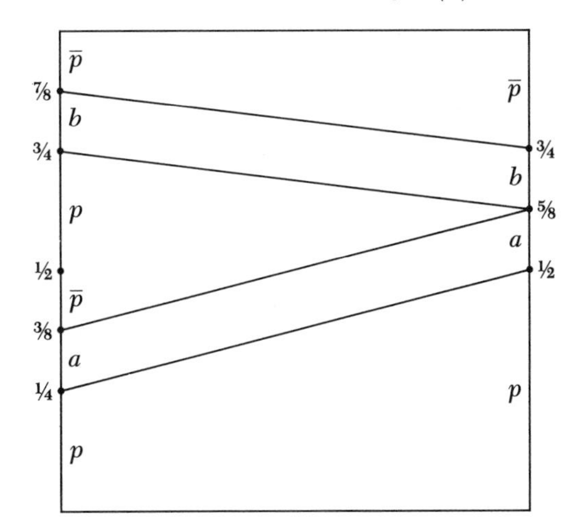

opy of $[p \cdot (a \cdot p)] \cdot [p \cdot (b \cdot \bar{p})]$ with $p \cdot [(a \cdot b) \cdot \bar{p}]$. Most of the details of this should be clear; the definition of the homotopy on the triangular region of its domain is left as an exercise.

It is obvious that if the roles of x_0 and p are interchanged with those of x_1 and $\bar{p}$, a morphism φ: $\pi_1(X,x_1) \to \pi_1(X,x_0)$ is defined just as θ was constructed. It is not hard to see that φ is a two-sided inverse for θ; by the symmetry of the constructions, it need only be shown that $\varphi \circ \theta$ is the identity on $\pi_1(X,x_0)$. The needed homotopy between the loops $\bar{p} \cdot ([p \cdot (a \cdot \bar{p})] \cdot p)$ and a at x_0 is suggested by a homotopy of $\bar{p} \cdot p$ with the constant loop at x_0. Make a diagram of this for yourself.

We have just shown that if x_0 and x_1 are in the same path component of X, then $\pi_1(X,x_0)$ is isomorphic to $\pi_1(X,x_1)$. If X is path connected, this means that $\pi_1(X,x)$ is independent (up to isomorphism) of the choice of the base point x. Hence, for a path-connected space X we shall denote by $\pi_1(X)$ any group isomorphic to some group $\pi_1(X,x)$, where $x \in X$. A group $\pi_1(X)$ is spoken of as the **fundamental group** (or "first homotopy group," or "Poincaré group") of X. This is, of course, an abuse of our language; $\pi_1(X)$ is really an isomorphism class of groups—all those groups isomorphic to $\pi_1(X,x)$ for some $x \in X$. But it will be convenient to say, for instance, when $\pi_1(X,x)$ is isomorphic to the integers, that $\pi_1(X)$ *equals* $\mathbf{Z}$.

Thus to each path-connected space X we attach a group $\pi_1(X)$. Further, a map f: $X \to Y$ from one such space to another induces a map $f^{\mathbf{I}}$: $X^{\mathbf{I}} \to Y^{\mathbf{I}}$, and $f^{\mathbf{I}}$ surely carries a loop at x in X to a loop at $y = f(x)$ in Y. Accordingly, a restriction, say $\tilde{f}$, of the map $f^{\mathbf{I}}$ carries the loop space at x in X into the loop space at y in Y, and thereby induces the function $\pi_0(\tilde{f})$ from the path components of one loop space to those of the other.

The latter function is usually denoted f_*; it is a function from $\pi_1(X,x)$ to $\pi_1(Y,y)$. If a is a loop at x in X, then $f^{\mathbf{I}}(a) = f \circ a$ is a loop at y in Y, and so $[f \circ a] = f_*[a]$ is the homotopy class of the loop $f \circ a$.

It is obvious that $f^{\mathbf{I}}$ preserves products of loops: $f^{\mathbf{I}}(a \cdot b) = f \circ (a \cdot b) = (f \circ a) \cdot (f \circ b)$. Consequently, f_* is a morphism: if we denote the product in $\pi_1(X,x)$ by juxtaposition, $[a \cdot b] = [a][b]$, then $f_*([a][b]) = f_*[a \cdot b] = [f \circ (a \cdot b)] = [(f \circ a) \cdot (f \circ b)] = f_*[a]\, f_*[b]$.

Moreover, if f and g are maps, $X \xrightarrow{f} Y \xrightarrow{g} Z$, $f(x) = y$ and $g(y) = z$, then $g \circ f$: $X \to Z$ and $(g \circ f)_*$: $\pi_1(X,x) \to \pi_1(Z,z)$. But $(g \circ f)_*[a] = [g \circ f \circ a] = g_*[f \circ a] = g_* \circ f_*[a]$, so $(g \circ f)_* = g_* \circ f_*$. Also, it is obvious that $(1_X)_*$ is the identity morphism of $\pi_1(X,x)$.

Finally, it should be clear what we mean by f_*: $\pi_1(X) \to \pi_1(Y)$ when X and Y are both path connected. Each of the above statements may be interpreted in terms of fundamental groups. If f: $X \to Y$ is a homeomorphism, then $(f^{-1})_* \circ f_* = (f^{-1} \circ f)_* = (1_X)_*$ is the identity morphism on $\pi_1(X)$, and by a similar argument, so is $f_* \circ (f^{-1})_*$. Hence f_*: $\pi_1(X) \to \pi_1(Y)$ has a two-sided inverse $(f^{-1})_* = (f_*)^{-1}$, and f_* is an isomorphism. With our earlier abuse of language, then, $\pi_1(X) = \pi_1(Y)$. Thus to each path-connected topological space X we have attached an algebraic object $\pi_1(X)$; $\pi_1(X)$ is a topological invariant, and maps of spaces induce morphisms of the attached invariant groups. [We remark that $\pi_1(X)$ is not a topological group; it has only algebraic structure.]

Exercises F and G

$\pi_1(\mathbf{R}^n)$, A TRIVIAL EXAMPLE

Consider a loop a: $\mathbf{I} \to \mathbf{R}^n$ at the origin 0 in the path-connected space $\mathbf{R}^n$. There is a homotopy h of a with the constant loop e at 0, defined by $h(u,t) = (1 - u)a(t)$. Since h is the product (scalar multiplication, pointwise) of maps, it is a map; $h(0,t) = a(t)$ and $h(1,t) = e(t) = 0$ for all t, and for each fixed $u \in \mathbf{I}$, $h(u,t)$ defines a loop at 0. Hence the loop space of $\mathbf{R}^n$ at 0 has just one path component, $[e]$, and $\pi_1(\mathbf{R}^n,0) \cong \pi_1(\mathbf{R}^n) = \{1\}$ (the singleton group).

Since each open ball in $\mathbf{R}^n$ is homeomorphic to $\mathbf{R}^n$ itself, the fundamental group of an open ball is trivial too. Furthermore, if the loop a above had its image in the closed unit ball centered at 0, then the image of the homotopy h also lay in that ball; this implies that every loop in the closed ball is **trivial** (that is, homotopic to the constant loop), and the fundamental

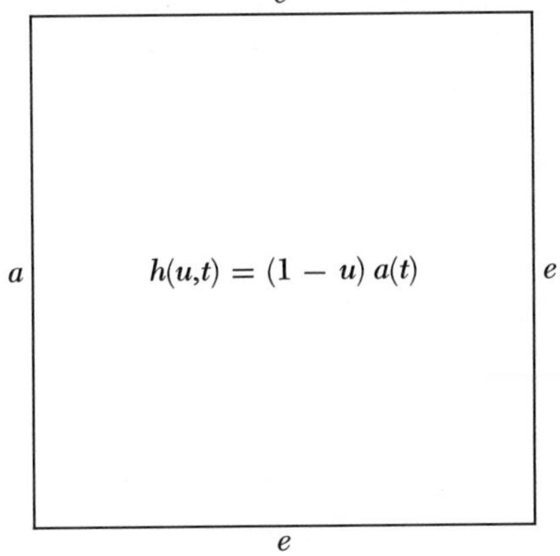
e
a
$h(u,t) = (1 - u)\,a(t)$
e
e

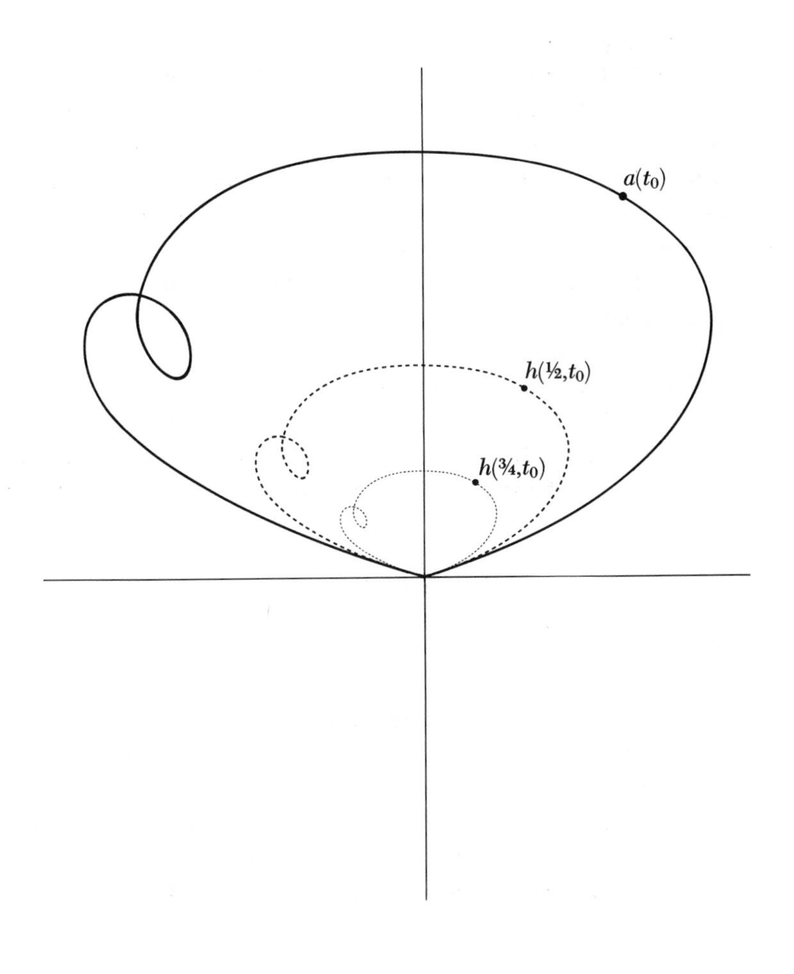
$a(t_0)$
$h(½,t_0)$
$h(¾,t_0)$

group of the closed ball is trivial. The same may be said for every closed ball in $\mathbf{R}^n$, as well as the n-cube $\mathbf{I}^n$, since these are homeomorphic to the closed unit ball.

Exercise H

FURTHER EXAMPLES

A connected space X such that $\pi_1(X) = \{1\}$ is called **simply connected;** hence we have just shown that for all positive integers n, $\mathbf{R}^n$ is simply connected. On the other hand, we asserted in the remarks introducing this chapter that the torus $S^1 \times S^1$ has as its fundamental group the gaussian integers; it is certainly not simply connected. The circle is perhaps an even more intuitive example of a not-simply connected space; its fundamental group will be seen in the next chapter to be the integers, $\pi_1(S^1) = \mathbf{Z}$. The proof of this fact is hard, however, despite its appeal to the imagination. Each nonnegative integer n of $\mathbf{Z}$ corresponds to a loop which winds itself counterclockwise around the circle n times, and $-n$ corresponds to a loop running clockwise n times around S^1. While this correspondence may require some thought, you will more readily agree that a loop going once around the circle is not homotopic to the constant loop e. (The proof of this is called for by Exercise J.)

Consider, in general, a loop a at a point x_0 in a space X. If a is homotopic to the constant loop e at x_0, then there exists a homotopy h: $\mathbf{I}^2 \to X$ of a with e, where $h(u,0) = h(u,1) = h(1,t) = x_0$ and $h(0,t) = a(t)$ for every t and u in $\mathbf{I}$. Now, a is a loop, $a(0) = a(1) = x_0$, so it factors through the quotient space S^1 of $\mathbf{I}$ which results from the identification of

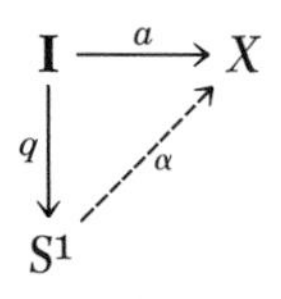

the end points of $\mathbf{I}$, to define a map α: $S^1 \to X$ such that $a = \alpha \circ q$. Thus every loop in X may be regarded as a "based" map of S^1 into X, one which carries $1 \in S^1$ to $x_0 \in X$. Further, the homotopy h is at each stage a loop, so h factors through the cylinder $\mathbf{I} \times S^1$, which is the quotient of $\mathbf{I} \times \mathbf{I}$ defined by identifying the point $(u,0)$ with $(u,1)$ for every $u \in \mathbf{I}$; this quotient map merely sews together two opposing edges of $\mathbf{I} \times \mathbf{I}$. On this cylinder, the appropriate factor of h takes each point of the top rim to the point $x_0 = h(1,t)$; hence we can further factor h through the quotient map on $\mathbf{I} \times S^1$ which identifies the circular top rim of the cylinder to a point. The quotient space is now a cone, which is clearly homeomorphic (by projection onto the plane of its base) to a closed disc D. Adding up these fac-

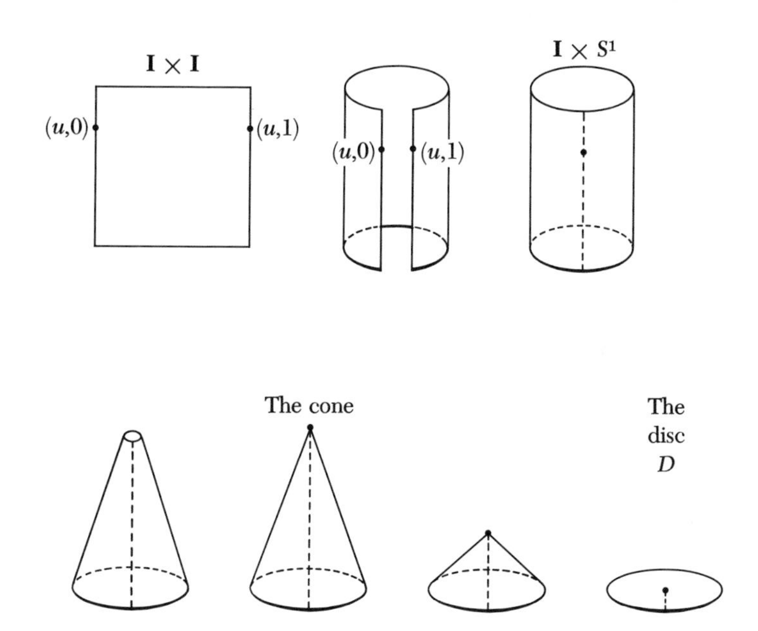

torizations and quotients, we get a factorization $h = H \circ Q$, where Q: $\mathbf{I}^2 \to D$ is a quotient map and H: $D \to X$. During all this, the bottom rim of the cylinder, which may be taken as the domain of the "loop" α that is like a, has become the outer circular edge of the disc. We conclude that if a map α: $S^1 \to X$ represents a loop a which is homotopic to the constant loop at x_0, then there exists a continuous extension H of α to the entire closed disc bounded by the circular domain of α. Conversely, if α: $S^1 \to X$ is a based map for which an extension H to the disc exists, then $h = H \circ Q$ is a homotopy of the loop $a = \alpha \circ q$ with the constant loop. Thus the loops homotopic to the constant loop are characterized as those loops represented by maps of the circle which can be extended to maps of the closed disc.

To return to our question about $\pi_1(S^1)$, do you think that the identity map on S^1 can be extended to the closed disc? That is, does there exist a map H: $D \to S^1$ with $H(x) = x$ for all $x \in S^1$? If not, then S^1 is not simply connected.

[In general, if $A \subset X$ and the identity map on A can be extended to a map $r: X \rightarrow A$ defined on all of X, then A is said to be a "retract" of X, and r is a retraction, $r(a) = a$ for each $a \in A$. Thus, if S^1 is simply connected, then S^1 is a retract of D.]

Exercises J and K

HOMOTOPIES OF MAPS

For convenience, we shall refer to a pair (X,x_0), where X is a space and x_0 a point of X, as a **based space.** The notation $f: (X,x_0) \rightarrow (Y,y_0)$ will be used for a **map of based spaces;** f is to be a continuous function from X to Y with $f(x_0) = y_0$ (it may occasionally be called a **based map**). If f and g are two based maps from (X,x_0) to (Y,y_0), a **homotopy of** f **with** g (or **from** f **to** g) is a map H: $\mathbf{I} \times X \rightarrow Y$ such that for all $x \in X$, $H(0,x) = f(x)$ and $H(1,x) = g(x)$; further, for each $u \in \mathbf{I}$, $H(u,x_0) = y_0$. If the Exponential Law applies (that is, if X is locally compact and Hausdorff), then $(Y^X)^{\mathbf{I}}$ is set isomorphic to $Y^{\mathbf{I} \times X}$, and two based maps f and g are homotopic iff their associated functions $\bar{f}$ and $\bar{g}$ both lie in the same path component of Y^X. The way in which this definition of homotopy generalizes that of loops will be clear from Exercise K; each loop is there corresponded to a based map on $(S^1,1)$.

As an example, we show that the identity map on $(\mathbf{R}^n,0)$ is homotopic to the constant map (which has zero as its only value). The homotopy is already familiar. Let $H(u,x) = (1 - u)x$, the multiple of the vector x by the scalar $1 - u$. Any space with this property, a based space (Y,y_0) such that its identity map is homotopic to its constant map, is called **contractible to the point** y_0 (and the homotopy of 1_Y is a "contraction"). A contractible space is clearly path connected; if $y \in Y$, then "the track left by y" under the homotopy of 1_Y with e is a path from y to y_0. Moreover, a contractible space is always simply connected; we shall show that, in fact, every two based maps f and g from a based space (X,x_0) into (Y,y_0) are homotopic if (Y,y_0) is contractible. Suppose H: $\mathbf{I} \times Y \rightarrow Y$ is a homotopy of 1_Y with e; we define a homotopy G of f with g by

$$G(u,x) = \begin{cases} H[2u, f(x)] & \text{if } 0 \leq u \leq \frac{1}{2} \\ H[2 - 2u, g(x)] & \text{if } \frac{1}{2} \leq u \leq 1. \end{cases}$$

As before, since the definitions of G agree on the intersection of the two closed domains of definition, on each of which G is a composite of continuous functions, the combined function G is continuous. But $G(0,x) = H[0,f(x)] = f(x)$, $G(1,x) = H[0,g(x)] = g(x)$, and for all u, $G(u,x_0) = H(v,y_0) = y_0$, where v is either $2u$ or $2 - 2u$. Hence, in particular, each pair of maps from $(S^1,1)$ into (Y,y_0), which may be considered as loops (see Exercise K), are homotopic; $\pi_1(Y) = \{1\}$.

(If a based map is homotopic to a constant map, it is called "inessential." Thus we have shown that every map into a contractible space is inessential.)

The importance of homotopies of maps is suggested by the fact that if f and g are homotopic maps from (X,x_0) to (Y,y_0), then $f_* = g_*: \pi_1(X,x_0) \to \pi_1(Y,y_0)$. To see this, remember that $f_*[a] = [f \circ a]$ and $g_*[a] = [g \circ a]$. If H is a homotopy of f with g, then the map $H \circ (1_I \times a)$, which sends (u,t) to $H[u,a(t)]$, is easily seen to be a homotopy of $f \circ a$ with $g \circ a$. Hence we could have said above that for contractible spaces X, $(1_X)_*$, an isomorphism of $\pi_1(X)$, is equal to the constant morphism whose only value is the identity $\{1\} = \pi_1\{x_0\}$ (why?); therefore $\pi_1(X) = \{1\}$.

Exercise L

HOMOTOPY TYPES

A map $f: (X,x_0) \to (Y,y_0)$ of based spaces is a **homotopy equivalence** iff there exists a based map $g: (Y,y_0) \to (X,x_0)$ such that $f \circ g$ is homotopic to 1_Y and $g \circ f$ is homotopic to 1_X. Clearly, in this case g is also a homotopy equivalence; then (X,x_0) and (Y,y_0) are said to be **homotopy equivalent,** or **of the same homotopy type.** As an example, a contractible space (Y,y_0) is of the same homotopy type as its one-point subspace $(X,x_0) = (y_0,y_0)$. Here we take f to be the inclusion map of (y_0,y_0) into (Y,y_0) and g to be the constant map of (Y,y_0) onto (y_0,y_0); $g \circ f$ thus equals 1_X, and $f \circ g$ is homotopic to 1_Y, since (Y,y_0) is contractible.

Another example is provided by the plane minus the origin, $(Y,y_0) = (\mathbf{C} - \{0\}, 1)$, and its subspace $(X,x_0) = (S^1,1)$, the unit circle. The based map f is to be the inclusion map of the subspace S^1 in the "deleted" plane. Since each nonzero complex number z has $|z| \neq 0$, we may define $g: \mathbf{C} - \{0\} \to S^1$ by $g(z) = z/|z|$. Clearly, $g \circ f = 1_X$. The map $f \circ g$ carries z to $z/|z| \in \mathbf{C} - \{0\}$; we must find a homotopy H of $f \circ g$ with 1_Y. Let $H(u,z) = (1 - u)z + uz/|z|$; as you check that H works, be sure to verify that its range is correct. That is, H must not take on the value 0.

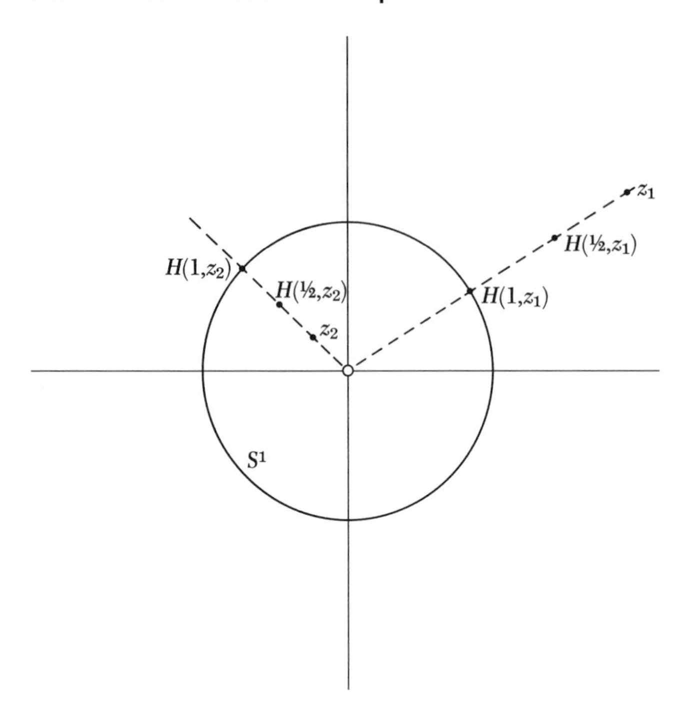

Granting this, we have shown that $(S^1,1)$ and $(\mathbf{C} - \{0\}, 1)$ are homotopy equivalent. The picture you should have in mind of this homotopy is that, under it, each point z of $\mathbf{C} - \{0\}$ moves along the ray from 0 to z toward the unit circle; if $|z| < 1$ the image of z moves outward, if $|z| > 1$ its image moves in toward the circle, and if $|z| = 1$ then z remains fixed during the entire homotopy.

Whenever two based spaces (X,x_0) and (Y,y_0) are of the same homotopy type, then $\pi_1(X,x_0) \cong \pi_1(Y,y_0)$ and their fundamental groups are equal. This is trivial; $(g \circ f)_* = g_* \circ f_*$ is the identity on $\pi_1(X,x_0)$, and $f_* \circ g_*$ is the identity on $\pi_1(Y,y_0)$, so f_* is an isomorphism. This implies, for instance, that $\pi_1(\mathbf{C} - \{0\}) = \pi_1(S^1)$ is a nontrivial group (see Exercises J and K).

Any property of spaces which, if held by a space, is held by each other space of the same homotopy type is called a **homotopy invariant.** Thus the cardinality (that is, the number of elements) of $\pi_0(X)$, or the fact that $\pi_1(X) = \mathbf{Z}$, for example, are homotopy invariants.

Exercises N and P

$\pi_1(S^n)$, A MORE DIFFICULT EXAMPLE

We shall now show that if $n \geq 2$ each loop $a\colon \mathbf{I} \to S^n$ in the n-sphere is homotopic to the constant loop at the distinguished point

$$N = (0, \ldots, 0, 1) \in S^n.$$

This is another strongly intuitive result. Visualize S^n as S^2 and imagine x to be a point of S^n which is not in the image $a(\mathbf{I})$ of the loop a; since $S^n - \{x\}$ is homeomorphic to $\mathbf{R}^n$, which is contractible, the loop a is homotopic to the constant loop in $S^n - \{x\}$, and therefore to the constant loop in S^n. However, this reasoning is falacious; there exist loops a in S^n for which $a(\mathbf{I}) = S^n$, for which there is no member x of $S^n - a(\mathbf{I})$ upon which to base this construction. Such "space-filling curves" were first discovered by Peano; often one is described in a course on real-variable theory.

There is a way around this problem, a device which breaks up the image of a into a finite number of pieces, each of which is a proper subset of S^n. For each point x of S^n let S_x be the open hemisphere centered at x, $S_x = \{y \in S^n\colon \|x - y\| < \sqrt{2}\}$. The family $\mathbb{S} = \{S_{a(t)}\colon t \in \mathbf{I}\}$ is an open cover of $a(\mathbf{I})$, so $a^{-1}(S_{a(t)})$ is a neighborhood of t in $\mathbf{I}$. Let T_t be the component of t in $a^{-1}(S_{a(t)})$; T_t is open, so $\{T_t\colon t \in \mathbf{I}\}$ is an open cover of $\mathbf{I}$. Let $T_{t_1}, T_{t_2}, \ldots, T_{t_j}$ be a finite subcover. We may choose an element T_1 of this subcover with $0 \in T_1$, then successively choose $T_2, T_3, \ldots, T_k$ so that, having determined T_i, we choose T_{i+1} such that $T_{i+1} \cap T_i \neq \emptyset$ and $lub\,(T_{i+1}) > lub\,(T_i)$. This process of choice stops when $1 \in T_k$ and $\{T_1, T_2, \ldots, T_k\}$ is a finite subcover of $\mathbf{I}$. Hence we can choose an increasing sequence $u_0 < u_1 < \cdots < u_k$ of points of $\mathbf{I}$ such that $u_0 = 0$, $u_i \in T_i \cap T_{i+1}$ for $i = 1, 2, \ldots, k - 1$, and $u_k = 1$; then the interval $[u_{i-1}, u_i]$ lies inside T_i for $i = 1, 2, \ldots, k$. This is the desired situation; the domain $\mathbf{I}$ of a has been divided into pieces each of which a carries into a proper subset, a hemisphere, of S^n. It is now no trick at all to smooth out the image of a, piece by piece.

Let $1 \leq i \leq k$; we propose to show now that a is homotopic to a loop a' in S^n such that $a' = a$ on $\mathbf{I} - [u_{i-1}, u_i]$ and $a'([u_{i-1}, u_i])$ is a great circle arc from $a(u_{i-1})$ to $a(u_i)$. The case where $a(u_{i-1}) = a(u_i)$ is trivial, since $S_{a(u_i)}$ is contractible; hence suppose $a(u_{i-1}) \neq a(u_i)$. Then $\{a(u_{i-1}), a(u_i)\}$ is an independent set of vectors in $\mathbf{R}^{n+1}$ and so determines a 2-dimensional subspace which is homeomorphic to $\mathbf{R}^2$; in this plane the two points $a(u_{i-1})$ and $a(u_i)$ lie in some open semicircle in the unit circle, and there is a unique

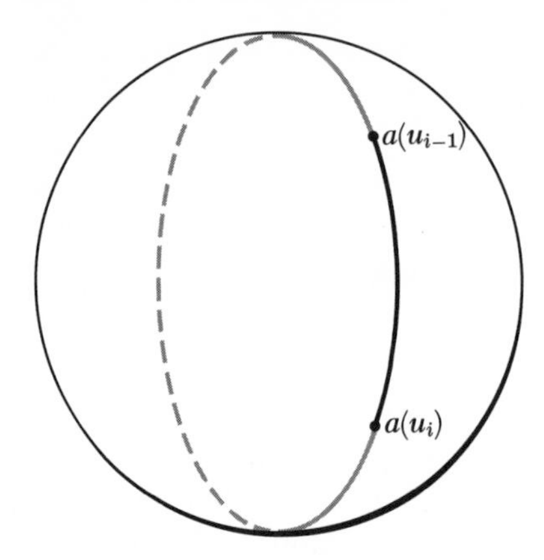

shorter arc of that circle from one to the other. This is the arc of a great circle on S^n which we meant; it is an exercise (see Exercise U) to exhibit a homotopy of a with an appropriate loop a'; the homotopy can be chosen so that the images of u_{i-1} and u_i are held fixed throughout.

By induction and the transitivity of the relation of homotopy, a is homotopic to a loop b whose image is the union of a finite set of arcs of great circles on S^n. Each of these arcs may be enclosed in an open subset

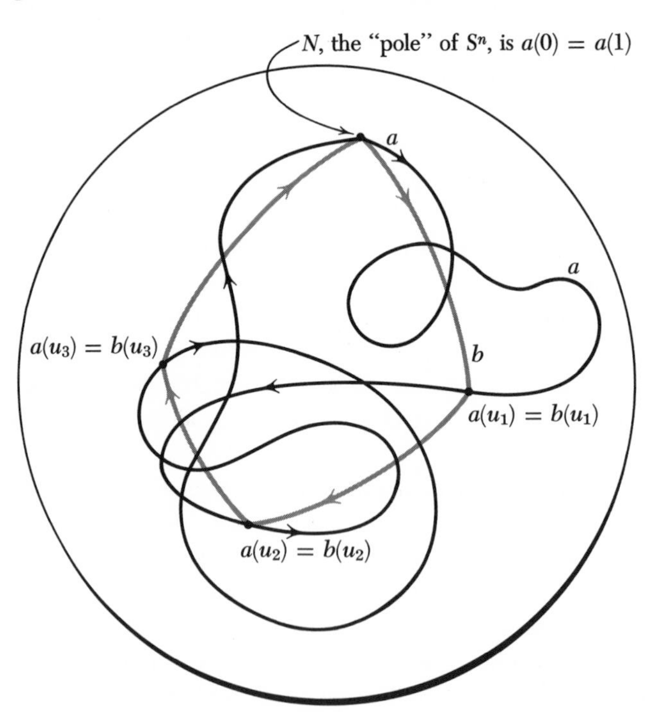

$A \subset S^n$ whose area is as small as we please; the area of a finite set of arcs is therefore zero, and the image of b is not all of S^n. But as we pointed out early in this example, b must be homotopic to the constant loop; this implies that a is also homotopic to a constant. Since a was an arbitrary loop in S^n, we conclude that $\pi_1(S^n) = \{1\}$.

It is an exercise (see Exercise W) to show that $\mathbf{R}^{n+1} - \{0\}$ is of the same homotopy type as S^n; hence $\pi_1(\mathbf{R}^{n+1} - \{0\}) = \{1\}$, $n \geq 2$.

Exercises U and W

$\pi_1(X \times Y)$

The **direct product of** two **based spaces** (X,x_0) and (Y,y_0) is defined to be the based space $[X \times Y, (x_0,y_0)]$, where $X \times Y$ means the direct product of topological spaces, of course. Let us denote the loop spaces by $\Omega(X)$, $\Omega(Y)$, and $\Omega(X \times Y)$, respectively, and let p and q be the projections of $X \times Y$ onto its factors X and Y. If $a \in \Omega(X \times Y)$, then $p \circ a \in \Omega(X)$ and $q \circ a \in \Omega(Y)$; this defines a function f: $\Omega(X \times Y) \rightarrow \Omega(X) \times \Omega(Y)$: $f(a) = (p \circ a, q \circ a)$. In the other direction, if $(b,c) \in \Omega(X) \times \Omega(Y)$, then there is a loop a in $X \times Y$ with $a(t) = [b(t),c(t)]$; let $a = g(b,c)$, so that g is a function from $\Omega(X) \times \Omega(Y)$ to $\Omega(X \times Y)$. Clearly, the composites $f \circ g$ and $g \circ f$ are both identity maps; it is an exercise (Exercise S) in the product of c-o topologies to show that f and g are homeomorphisms. Further, both f and g preserve products of loops; that is, $f(a \cdot a') = f(a) \cdot f(a')$, etc., where the product in $\Omega(X) \times \Omega(Y)$ is defined just as for the direct product of groups. (In other words, if (b,c) and (b',c') are elements of $\Omega(X) \times \Omega(Y)$, then $(b,c) \cdot (b',c') = (b \cdot b', c \cdot c')$. The proof of continuity of multiplication which was given for the direct product of groups works quite as well here.)

Now, f induces a function from $\pi_0[\Omega(X \times Y)]$ to $\pi_0[\Omega(X) \times \Omega(Y)]$, which latter set is isomorphic to $\pi_0[\Omega(X)] \times \pi_0[\Omega(Y)]$; putting this all together, we have a function from $\pi_1[X \times Y, (x_0,y_0)]$ to $\pi_1(X,x_0) \times \pi_1(Y,y_0)$. It is clear, moreover, that the inverse of this function is the corresponding function induced by g, and that both of these induced functions are morphisms, since f and g preserve the loop product. Hence we have an isomorphism

$$\pi_1[X \times Y, (x_0,y_0)] \cong \pi_1(X,x_0) \times \pi_1(Y,y_0);$$

or more succinctly,

$$\pi_1(X \times Y) = \pi_1(X) \times \pi_1(Y)$$

(if X and Y are path connected).

For an example, notice that $\mathbf{C} - \{0\}$ is homeomorphic to $S^1 \times \mathbf{R}$; a homeomorphism might carry each nonzero complex number z to $(z/|z|, x)$, where $x = |z| - 1$ if $|z| \geq 1$, and $x = (|z| - 1)/|z|$ if $0 < |z| \leq 1$. But we have seen that $\pi_1(\mathbf{R}) = \{1\}$, since $\mathbf{R}$ is contractible, so $\pi_1(\mathbf{C} - \{0\}) = \pi_1(S^1) \cong \pi_1(S^1) \times \{1\}$. This is a new derivation of this result; we have already seen that it is a consequence of the homotopy equivalence of $\mathbf{C} - \{0\}$ and S^1.

Another example is the 2-torus $T^2 = S^1 \times S^1$ (the surface of an inner tube): assuming that $\pi_1(S^1) = \mathbf{Z}$, we have $\pi_1(T^2) = \mathbf{Z} \times \mathbf{Z}$.

Exercise S

REFERENCES

There are several treatments of the fundamental group which are at the same level as this chapter. You should find the following quite digestible:

Crowell and Fox, *Introduction to Knot Theory*, chap. II (Boston: Ginn, 1963).

P. Hilton, *Introduction to Homotopy Theory* (New York: Cambridge University Press, 1953).

S. -T. Hu, *Elements of General Topology*, chap. VI (San Francisco: Holden-Day, 1964).

L. Pontrjagin, *Topological Groups*, pp. 217–225 (Princeton, N.J.: Princeton University Press, 1939).

Although the only fundamental groups we have actually calculated thus far have all been trivial, the groups for the circle and the torus were announced to be $\mathbf{Z}$ and $\mathbf{Z} \times \mathbf{Z}$, respectively. You may by now surmise that fundamental groups are always abelian. This is false; as an example, the fundamental group of a pretzel having two or more holes is not abelian. In fact, if G is any group whatsoever, there exists a space X for which $\pi_1(X) = G$. Thorough treatments of these and broader questions are embedded in:

Hilton and Wylie, *Homology Theory* (New York: Cambridge University Press, 1960).

S. -T. Hu, *Homotopy Theory* (New York: Academic Press, 1959).

EXERCISES

A Give a complete description of the loop $a \cdot (b \cdot c)$, where a, b, and c are loops at x_0 in a space X.

B Let X and X' be spaces with loop spaces Ω and Ω' at the distinguished points x_0 and x_0', respectively. If $h\colon X \to X'$ is a homeomorphism and $h(x_0) = x_0'$, show that there is a homeomorphism of Ω with Ω' which preserves the products of loops.

C Let a be the loop at 1 in the circle S^1 given by $a(t) = (cos\ 2\pi t, sin\ 2\pi t)$. Define an inverse $[\bar{a}]$ for $[a]$, the component of a in the space of loops at 1 in S^1, and give a detailed proof that your inverse works.

D Complete the discussion and the diagram in the text to show that for each loop $a \in \Omega$ there exists a loop $\bar{a}$ such that *both* $a \cdot \bar{a}$ and $\bar{a} \cdot a$ are loops homotopic to the constant loop.

E Exhibit in detail a homotopy between the loops a and $\bar{a}$ of Exercise C when they are both regarded as loops at 1 *in the plane* **C**.

F Complete the discussion and the diagram in the text to show that the isomorphism $\theta\colon \pi_1(X,x_0) \cong \pi_1(X,x_1)$ is *well defined* by a path p from x_1 to x_0. That is, if a and b are homotopic loops at x_0, prove that $p \cdot (a \cdot \bar{p})$ is homotopic to $p \cdot (b \cdot \bar{p})$, where $\bar{p}(t) = p(1 - t)$; thus it makes sense to say $\theta[a] = [p \cdot (a \cdot \bar{p})]$.

G Prove in detail that the function θ, defined in the text and Exercise F above, is a morphism. [*Hint:* A modification of the homotopy in $\pi_1(X,x_0)$ of $\bar{p} \cdot p$ with e is what is needed.]

H Give a proof that scalar multiplication of vectors is a continuous function from $\mathbf{R} \times \mathbf{R}^n$ into $\mathbf{R}^n$. Then explain why the pointwise scalar multiple fg of two maps $f\colon S \to \mathbf{R}$ and $g\colon S \to \mathbf{R}^n$, where $fg(s) = f(s)g(s)$, is necessarily continuous.

J Attempt to prove that the loop a at 1 in S^1 which has values $a(t) = (cos\ 2\pi t, sin\ 2\pi t)$ is *not* homotopic to the constant loop at 1. The lesson here is that this is intuitively obvious yet quite difficult to prove; do not give up without a struggle.

K Argue that the result of Exercise J is equivalent to the fact that the identity map on the circle S^1 is not homotopic to the constant map sending all of S^1 to its base point 1 (both are maps of based spaces). This means that for S^1 contractibility is *equivalent* to the triviality of the fundamental group. Does this equivalence hold for every space?

L Prove that homotopy of based maps defines an equivalence relation on the set of all based maps from one given based space to another.

M Show that if (X,x_0) is contractible, then its loop space (Ω,e) is contractible. How does this show that X is simply connected?

N Divide the following alphabet into sets of letters of the same homotopy type:

A B C D E F G H I J K L M N O P Q R S T U V W X Y Z.

Discuss the relation of the result to that of Exercise IV.U (division of the alphabet into homeomorphism classes).

P Prove that a homotopy equivalence $f: X \to Y$ induces a set isomorphism $\pi_0(f): \pi_0(X) \to \pi_0(Y)$; in other words, the cardinality of $\pi_0(X)$ is a homotopy invariant.

Q A subspace S of $\mathbf{R}^n$ is called a "star about a point $x \in S$" iff for every $y \in S$ the straight-line segment from x to y lies in S. Show that if S is

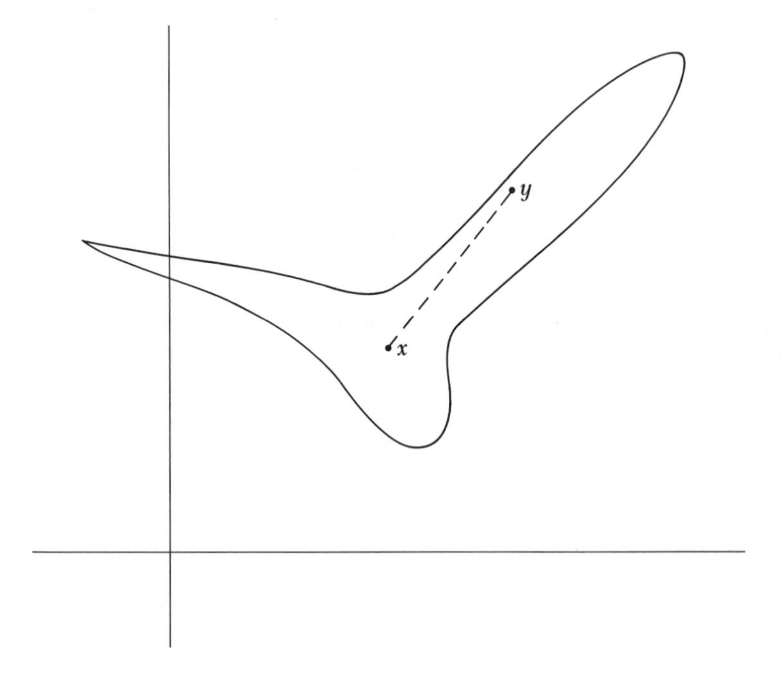

a star about x, then (S,x) is contractible. (A subspace S of $\mathbf{R}^n$ is "convex" iff it is a star about each of its points; convex spaces are therefore contractible.)

R Assume the results of Exercises J and K and show that, even if a based map f is 1-1 (or onto), f_* need not be 1-1 (or onto).

S Show that the map $f: \Omega(X \times Y) \to \Omega(X) \times \Omega(Y)$, which projects loops onto their "factor loops," is a homeomorphism. The text argues that f is a set isomorphism; you need only show that it is continuous and open.

T Prove that the property of contractibility is a homotopy invariant. Is every homotopy invariant also a topological invariant? Is every topological invariant also a homotopy invariant?

U Prove that if a is a path whose image lies in an open hemisphere D of S^n, and $A \subset D$ is the unique arc of a great circle which lies between

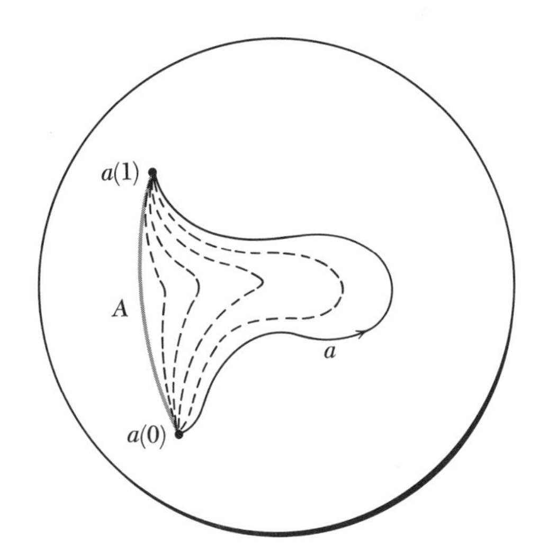

$a(0)$ and $a(1)$, then there is a homotopy of a with a path whose image is A, each stage of which homotopy is a path from $a(0)$ to $a(1)$. More arithmetically, the problem is to find a map $H: \mathbf{I}^2 \to S^n$ such that $H(0,t) = a(t)$, $H(1 \times \mathbf{I}) \subset A$ and $H(\mathbf{I} \times 0) = a(0)$, $H(\mathbf{I} \times 1) = a(1)$.

W Prove that if Y is contractible, then the projection of $X \times Y$ onto X is a homotopy equivalence. Next establish the fact that $\mathbf{R}^{n+1} - \{0\}$ is homeomorphic to $S^n \times \mathbf{R}$; then show that $\mathbf{R}^{n+1} - \{0\}$ is simply connected, $n > 1$.

PROBLEMS

AA Automorphisms of $\pi_1(X)$ In the definition of the isomorphism θ: $\pi_1(X,x_0) \cong \pi_1(X,x_1)$ by a path p from x_1 to x_0, let a different path p' from x_1 to x_0 be used to define an isomorphism θ': $\pi_1(X,x_0) \cong \pi_1(X,x_1)$. Show that $\theta = \theta'$ iff $p' \cdot \bar{p}$ lies in the center of the group $\pi_1(X,x_1)$. Here, of course, $p'(1) = p(0) = p(1)$, so $p' \cdot p$ is defined; its values are given by

$$(p' \cdot \bar{p})(t) = \begin{cases} p'(2t) & \text{if } 0 \leq t \leq \frac{1}{2} \\ p(2t - 1) = p(2 - 2t) & \text{if } \frac{1}{2} \leq t \leq 1. \end{cases}$$

Hence $(p' \cdot p)(0) = (p' \cdot p)(1) = x_1$, so this is a loop at x_1.

In the special case where $x_0 = x_1$, each loop p at x_0 is a path from x_0 to x_0; thus p defines an automorphism θ of the group $\pi_1(X,x_0)$ with itself (see Prob. II.AA). It is obvious that the constant loop e at x_0 yields the identity automorphism by this construction; hence the loop p defines the identity automorphism of $\pi_1(X,x_0)$ with itself iff $[p]$ lies in the center of $\pi_1(X,x_0)$. Is it clear that translation along p constructs the same (inner) automorphism of $\pi_1(X,x_0)$ as does conjugation by $[p]$?

BB A Fixed-point Theorem You have seen that the results of Exercises J and K are equivalent to the fact that S^1 is not a retract of the disc D. Show that the "fixed-point theorem" for D is another equivalent statement: if g: $D \to D$ is a map, then there is a point $d \in D$ with $g(d) = d$. A clue to the trick needed here is contained in Prob. III.BB; the analogy is that the 0-sphere S^0 is the set of unit vectors in $\mathbf{R}$, $\{-1,1\}$, with the relative (that is, discrete) topology. Thus a space X is path connected iff every map from S^0 to X can be extended to the "1-disc" $[-1,1]$; X is simply connected iff every map from S^1 to X can be extended to the "2-disc" D.

The Brouwer fixed-point theorem, which we shall not prove, states more generally that every map from the closed ball D^n of unit radius centered at the origin in $\mathbf{R}^n$ into D^n itself has a fixed point. Using a technique quite like the one used in this problem, we can show that this theorem is equivalent to the fact that there is no map on D^n to S^{n-1} which extends the identity map on S^{n-1}. Can you state some other equivalent facts, and prove that they are equivalent?

CC The Möbius Strip Let S be the strip $\{(x,y) \in \mathbf{R}^2: 0 \leq x \leq 1$ and $0 \leq y \leq 1\}$ in the plane, and let an equivalence relation be defined on S with equivalence classes of the form, for each $(x,y) \in S$, $\{(z,w) \in S: (z,w) = (x,y)$ or $(z,w) = (x \pm 1, 1 - y)\}$. Thus the equivalence classes are singletons except for points along the left and right edges of S, and there the classes are doubletons, one member from each edge.

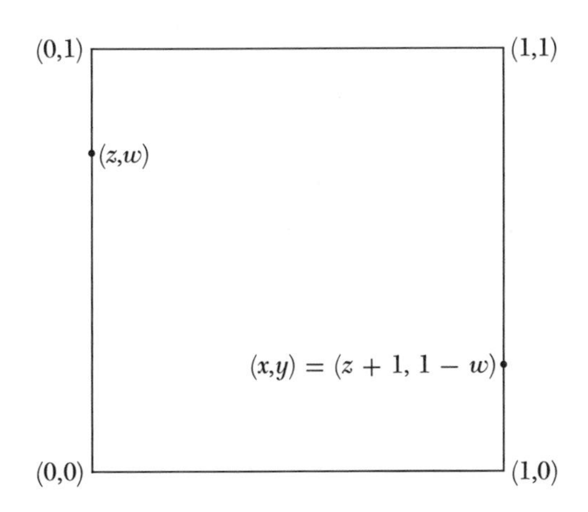

The two edges are identified with a twist. The quotient space M so defined is called a "Möbius strip."

It is intuitively clear that M is not homeomorphic to a cylinder; that is, M is not a homeomorph of the product $S^1 \times \mathbf{I}$. Show that M is of the same homotopy type as S^1 (and, therefore, as $S^1 \times \mathbf{I}$).

DD The Cone Construction Let X be a space; the "cone $K(X)$ on X" is constructed by the identification of the subset $X \times 1$ of the "cylinder" $X \times \mathbf{I}$ to a point,

$$K(X) = \frac{X \times \mathbf{I}}{X \times 1}.$$

Show that if X is a compact subset of $\mathbf{R}^n$, where

$$\mathbf{R}^n = \{(x_1, x_2, \ldots, x_n, x_{n+1}): x_{n+1} = 0\}$$

is regarded as a subset of $\mathbf{R}^{n+1}$, then $K(X)$ is homeomorphic to the set of points in $\mathbf{R}^{n+1}$ which lie on the line segment between $z = (0, 0, \ldots, 0, 1)$ and some point x of X; $K(X) \cong \{y \in \mathbf{R}^{n+1}$: there exist $t \in \mathbf{I}$ and $x \in X$ such that $y = tz + (1 - t)x\}$. (This subset of $\mathbf{R}^{n+1}$ is sometimes called the "euclidean cone on X.") The space X is homeomorphic to the subset $X \times 0$ of $X \times \mathbf{I}$, and the quotient map q: $X \times \mathbf{I} \to K(X)$, when restricted to $X \times 0$, defines a homeomorphism of $X \times 0$ with a subspace (the "base") of the cone. Thus X may be regarded as a subspace of $K(X)$. However, if $A \subset X$, it is not necessarily true that $K(A)$ is a subspace of $K(X)$; that is, the map $K(i)$: $K(A) \to K(X)$ which the inclusion i: $A \to X$ induces is not necessarily open. [*Hint:* Try the noncompact subspace $(0,1)$ of $\mathbf{I} = X$.]

If (X,x_0) is a based space, the cone construction is customarily altered to define $K(X,x_0)$ as the quotient space of $K(X)$ resulting from the identification of $q(x_0 \times \mathbf{I})$ to a point; if $S = X \times 1 \cup x_0 \times \mathbf{I}$, then

$$K(X,x_0) = \frac{X \times \mathbf{I}}{S},$$

often called the "reduced cone on (X,x_0)." The point S of this quotient is the base point of $K(X,x_0)$. A map f: $(X,x_0) \to (Y,y_0)$ is "inessential" (that is, homotopic to the constant map) iff there exists a map F: $K(X,x_0) \to (Y,y_0)$ whose restriction to (X,x_0) is the map f. In particular, X is contractible to the point x_0 iff f extends to all of $[K(X,x_0),S]$.

EE Higher Homotopy Groups The "n-th homotopy group" $\pi_n(X,x_0)$ of a based space (X,x_0) may be defined inductively by

$$\pi_n(X,x_0) = \pi_{n-1}[\Omega(X,x_0),e]$$

(where e is the constant loop). A map a representing an element $[a]$ of $\pi_n(X,x_0)$ defines a unique map $\tilde{a}$ from the n-cube $\mathbf{I}^n$ into X which is constant on the set of those points of $\mathbf{I}^n$ which have at least one coordinate equal to 0 or 1; this is exactly the topological boundary of $\mathbf{I}^n$ in $\mathbf{R}^{n+1}$. Hence the members of $\pi_n(X,x_0)$ are in a natural 1-1 correspondence with the equivalence classes of maps from an n-sphere (S^n,N) into (X,x_0) under the equivalence relation of homotopy. [The base point of S^n is $N = (0, \ldots, 0, 1) \in \mathbf{R}^{n+1}$.] Under this correspondence the product of two elements $[a]$ and $[b]$ of $\pi_n(X,x_0)$ goes over to the homotopy class of a map on S^n described as follows. Let $\bar{a}$ and $\bar{b}$ be maps of S^n into X whose homotopy classes correspond to the elements $[a]$ and $[b]$ of $\pi_n(X,x_0)$. Form a quotient of S^n by identifying points on the "equator" $E = \{(x_1, \ldots, x_{n+1}) \in S^n\colon x_1 = 0\}$; this quotient is, topologically, the union of two n-spheres with a single distinguished point in common. The map on S^n corresponding to the product of $[a]$ and $[b]$ in $\pi_n(X,x_0)$ is the homotopy class of the map

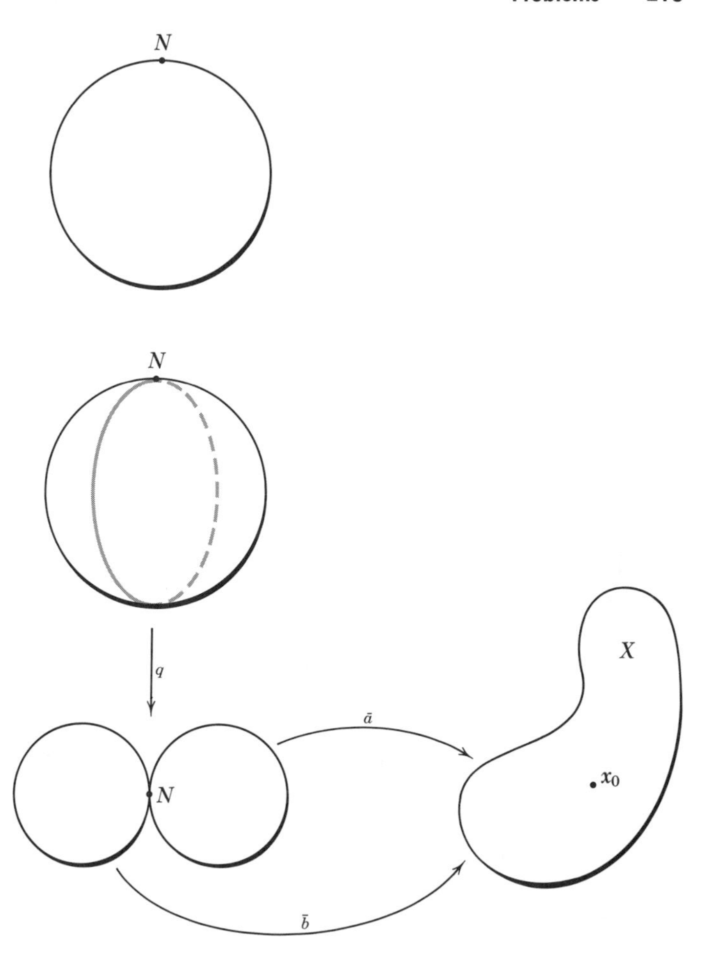

obtained by composing the above quotient map on S^n with a map on S^n/E which "looks like" $\bar{a}$ on one of the two n-spheres which compose S^n/E and looks like $\bar{b}$ on the other n-sphere.

[If n is at least 2, a path of rotations of (S^n,N) can be used to define a homotopy of the map described above with another map on S^n which, when factored through S^n/E, interchanges the roles of $\bar{a}$ and $\bar{b}$ on the two n-spheres of S^n/E. This yields a proof that $\pi_n(X,x_0)$ is abelian when $n \geq 2$.]

The group $\pi_n(X,x_0)$ is a homotopy invariant of (X,x_0) for each integer n, and if X is contractible to x_0, then $\pi_n(X,x_0) = \{1\}$ for each n. [A feeling for the n-th homotopy group may be obtained from the fact that for each n, $\pi_n(S^n) = \mathbf{Z}$, but this is difficult to prove now.]

FF Products of Loop Spaces For every indexed family $\{X_\lambda\colon \lambda \in L\}$ of spaces, $\Omega(\times\{X_\lambda\colon \lambda \in L\})$ is homeomorphic to $\times\{\Omega X_\lambda\colon \lambda \in L\}$. (Here notation fails to reflect our choice of a distinguished point x_λ for each X_λ, and of the distinguished point $\{x_\lambda\colon \lambda \in L\}$ of the product.)

GG Homotopy Equivalence Let $A \xrightarrow{f} B \xrightarrow{g} C \xrightarrow{h} D$ be a sequence of maps, with $g \circ f$ and $h \circ g$ both homotopy equivalences. Then f, g, and h are all homotopy equivalences.

The Fundamental Group of the Circle

CHAPTER IX

In the last chapter we computed the fundamental groups of a few types of spaces. For instance, the contractible spaces, which include all euclidean n-spaces and their starlike subsets, were shown to have trivial fundamental groups. The n-spheres for $n \geq 2$ were also found to be simply connected. In fact, we have not yet proved that any space whatsoever has a nontrivial fundamental group. This is surprising; an intuitive grasp of the definition convinces us immediately that there are lots of loops in spaces which cannot be deformed continuously into constant loops, but up to this point, we have no firm reason to believe that this notion of fundamental group is not empty of significance, that not every space is assigned the group $\{1\}$.

We have seen that a loop in a space may be regarded as a map of a circle into the space; accordingly, loops are in some sense generalized circles. This suggests that the crucial test space for the usefulness of this notion is the circle. We have seen that if the identity map on the circle is homotopically trivial, then its domain can be extended to the whole closed unit disc in the plane. Then each given map from the circle into an arbitrary space may be composed with this extension of the identity map on the circle to yield an extension of the given map to the whole disc; this in turn implies that the given map is homotopic to a constant map. Therefore the triviality of the fundamental group of the circle would imply the triviality of every fundamental group.

We shall now show, as advertised in the previous chapter, that $\pi_1(S^1) = \mathbf{Z}$, the integers. Our proof will begin with the construction of a topological group called the universal covering group of the circle, and a morphism of this group onto the circle; the kernel of this morphism will be the fundamental group of the circle! We shall then discover that there is an isomorphism of the universal covering group of the circle with the real line, and that this isomorphism carries the fundamental group of the circle onto the integers.

THE PATH GROUP OF A TOPOLOGICAL GROUP

The first part of our program to compute $\pi_1(S^1)$ is a construction which is valid for a broad class of topological groups. For a group G of this class there are special techniques available for the study of the fundamental group $\pi_1(G)$ of the underlying topological space of G (for which we use the same symbol "G"). Accordingly, we base the constructions of the path group and the universal covering group upon a group G which has the following three properties:

- **i** G is path connected.
- **ii** G is **locally path connected.** This means that if g is a point of an open set V of G, then the path component V_0 of g in V is also open.
- **iii** G is **semilocally simply connected.** This means there is at least one neighborhood N of each point g of G such that every loop at g in G whose image lies in N is homotopic, via loops at g in G, to the constant loop at g.

The new definitions in properties ii and iii apply to an arbitrary space; since G is a group, it is clear that the demands need be met only at the single point e, the identity of G. Thus, for a group, property ii is equivalent to the demand that inside each nucleus there lie a path-connected nucleus; property iii is equivalent to the demand that there exist a nucleus N such that the c-o subbasic set $(\mathbf{I},N)$ is contained in $\Omega_0(G,e)$, the set of trivial loops at e in G (why?). The circle has these three properties; in fact, every group manifold has the latter two properties, since inside each nucleus lies a nucleus which is a homeomorph of an ε-ball in some $\mathbf{R}^n$, and every such ε-ball is both path connected and simply connected. Hence each path-connected group manifold satisfies our conditions.

A comment about the definitions of "local" properties in topology is in order here. Our statement of condition ii may be immediately rephrased to demand that there exist at each point a local base of path-connected sets.

This latter statement resembles the requirement, at least in Hausdorff spaces, of local compactness: a T_2-space is locally compact iff there exists at each point a local base of compact sets. But a more usual definition says that a space is locally path connected iff for each open set V and each point $g \in V$ there is a neighborhood N of g such that for every point $x \in N$ there exists a path in V from x to g. This seems more complicated than condition ii, but a moment's reflection will show the two definitions to be equivalent. The latter definition of local path-connectedness has a form resembling statement iii; the precise relation of the two is examined in Exercise E, which may justify the bizarre definition of semilocal simple connectivity.

An example of a space which is not semilocally simply connected is the subspace of the plane which is the union of the countable collection of circles $\{S_i: i = 1, 2, \ldots\}$, where each S_i has center $(1/i,0)$ and radius $1/i$;

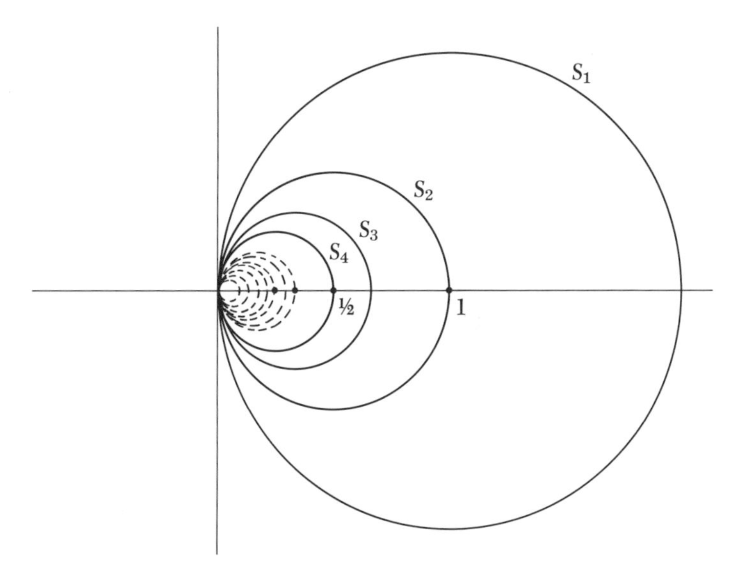

they are all tangent at the origin. Every neighborhood of the origin contains loops around little circles which are not homotopically trivial (this is not a formal proof; see Exercise VIII.J). To find a space which is not locally path connected is an exercise.

It is henceforth assumed, then, that the topological group G which we shall consider has properties i, ii, and iii. We begin our study of the fundamental group of the group G by constructing the **path group** E of (G,e): E is the set of based maps of $(\mathbf{I},0)$ into (G,e), that is, the set of paths in G which begin at e, with the c-o topology. If a and b are members of E, there is a pointwise-product path ab in E, defined by $ab(t) = a(t)b(t)$ for all $t \in \mathbf{I}$, the latter product being in the group G. (Note that this product "ab" is denoted by juxtaposition, as opposed to a product "$a \cdot b$," previously defined for some pairs of paths.) This pointwise product of paths is a group operation for E; a^{-1} is the path with values $a^{-1}(t) = [a(t)]^{-1}$, the constant path e is the identity of E (which explains that choice of notation), and associativity in E is equivalent to that of G itself. We assert that E is a topological group; inversion is continuous, since each c-o-subbasis element (K,W) has as its inverse (K,W^{-1}) (here recall that $W^{-1} = \{w^{-1}: w \in W\}$ is open in G). The set $E - \{e\}$ is open, since if $a \in E - \{e\}$, say $a(t_0) \neq e \in G$, then the set of paths which map the compact set $\{t_0\}$ into the open set $G - \{e\}$ is a neighborhood of a which does not contain the constant path. The proof that the product on E is continuous will use only the fact that $\mathbf{I}$ is regular; we shall show that if (K,W) is a subbase element for E and $ab \in (K,W)$, then there are basic neighborhoods A and B of a and b with the product $AB \subset (K,W)$. For each $t \in K$ choose a neighborhood $U_t \times V_t$ of $[a(t),b(t)]$ in $G \times G$ so that $U_tV_t \subset W$ (remember that multiplication in G is continuous), and thereby get the neighborhood $a^{-1}(U_t) \cap b^{-1}(V_t)$ of t in $\mathbf{I}$. Since $\mathbf{I}$ is regular, there is an open neighborhood T_t of t with its closure $T_t^- \subset a^{-1}(U_t) \cap b^{-1}(V_t)$, and $T_t^- \cap K$ is a closed subset of the compact set K, and is thus compact. Now, $\{T_t: t \in K\}$ is an open cover of K; select a finite subcover $T_1, T_2, \ldots, T_n$. Take A to be $\bigcap\{(T_i^- \cap K, U_i): i = 1, 2, \ldots, n\}$ and B to be $\bigcap\{(T_i^- \cap K, V_i): i = 1, 2, \ldots, n\}$, where U_i and V_i correspond to the chosen set T_i for each i. It is easy to check that $(a,b) \in A \times B$ and $AB \subset (K,W)$; hence E is a topological group. We have shown, in fact, that *if X is regular, then G^X is a topological group with the pointwise multiplication.* ■

There is a projection $p: E \to G: p(a) = a(1)$ which sends each path to its end point, p is continuous (why?), and it preserves products: $p(ab) =$

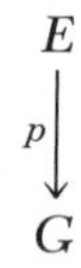

$p(a)p(b)$. Since G is locally path connected, p is also open; we need only check this at $e \in E$. Let (K,W) be a subbasic element for the c-o topology of E with $e(K) \subset W$. If $K \neq \emptyset$, then W is a nucleus of G, and so the path component W_0 of e in W is also open in G, and $e \in (\mathbf{I},W_0) \subset (K,W)$. But $p(\mathbf{I},W_0) = W_0$ is contained in $p(K,W)$, which is therefore a nucleus; p is an open map and a morphism of topological groups. It is an epimorphism, since G is path connected. Taken together, these facts mean that G is (isomorphic to) a quotient group of E.

Exercises B, C, and D

THE UNIVERSAL COVERING GROUP

The kernel of p is Ω, the loop space of (G,e), and thus $G \cong E/\Omega$; Ω is a closed normal subgroup of E. Now consider the path component Ω_0 of e in Ω. As we saw in Chap. VI, it is normal in Ω. Further, since G is semi-locally simply connected, there is a nucleus $(\mathbf{I},N)$, a subbasic element of Ω, which lies entirely in Ω_0. Because the subgroup Ω_0 contains a nucleus, it is open, and thus closed, in Ω. Being a closed subset of the closed subset Ω of E, Ω_0 is a closed subgroup of E. But if $a \in \Omega_0$ with P a path in Ω from a to e, and $b \in E$, then for each $u \in \mathbf{I}$ the conjugate $bP(u)b^{-1}$ of the loop $P(u)$ is in Ω (since Ω is normal in E). This yields a path (why?) in Ω from $bP(0)b^{-1} = bab^{-1}$ to $bP(1)b^{-1} = beb^{-1} = e$, so $bab^{-1} \in \Omega_0$ and Ω_0 is normal in E:

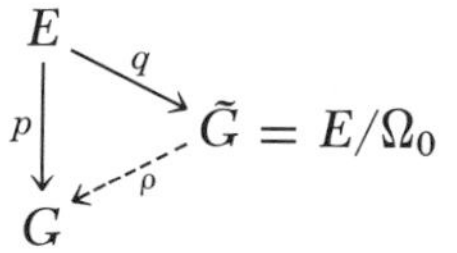

$$\begin{array}{ccc} E & \xrightarrow{q} & \tilde{G} = E/\Omega_0 \\ {\scriptstyle p}\downarrow & \swarrow{\scriptstyle \rho} & \\ G & & \end{array}$$

The quotient $\tilde{G} = E/\Omega_0$ is called the **universal covering group**† of G; it is a topological group, and if q: $E \to \tilde{G}$ is the quotient morphism, then p factors through q; that is, there is an epimorphism ρ: $\tilde{G} \to G$ such that $p = \rho \circ q$ (see Chap. V). The kernel of the **covering map**† ρ is Ω/Ω_0, since Ω is the kernel of p, and Ω/Ω_0 is just the group of path components of the topological group Ω, which has the pointwise multiplication of its member loops. Of course, the elements of Ω/Ω_0 are exactly the elements of $\pi_1(G,e)$. We claim that the group operations are the same, as well; that is, the pointwise product ab of two loops is homotopic to the loop product $a \cdot b$. Since $a \cdot e$ is in the same path component $[a]$ as is a ($a \cdot e$ goes around a at double speed and then stands still at e) and $e \cdot b \in [b]$, we have $[ab] = [(a \cdot e)(e \cdot b)]$. But clearly, $(a \cdot e)(e \cdot b)$ is exactly the loop $a \cdot b$; it goes first around a and then around b. Hence $[ab] = [a \cdot b]$ and $\Omega/\Omega_0 = \pi_1(G,e)$.

† Covering groups and covering maps have no direct connection with the earlier defined notion of a cover of a space. Naturally enough, there is some grammatical confusion of these words in the literature, but the context usually prevents mathematical confusion.

Incidentally, $[ab] = [(e \cdot a)(b \cdot e)] = [b \cdot a]$ as well; this shows that *the fundamental group of a topological group is abelian.* ∎

As another corollary, we have the result that *G is simply connected iff ρ is an isomorphism of $\tilde{G}$ with G.* ∎

Exercise F

THE PATH GROUP OF THE CIRCLE

What is the interpretation of all this theoretical structure in case the group G involved is our old friend the circle? The theory suggests that one way we can compute the fundamental group of the circle is first to consider the group E of all paths at $1 \in S^1$; then we must decide which paths are loops and which loops are trivial. The universal covering group $\widetilde{S^1}$ must be constructed next, and the covering morphism ρ; finally, we realize $\pi_1(S^1)$ as the kernel of ρ. This may at first blush seem to be a needless complication of our problem; why not just compute the fundamental group directly? But exactly what does this entail? A direct attack must consider the set Ω of loops. Since these are paths, we may as well admit that we look first at the set of paths on the circle and decide next which paths are loops and then which loops are trivial. Finally, the relation of homotopy between loops must be clarified in order to establish the group $\pi_1(S^1)$. This is easily seen to be a restatement of the theoretical program, except that the theory offers a simplification: the group structure of E and $\widetilde{S^1}$ can be used to simplify our examination of the relationship of homotopy. That is, we need only find the subgroup Ω_0 of trivial loops; the equivalence classes of mutually homotopic loops which are the members of $\pi_1(S^1)$ are then the cosets of Ω_0 in Ω. Thus the path group and the universal covering group are not new constructs, but are entirely implicit in the definition of the fundamental group itself (for topological groups). They are intimately bound up in any computation or theoretical discussion of the fundamental group.

We shall now exploit this structure to compute $\pi_1(S^1)$. Our first result, the theorem of which this section constitutes a proof, is as follows.

THEOREM *The exponential map $exp\colon \mathbf{R} \to S^1$ induces an isomorphism $exp^{\mathbf{I}}$ of the path group E' of the real line with the path group E of the circle,*

$$exp^{\mathbf{I}}\colon E' \cong E.$$

Specifically, the map $exp\colon \mathbf{R} \to S^1$ wraps the line around the circle; its values are $exp\ (r) = (cos\ r, sin\ r) \in S^1$. It is a morphism of topological groups, but we have seen that it is not 1-1, although it is onto; in fact, it is not even homeomorphic to a direct-product projection (see Chap. V).

$$\begin{array}{ccc} E' & \xrightarrow{exp^{\mathbf{I}}} & E \\ {\scriptstyle p'}\downarrow & & \downarrow{\scriptstyle p} \\ \mathbf{R} & \xrightarrow{exp} & S^1 \end{array}$$

Nevertheless, the induced map $exp^{\mathbf{I}}$, which carries $\mathbf{R}^{\mathbf{I}}$ into $(S^1)^{\mathbf{I}}$ by composition, $exp^{\mathbf{I}}(a) = exp \circ a$, has a restriction to the path group E' of $\mathbf{R}$ (we use the symbol $exp^{\mathbf{I}}$ for this restriction, too) which is an isomorphism of E' with E. Picture $exp^{\mathbf{I}}$ as projecting a path at the origin of the line by wrapping that path around the circle. The function $exp^{\mathbf{I}}$ is trivially a morphism of topological groups, since *exp* is (Exercise J asks for details).

LEMMA *There exists an inverse function for* $exp^{\mathbf{I}}$*; that is,* $exp^{\mathbf{I}}$ *is an algebraic isomorphism.*

Proof We begin the construction of the inverse of $exp^{\mathbf{I}}$ by showing that for each path a of E there exists a unique path a' in $\mathbf{R}$ which begins at 0 and for which $exp \circ a' = a$. The path a' will be built a piece at a time; note that *exp* is 1-1 on each open interval of length π in $\mathbf{R}$. Thus *exp*, restricted to an open interval of length π, is a homeomorphism onto an open arc of length π (*exp* is a "local homeomorphism"). For each $t \in \mathbf{I}$, the do-

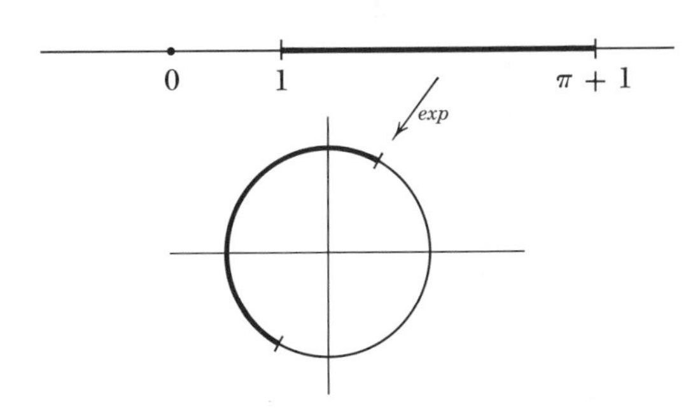

main of a, let N_t be the component of t in the open set which is the inverse image under a of the open arc of length π centered at $a(t)$; N_t is an open set containing t (why?). Select a finite subcover $\{N_1, N_2, \ldots, N_k\}$ of the cover $\{N_t: t \in \mathbf{I}\}$ of $\mathbf{I}$; reindex these if need be, so that their left end points $u_1, u_2, \ldots, u_k$ form an increasing sequence. Clearly, $u_1 = 0$, $u_i \in N_{i-1}$ for $i = 2, 3, \ldots, k$, and $a(N_i)$ is a connected subset (an "arc") of S^1 of length no greater than π. The path a' is first defined on N_1. There is a unique interval J_1 of length π which contains 0 and which *exp* maps homeomorphically onto the arc A_1 of length π which is centered at t_1 (t_1 is the point of **I** corresponding to $N_1 = N_{t_1}$); A_1 contains $a(N_1)$. If f_1: $A_1 \to J_1$ is the inverse of the homeomorphism defined by *exp* of J_1 with A_1, let $a'(t) = f_1[a(t)]$ for $t \in N_1$. Inductively, if a' has been defined on $N_1 \cup N_2 \cup \cdots \cup N_i$, find the unique interval J_{i+1} of length π which contains $a'(u_i)$ and which *exp* maps onto the arc A_{i+1} centered at $a(t_{i+1})$; use the homeomorphism f_{i+1}: $A_{i+1} \to J_{i+1}$ to define $a'(t) = f_{i+1}[a(t)]$ on N_{i+1}. This extension of the definition of a' to N_{i+1} can be made to agree with the definition at the previous stage on $N_i \cap N_{i+1}$. Our ability to "paste

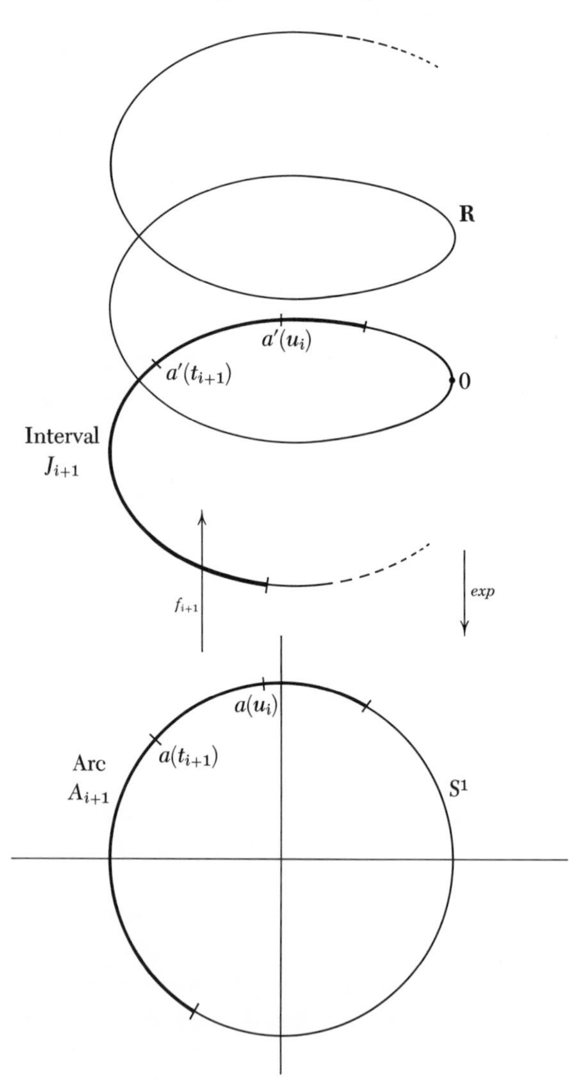

together" the homeomorphisms f_i and f_{i+1} depends crucially on the fact that two overlapping open arcs, each of length π, do not together cover S^1. This explains the choice of the length π for our homeomorphic intervals and arcs. Since $\{N_1, N_2, \ldots, N_k\}$ forms a cover of **I**, a' is defined and continuous on all of **I**, and $exp \circ a' = a$.

Picture a path in the circle as a string laid on the circle; the real line is the helix which *exp* projects downward onto the circle. We have shown that if the string a is laid down gradually, starting at $1 \in S^1$, we can follow with another string a' up above on the helix, starting at $0 \in \mathbf{R}$, so that at all times t the shadow on the circle (via *exp*) of the point $a'(t)$ being laid onto the helix is exactly the point of string $a(t)$ being at that moment laid down on the circle.

We have shown the existence of a **lift** a' of a path a in S^1 to a path in **R**; now suppose that both a' and a'': $\mathbf{I} \to \mathbf{R}$ are lifts of a which begin at 0 (so that $exp \circ a' = exp \circ a'' = a$). Consider the subset L of **I** on which $a' = a''$: $0 \in L$, and L is open, since if $t \in N_i$ (the choice of N_i depended only on a) and $a'(t) = a''(t)$ lies in a set J_i, then the homeomorphism f_i: $A_i \to J_i$ guarantees that $a' = a''$ on all of N_i (f_i is the functional inverse of the restriction of *exp* to J_i). But the difference $a' - a''$ is a continuous function, and $L = (a' - a'')^{-1}(0)$ is therefore closed. Hence L is a nonempty open and closed subset of **I**; so $L = \mathbf{I}$, $a' = a''$, and the lift of a is uniquely defined. ■ A reader who understands the definition of the multiple-valued natural logarithmic "function" on the complex plane will realize that we have just shown that it is possible to make a unique continuous choice of values of the log_e function on the values of a: $a' = (1/i)\, log_e \circ a$.

The assignment of a' to a is a function ℓ: $E \to E'$ such that $exp^{\mathbf{I}} \circ \ell = 1_E$ and $\ell \circ exp^{\mathbf{I}} = 1_{E'}$. [We might well have used the notation $(log_e)^{\mathbf{I}}$ instead of ℓ.] The group operation in E is the pointwise addition of angles and in E' the pointwise addition of reals. The function *exp* is a morphism, so $exp \circ (a' + b') = (exp \circ a')(exp \circ b') = ab$, and $a' + b'$ must be the unique path $(ab)'$ which "lifts" ab; thus ℓ preserves products. (More generally, the functional inverse of a morphism is always an algebraic morphism.)

Now, $exp^{\mathbf{I}}$ is continuous because *exp* is, and it is an algebraic isomorphism. The following lemma will conclude the proof of our theorem that $exp^{\mathbf{I}}$ is a topological isomorphism.

LEMMA *The map* $exp^{\mathbf{I}}$ *is open.*

Proof Let (K, W) be a subbasic nucleus of E'; clearly, $0 \in W \subset \mathbf{R}$, and K is compact in **I**. The map *exp* is a homeomorphism, which we shall denote by f, when restricted to $(-\pi/2, \pi/2) \cap W = V$, and $(\mathbf{I}, V) \subset (K, W)$.

The map $exp^{\mathbf{I}}$ carries $(\mathbf{I},V)$ into $[\mathbf{I}, exp\,(V)]$, the set of paths in S^1 whose images lie entirely inside $exp\,(V)$, and each member a of $[\mathbf{I}, exp\,(V)]$ has a lift a' which is a member of $(\mathbf{I},V)$; hence $exp^{\mathbf{I}}\,(K,W) \supset f^{\mathbf{I}}(\mathbf{I},V) = [\mathbf{I}, exp\,(V)]$, which is a nucleus of E. Thus $exp^{\mathbf{I}}\,(K,W)$ is a nucleus and $exp^{\mathbf{I}}$ is open at the identity e of E'. This implies (see Exercise V.Q) that $exp^{\mathbf{I}}$ is open, and E' is isomorphic to E. ■

Exercises G and J

THE UNIVERSAL COVERING GROUP OF THE CIRCLE

Now that we have the somewhat surprising result that the circle and the line have isomorphic path groups, we can sketch our plan of attack on $\pi_1(S^1)$. It amounts to this: The subgroup Ω_0' of E' consisting of all trivial loops in the line (so $\Omega_0' = \Omega'$) is carried by the isomorphism $exp^{\mathbf{I}}$ onto Ω_0, the subgroup of E of all loops in the circle which are homotopically trivial. This immediately implies that E'/Ω_0' is isomorphic to E/Ω_0; that is, the universal covering groups for the line and the circle are isomorphic. Because we already know that the simply connected group $\mathbf{R}$ is isomorphic to its universal covering group, $\mathbf{R} \cong E'/\Omega' = E'/\Omega_0'$, this means that there is an isomorphism of $\widetilde{S^1}$ with $\mathbf{R}$! This section, then, is devoted to a proof of the following theorem.

THEOREM *There exists an isomorphism h: $\mathbf{R} \to \widetilde{S^1}$ such that $\rho \circ h = exp$:*

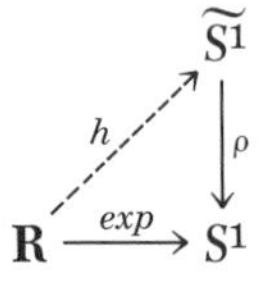

$$\begin{array}{ccc} & & \widetilde{S^1} \\ & \overset{h}{\nearrow} & \downarrow \rho \\ \mathbf{R} & \xrightarrow{exp} & S^1 \end{array}$$

We begin with the construction of an isomorphism of $\mathbf{R}$ with a certain subgroup of the real path space E'. For each $r \in \mathbf{R}$ let $\sigma(r)$ be the path in $\mathbf{R}$ at 0 which has the value tr for each $t \in \mathbf{I}$ so that $\sigma(r)$ is a line from 0 to r. Thus σ is a function from $\mathbf{R}$ to E'; it is an easy exercise to check that it is an isomorphism of $\mathbf{R}$ with $\sigma(\mathbf{R}) \subset E'$. Thus the subgroup $exp^{\mathbf{I}} \circ \sigma(\mathbf{R})$ of E is isomorphic to $\mathbf{R}$; it is the subgroup of all those paths beginning at 1 in S^1 which "have constant speed." Furthermore, if a is a member of this subgroup, say $a = exp^{\mathbf{I}}\,[\sigma(r)]$, then it is clear from the step-by-step construction of $a' \in E'$ that a' also has constant speed, and $a'(1) = r$.

We can now define the isomorphism h of $\mathbf{R}$ with S^1 which was promised above: set $h = q \circ exp^{\mathbf{I}} \circ \sigma$. That h is a morphism is clear; it is a composite of morphisms. Specifically, h assigns to each real number r the coset of $\Omega_0(\mathrm{S}^1)$ determined by the path in S^1 with constant counterclockwise speed r. The definition of h satisfies the requirement we set that the diagram

$$\begin{array}{ccc} & & \widetilde{\mathrm{S}^1} \\ & {}^{h}\nearrow & \downarrow \rho \\ \mathbf{R} & \xrightarrow{exp} & \mathrm{S}^1 \end{array}$$

be commutative:

$$\begin{aligned} \rho \circ h(r) &= \rho \circ q \circ exp^{\mathbf{I}} \circ \sigma(r) \\ &= p \circ exp^{\mathbf{I}} \circ \sigma(r) \\ &= exp\ [\sigma(r)(1)] \\ &= exp\ (r). \end{aligned}$$

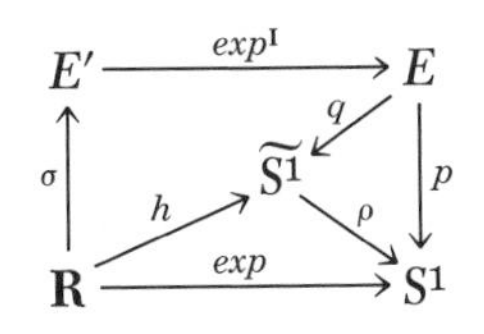

It is an exercise to show that h is open. The problem now is to show that h is 1-1 and onto, or equivalently, to show that each coset of Ω_0 in E contains exactly one path in S^1 of constant speed.

LEMMA *The morphism h is onto.*

Proof If $a \in E$, then there is a lift of a to a path a' at $0 \in \mathbf{R}$. There is a loop $\bar{a}$ at $0 \in \mathbf{R}$ which measures the deviation of a' from a constant-speed path, $\bar{a} = \sigma[a'(1)] - a'$, so $\bar{a}(t) = ta'(1) - a'(t)$. Since $\mathbf{R}$ is simply connected, there is a path A' in Ω' which begins at the constant loop and ends at $\bar{a}$; for each $u \in \mathbf{I}$ we define $A'(u)$ to be the loop in $\mathbf{R}$ which has values for each $t \in \mathbf{I}$: $A'(u)(t) = u[ta'(1) - a'(t)]$. The composite $exp^{\mathbf{I}} \circ A' = A$ is a path in E which begins at the constant path in S^1; each of its values $exp^{\mathbf{I}} \circ A'(u)$ is a path in S^1 beginning at $exp\ (u[0a'(1) - a'(0)]) = exp\ (0) = 1$ and ends at $exp\ [ua'(1) - ua'(1)] = 1$. Hence $exp^{\mathbf{I}} \circ A'$ is a path of loops in S^1, that is, a path in Ω. It begins at the constant loop e and ends at

$$\begin{aligned} exp^{\mathbf{I}}\ [A'(1)] &= exp^{\mathbf{I}}\ (\bar{a}) \\ &= exp^{\mathbf{I}}\ [\sigma \circ a'(1) - a'] \\ &= [exp^{\mathbf{I}} \circ \sigma \circ a'(1)](exp^{\mathbf{I}} \circ a')^{-1} \\ &= [exp^{\mathbf{I}} \circ \sigma \circ a'(1)]a^{-1} \end{aligned}$$

The multiplication and inversion in the last line above are those of the group E; thus $exp^{\mathbf{I}} \circ \sigma \circ a'(1)$ and a are two paths which lie together in the same coset of Ω_0 in E. Since $exp^{\mathbf{I}} \circ \sigma \circ a'(1)$ is just the path in S^1 with constant speed $a'(1)$, we have shown that each coset $a\Omega_0$ of Ω_0 in Ω contains at least one path of constant speed; h is onto. ∎

LEMMA *The morphism h is one-to-one.*

Proof That h is monic is equivalent to the fact that the kernel of q contains only one member, e, of $exp^{\mathbf{I}} \circ \sigma(\mathbf{R})$. That is, the only path in $\Omega_0(S^1)$ having constant speed is the constant path (with speed 0). To see this, assume $a \in \Omega_0$, and let H be a homotopy of a with e; H is a map from $\mathbf{I}^2$ to S^1 with $H(0,t) = a(t)$ and $H(1,t) = 1$ for each $t \in \mathbf{I}$. Suppose for a moment that there exists a map $H'\colon \mathbf{I}^2 \to \mathbf{R}$ such that $exp \circ H' = H$ and $H'(\mathbf{I} \times 0) = 0$. Then, as the diagram indicates, along the right edge of its domain H' is the *unique* lift of a, so $H'(1,t) = a'(t)$. On the other three edges of its domain H' is the constant path, so $H'(1,t) = H'(u,0) = H'(u,1) = 0$ for all t

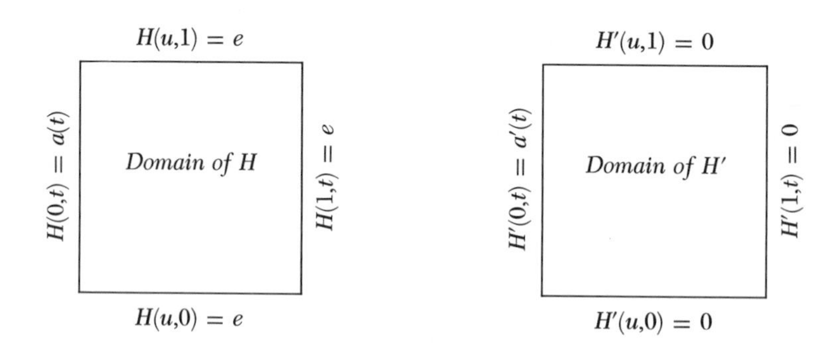

and u. However, a' begins and ends at 0; if a is a constant speed path in S^1, then a' has constant speed, which must be zero, and $a = exp^{\mathbf{I}} \circ \sigma \circ a'(1) = e$, the constant path in S^1. This will show that h is a monomorphism, and therefore an isomorphism.

All that remains is to show the existence of a lift H' of a homotopy H in Ω_0. The following lemma will substantiate our supposition.

LEMMA *There exists a lift $H'\colon \mathbf{I}^2 \to \mathbf{R}$ of the homotopy $H\colon \mathbf{I}^2 \to S^1$ such that $exp \circ H' = H$. That is, a path at e in Ω_0 can be lifted to a unique path at e in Ω'_0.*

Proof This construction is quite similar to our previous lifting of a path in S^1. In fact, the exponential law states that for each fixed u the function

$H(u,t)$ of t defines a path a_u in S^1, and we may get a lifted path a'_u at 0 in $\mathbf{R}$. If we define $H'(u,t) = a'_u(t)$, then

$$exp \circ H'(u,t) = exp\,[a'_u(t)] = a_u(t) = H(u,t).$$

But is H' continuous on $\mathbf{I}^2$? Well, fix an element u_0 of $\mathbf{I}$ in mind, and for each $t \in \mathbf{I}$ choose a rectangular subbasic neighborhood $M_t \times N_t$ of (u_0,t) in $\mathbf{I} \times \mathbf{I}$ such that $H(M_t \times N_t)$ lies in the arc of length π centered at $H(u_0,t)$. Just as before, choose a finite set $M_1 \times N_1, M_2 \times N_2, \ldots, M_k \times N_k$ with $u_0 \times \mathbf{I} \subset \bigcup\{M_i \times N_i: i = 1, \ldots, k\}$; clearly, $\mathbf{I} = \bigcup\{N_i: i = 1, \ldots, k\}$. Define $M_0 = \bigcap\{M_i: i = 1, \ldots, k\}$; M_0 is a neighborhood of u_0 in $\mathbf{I}$, so $M_0 \times \mathbf{I}$ is a neighborhood of $u_0 \times \mathbf{I}$ in $\mathbf{I}^2$. As before, on each successive set $M_0 \times N_1, M_0 \times N_2, \ldots$ we may define inductively a unique continuous lift of H which maps $M_0 \times 0$ to 0; this lift obviously must agree with (that is, have the same values as) the H' defined above on $M_0 \times \mathbf{I}$, since for each fixed $u \in M_0$ it lifts the path $H(u,t)$.

However, H' is continuous on a neighborhood of the line $u_0 \times \mathbf{I}$ for each $u_0 \in \mathbf{I}$; H' is the unique map of $\mathbf{I}^2$ into $\mathbf{R}$ for which $H'(\mathbf{I} \times 0) = 0$ and $exp \circ H' = H$. This concludes the proof that h is monic; thus h is an isomorphism. ■

The inverse morphism to h assigns to each coset $a\Omega_0 \in S^1$ the real number $a'(1)$. We have seen that this is a well-defined function [that is, if $b \in a\Omega_0$, then $b'(1) = a'(1)$] which is also an isomorphism. Further, the commutativity requirement is satisfied; $exp \circ h^{-1}(a\Omega_0) = exp \circ a'(1) = a(1) = \rho(a\Omega_0)$, so $exp \circ h^{-1} = \rho$.

Exercises H and L

SOME NONTRIVIAL FUNDAMENTAL GROUPS

Now, the kernel of ρ is Ω/Ω_0, the fundamental group of S^1; since h^{-1} is an isomorphism, $\pi_1(S^1,1)$ is isomorphic to the kernel of exp (see Exercise K). This isomorphism is determined by h^{-1}; the coset $a\Omega_0$ is in the kernel of ρ iff $p(a) = a(1) = 1$, the identity of S^1. This happens iff $exp[a'(1)] = 1$, that is, iff $a'(1)$ is an integral multiple $2k\pi$ of 2π. The integer k is called the **degree** of a, denoted **deg** (a); it is, suggestively, the "net number of times" the loop a wraps itself around the circle S in a counterclockwise direction (if k is negative, the "net" wrapping is in the clockwise direction). The loop products ab and $a \cdot b$ in Ω clearly both correspond to the addition of degrees; that is, *deg*: $\pi_1(S^1,1) \rightarrow \mathbf{Z}$ is an isomorphism onto the additive group of integers. We have proved that

$$\pi_1(S^1) = \mathbf{Z}.$$

Another immediate result is that the fundamental group of the 2-torus $T^2 = S^1 \times S^1$ is the direct product of two copies of $\mathbf{Z}$,

$$\pi_1(T^2) = \mathbf{Z} \times \mathbf{Z}.$$

Also,
$$\pi_1(S^1 \times \mathbf{R}) = \pi_1(\mathbf{C} - \{0\}) = \mathbf{Z}.$$

Of course, this last statement should be generalized: any space having the same homotopy type as the circle has the integers as its fundamental group. An "annulus," for instance, which is the set (closed or open) of plane points lying between two concentric circles of differing radii, is homotopy equivalent to the circle, as is the Möbius strip (see Prob. VIII.CC).

Exercise K

THE FUNDAMENTAL THEOREM OF ALGEBRA

Since you first met the quadratic formula in high school, you have been convinced that a quadratic polynomial which has real or complex coefficients must always have a complex "zero" or "root"; that is, if the polynomial is $ax^2 + bx + c$, $a \neq 0$, then there exists a complex number z with $az^2 + bz + c = 0$. There are always exactly two such roots, in fact, if you agree to count z a double root in case $ax^2 + bx + c = a(x - z)^2$. There are similar but more complicated formulas which give roots for arbitrary cubic or quartic polynomials; you may not have seen these, since it is not fashionable nowadays to include them in elementary courses. But it is a theorem of abstract algebra that such formulas do not exist to find roots for polynomials of degree 5 or more. To be sure, you might be lucky or skillful enough to factor such a polynomial and thereby find a root for it; this is easy for the polynomial $(x - 3)^{17}$ of degree 17. But there is no guarantee of eventual success in such an endeavor, unless the polynomial is a made-up problem in a textbook. Hence you have no algebraic reason to believe that a polynomial of, say, degree 1429 necessarily has any roots at all.

The fundamental theorem of algebra states that *each nonconstant polynomial with complex coefficients has at least one complex root.*

Curiously, there is no known proof of this fact which uses only algebraic manipulations; we now offer a topological proof.

To better understand this proof, you will find it helpful to consider first a simple example in some detail. The polynomial $p = x^2 - 4x + 15/4 = (x - 3/2)(x - 5/2)$ has two real roots, at $x = 3/2$ and $x = 5/2$. Its constant term $15/4$ is its value at the origin 0 of the plane. Since p: $\mathbf{C} \to \mathbf{C}$ is continuous, the image under p of the unit circle of its domain is a closed curve in the range, and that curve does not encircle the origin (it is the curve labeled $r = 1$ in the diagram). Since $3/2$ is a root of p, the domain circle of radius $3/2$ must be carried by p to an image curve which goes through the origin

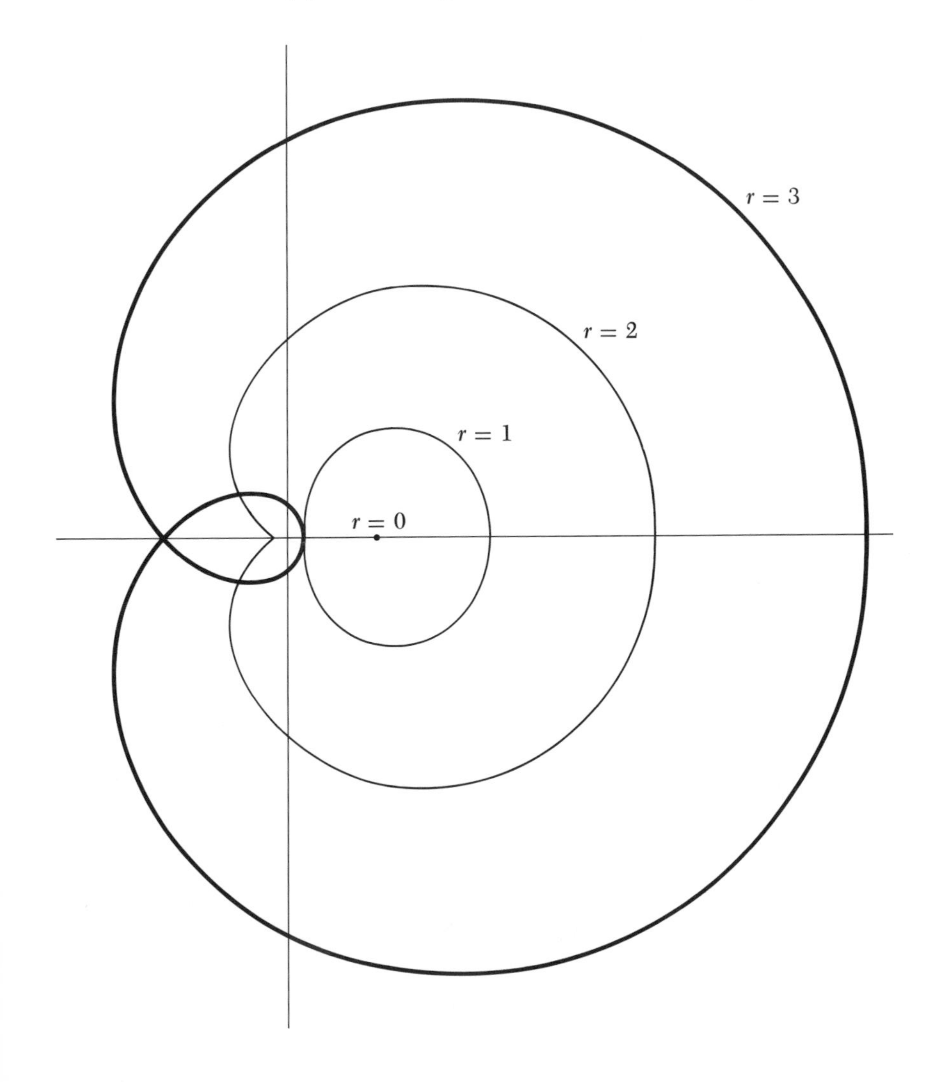

(this curve is not shown). If we think of the domain circle centered at 0 as expanding, the image curve also swells outward; at radius 2 in the domain, the image curve (labeled $r = 2$) encircles the origin. This latter curve has a sharp cusp on it; as the radius of the domain circle grows still more, the cusp becomes a crossing point and there is a second little loop of the image curve which just touches the origin when r reaches $5/2$ (not shown). When the domain circle has radius 3 the image curve ($r = 3$) loops around the origin twice. It is easy to imagine that as the domain circle gets still larger without limit, the image curve gets larger and larger, always looping twice around the origin.

The arithmetic of this proof is carried out on the circle and in $\pi_1(S^1)$. It depends heavily on the continuity of division in the group of nonzero complex numbers, but an informal sketch of it is better presented in the plane. There we think of a polynomial p of degree n as a function which carries each circle of radius r and center 0 to a closed path in $\mathbf{C}$; there is a continuous deformation of such a path as r varies. When $r = 0$ the path is constant, and when r is very large the path winds around the origin n times. At no stage does the path go through the origin if p has no complex root; in that case every one of these paths must lie in the same element of $\pi_1(\mathbf{C} - \{0\})$. Since the path is constant when $r = 0$, this means $n = 0$, and the polynomial p with no complex roots must be a constant polynomial.

Now for the proof of the theorem. Suppose p is a polynomial of degree n with complex coefficients, say $p = a_n x^n + a_{n-1}x^{n-1} + \cdots + a_1 x + a_0$; since we presume $a_n \neq 0$, we can divide through by a_n to get a new polynomial whose leading coefficient a_n is 1, and we suppose this done. We shall denote the "value" of p at a complex number z by $p(z)$, $p(z) = z^n + a_{n-1}z^{n-1} + \cdots + a_0$. Now, we assume that p has no complex roots, that is, that there is no complex number z such that $p(z) = 0$. For each real number $r \geq 0$ let $P(r)$ be the loop at 1 in S^1 whose value at $t \in \mathbf{I}$ is

$$P(r)(t) = \frac{|p(r)|}{p(r)} \frac{p[r\, exp\, (2\pi t)]}{|p[r\, exp\, (2\pi t)]|}.$$

Since p has no roots, each of the two complex fractions on the right-hand side is well defined and has absolute value 1. Therefore $P(r)(t) \in S^1$ for each $r \geq 0$ and each $t \in \mathbf{I}$, and it is clear that $P(0)$ is the constant loop in S^1. The first fraction in the definition of $P(r)$ is there to adjust the starting point of the loop; the second fraction corresponds to a projection into S^1 of the loop $p[r\, exp\, (2\pi t)]$ in $\mathbf{C} - \{0\}$, an adjustment of absolute values. In general, the degree of the loop $P(r)$ is called the "winding number" of the closed path in the plane with values $p[r\, exp\, (2\pi t)]$.

Since the values $P(r)(t)$ are obviously continuous simultaneously in r and t, $P(0)$ is homotopic to each loop $P(r)$; a homotopy H_r is defined by

$$H_r(u,t) = P(u,r)(t).$$

This says that each loop $P(r)$ has degree 0, since $P(0)$ does. We propose to show that there is a positive real r_0 with $deg\,[P(r_0)] = n$. This will finish our proof, since then $n = 0$ and this polynomial p of degree n without roots has degree 0; it is a constant polynomial.

Take r_0 to be $(n + 1)a$, where a is the largest of the numbers 1, $|a_{n-1}|, \ldots, |a_0|$, and let $A(t) = r_0\, exp\,(2\pi t)$ be the closed path in the plane describing the circle of radius r_0 centered at the origin. If q is the polynomial z^n, then $q \circ A$ is a closed path with values $q \circ A(t) = r_0{}^n$ $exp\,(2\pi nt)$; you may see that its winding number is n. We now claim that for each $t \in \mathbf{I}$, $p \circ A(t)$ lies inside the ball of radius $r_0{}^n$ centered at $q \circ A(t)$, since

$$\begin{aligned}
&|q \circ A(t) - p \circ A(t)| \\
&\quad = |A(t)^n - [A(t)^n + a_{n-1}A(t)^{n-1} + \cdots + a_0]| \\
&\quad \leq |a_{n-1}A(t)^{n-1}| + |a_{n-2}A(t)^{n-2}| + \cdots + |a_0| \\
&\quad \leq |a_{n-1}|\,|A(t)^{n-1}| + |a_{n-2}|\,|A(t)^{n-2}| + \cdots + |a_0| \\
&\quad \leq |a_{n-1}|r_0{}^{n-1} + |a_{n-2}|r_0{}^{n-2} + \cdots + |a_0| \\
&\quad \leq \frac{r_0}{n+1}r_0{}^{n-1} + \frac{r_0}{n+1}r_0{}^{n-2} + \cdots + \frac{r_0}{n+1} \leq \frac{n}{n+1}r_0{}^n \\
&\quad < r_0{}^n.
\end{aligned}$$

This result implies that the line segment lying between $q \circ A(t)$ and $p \circ A(t)$ does not cross the origin; that is, if $0 \leq u \leq 1$, then the point $H(u,t) = u[q \circ A(t)] + (1 - u)[p \circ A(t)]$ is not zero for any $t \in \mathbf{I}$. In an informal sense, the function H is a deformation of the closed path $q \circ A$(t) into $p \circ A(t)$. Formally, let

$$J(u,t) = \frac{|H(u,0)|}{H(u,0)}\frac{H(u,t)}{|H(u,t)|};$$

J is a map from $\mathbf{I}^2$ into S^1,

$$\begin{aligned}
J(0,t) &= \frac{|p(r_0)|}{p(r_0)}\frac{p \circ A(t)}{|p \circ A(t)|} = P(r_0)(t), \\
J(1,t) &= \frac{|q(r_0)|}{q(r_0)}\frac{q \circ A(t)}{|q \circ A(t)|} = \frac{r_0{}^n}{r_0{}^n}\frac{r_0{}^n\, exp\,(2\pi nt)}{r_0{}^n} = exp\,(2\pi nt).
\end{aligned}$$

This map J is a homotopy in S^1 between the loops $J(0,t)$ and $J(1,t)$, since at each stage u

$$J(u,0) = \frac{|H(u,0)|}{H(u,0)} \frac{H(u,0)}{|H(u,0)|} = 1,$$

and $J(u,1) = J(u,0) = 1$, since $A(1) = A(0)$. Thus J is a homotopy of $P(r_0)$ with the loop whose values are $exp\ (2\pi nt)$. This latter loop clearly has degree n; it wraps the interval around the circle counterclockwise n times. ■

Exercises M and N

REFERENCES

A construction of the universal covering group is made in

P. M. Cohn, *Lie Groups,* chap. VII (New York: Cambridge University Press, 1957).

A more general construction, valid for any space satisfying our requirements i, ii, and iii, is offered in

L. Pontrjagin, *Topological Groups,* chap. VIII (Princeton, N.J.: Princeton University Press, 1958).

Several very readable computations of the fundamental group of the circle exist; you might enjoy

Chinn and Steenrod, *First Concepts of Topology* (New York: Random House, 1966).

Crowell and Fox, *Introduction to Knot Theory,* chap. II (Boston: Ginn, 1963).

S. -T. Hu, *Elements of General Topology,* pp. 193–200 (San Francisco: Holden-Day, 1964).

The first of these is a paperback; it discusses winding numbers (secs. 20–27) and the fundamental theorem of algebra (sec. 36) in detail.

A more sophisticated treatment of fundamental groups and $\pi_1(S^1)$ may be found in

E. H. Spanier, *Algebraic Topology,* chap. 1 (New York: McGraw-Hill, 1966).

In the next chapter we shall continue our study of the fundamental groups of topological groups in general; suggestions of further topics are postponed until then.

EXERCISES

A Supply complete details for the argument (sketched in the introduction to this chapter) that if the fundamental group of the circle were trivial, then the fundamental group of every space would be trivial.

B Show that a connected and locally path-connected space is necessarily path connected. This means that these two global notions of connectivity coincide for locally path-connected spaces (but not for their subsets, to be sure).

C Exhibit a space which fails to be locally path connected.

D Show that the path space $E = \{f: (\mathbf{I},0) \to (X,x_0)\}$ of a space (X,x_0) is always contractible.

E Argue that the openness of Ω_0 in Ω is equivalent to the semilocal simple connectivity of G. A stronger condition is that Ω be locally path connected; phrase an equivalent condition on G. (A group is called "locally simply connected" or "locally 1-connected" if its loop space is locally path connected.)

F An identity e for a space X with a multiplication $m: X \times X \to X$ is a point e of X such that $m(e,x) = m(x,e) = x$ for all $x \in X$. If X has a multiplication and an identity e, then (X,e) is called an "H-space" (or "Hopf space"). All topological groups are H-spaces; an example which is not a group is the complement of the open-unit disc in the plane, with complex multiplication.

Show that the fundamental group of an H-space is necessarily abelian.

G The example of the circle in this chapter contains an argument that the set of points on which two real-valued maps agree is closed. Prove that, more generally, if f and g are maps from X to Y and Y is Hausdorff, then $\{x: f(x) = g(x)\}$ is a closed subset of X (sometimes called the "equalizer" of f and g).

H Show in detail that the function σ: $\mathbf{R} \to E'$ which is described in the example of the circle, $\sigma(r)(t) = tr$, is a topological isomorphism of $\mathbf{R}$ with $\sigma(\mathbf{R})$. In fact, σ is a cross section to the map p': $E' \to \mathbf{R}$ and also a morphism; it is called a "cross section to the morphism" p' of topological groups.

J Prove in detail that $exp^{\mathbf{I}}$: $E' \to E$ is a morphism of topological groups.

K Let f: $G_1 \cong G_2$ be an algebraic isomorphism of groups and m_i: $G_i \to H$ be a morphism for $i = 1$ and 2, so that $m_1 = m_2 \circ f$:

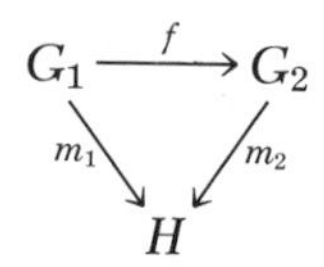

Show that the restriction of f to $Ker\,(m_1)$ is an isomorphism of $Ker\,(m_1)$ with $Ker\,(m_2)$.

L Prove that the morphism $h = q \circ exp^{\mathbf{I}} \circ \sigma$: $\mathbf{R} \to \widetilde{S^1}$ (described in the text) is an open function.

M Supply the missing details to show that the function H used in the proof of the fundamental theorem of algebra is continuous simultaneously in t and u.

N Use the division algorithm for polynomials to show that if p_0 is a polynomial and $p_0(z_1) = 0$, then there is a polynomial p_1 such that $p_0 = (x - z_1)p_1$. Use this fact to make an inductive proof that $p_0 = (x - z_1)(x - z_2) \cdots (x - z_n)$, where n is the degree of p_0 and each z_i is a root of p_0; this says that p_0 has exactly n roots (some may be multiple roots, counted several times).

PROBLEMS

AA Free Homotopy Two loops a_0 and a_1 at the points x_0 and x_1 in X are "freely homotopic" if they are connected by a "free homotopy" H: $\mathbf{I}^2 \to X$; that is, there exists a map H such that $H(0,t) = a_0(t)$ and $H(1,t) = a_1(t)$ for all t and $H(u,0) = H(u,1)$ for all u. This means that a_0 and a_1 are joined together by a path of loops, but that the base point of these loops is allowed to wander around in X during

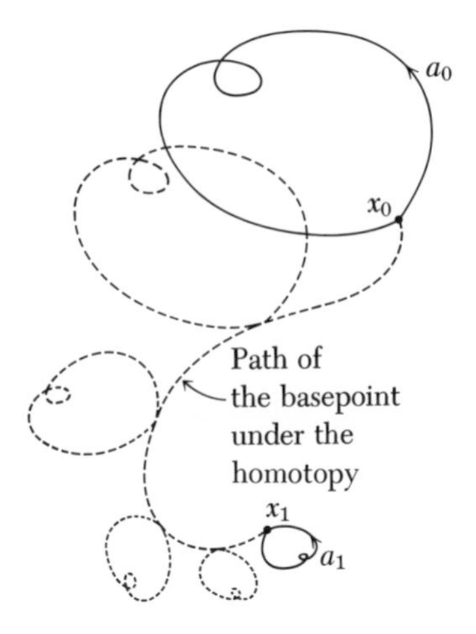

the homotopy. Show that if X is the circle, or any other group space, then two loops at e are freely homotopic iff they are homotopic.

Can you see a free homotopy lurking inside our proof of the fundamental theorem of algebra?

BB Only Separately Continuous Homotopies Exhibit a function $\mathfrak{k}: \mathbf{I}^2 \to S^1$ which satisfies the requirements for a homotopy of the loop a, $a(t) = exp\,(2\pi t)$, with the constant loop, except that $\mathfrak{k}$ is to be continuous only *separately* in each variable. That is, $\mathfrak{k}(u,t)$ is to be continuous in u for each fixed t and continuous in t for each fixed u. Then show that the equivalence relation between loops which is defined by this notion of separately continuous homotopy makes every loop trivial in an arbitrary space X.

CC The Figure 8 Let X be the figure 8, two circles tangent at a single point, and let a be a loop at the crossover point x in X which goes once around one circle, while b loops once around the other. Show that a and b together generate $\pi_1(X,x)$, that is, that the only sub-

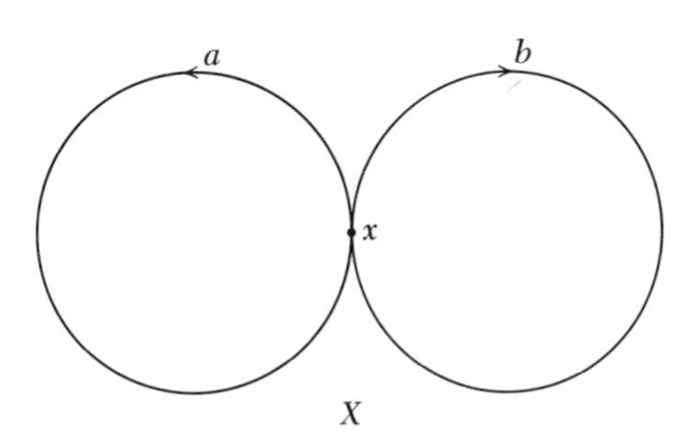

group containing both a and b is $\pi_1(X,x)$ itself. Then use the quotient maps which collapse just one of the circles to a point to define an epimorphism of $\pi_1(X,x)$ onto $\mathbf{Z} \times \mathbf{Z}$. The existence of such a morphism implies that $\pi_1(X,x)$ is not isomorphic to the integers. [In fact, $\pi_1(X,x)$ is not abelian. It is the "free group" generated by a and b; we have not defined this concept.]

DD The Homotopy Group $\pi_2(S^1)$ The second homotopy group of the circle, $\pi_2(S^1)$, is the group of homotopy classes of maps of $\mathbf{I}^2$ into S^1, where each map and each stage of a homotopy are required to send the whole boundary $\partial\mathbf{I}^2$ to $1 \in S^1$ (see Prob. VIII.EE). Put another way, it is the group of path components of $\Omega[\Omega(S^1,1),e]$, the group of path components of the subspace of $(S^1)^{(\mathbf{I}^2)}$ consisting of those based maps which are constant on $\partial\mathbf{I}^2$.

Use the technique developed in this chapter for lifting homotopies to show that $\pi_2(S^1)$ is trivial. Can you generalize this result to prove that for all $n > 1$, $\pi_n(S^1) = \{1\}$?

Locally Isomorphic Groups

CHAPTER X

Armed with an intimate knowledge of the relationship expressed by *exp* between the line and the fundamental group of the circle, we return now to the investigation of topological groups in general which we began in the last chapter. We discovered there that, at least in the case of the circle, the covering map from its universal covering group to a given group has a restriction which is a homeomorphism from one nucleus to another. This turns out to be true in general. That the universal covering group is simply connected is also always true.

We begin our work by isolating those properties of the morphism *exp* which enabled us to show that $\mathbf{R} \cong \widetilde{S^1}$. Although in our proofs of the last chapter we did use some special properties of **R**, we shall soon see that the construction of this isomorphism need depend only on the following facets of the situation: **R** is simply connected and there exists a homeomorphism from a nucleus of **R** to a nucleus of S^1 which preserves the group operation in so far as that operation can be carried out in the domain nucleus.

THE DEFINITION

If N and N' are nuclei of two topological groups G and G', respectively, for which there exists a homeomorphism $f: N \to N'$ such that

i If g, h, and gh are all in N, then $f(g)f(h) = f(gh)$,
ii If g', h', and $g'h'$ are all in N', then $f^{-1}(g')f^{-1}(h') = f^{-1}(g'h')$,

then G and G' are said to be **locally isomorphic.** We have just observed an example of this: $\mathbf{R}$ and S^1 are locally isomorphic. The restriction of the exponential map to the nucleus $(-\pi/2,\pi/2)$ of $\mathbf{R}$ is the **local isomorphism** f.

This example may be generalized considerably. Suppose G is a topological group with S a normal subgroup which is discrete in its relative topology. Then there is a nucleus M of G with $M \cap S = \{e\}$. Choose a nucleus N such that $NN^{-1} \subset M$, and let f be the restriction of the quotient map $\varphi: G \to G/S$ to the nucleus N. The map f is an open 1-1 map onto the nucleus $\varphi(N)$; since if $f(g) = f(h)$, then $\varphi(gh^{-1}) = e$, so $gh^{-1} \in S$. But $g, h \in N$ implies $gh^{-1} \in M$, and because $M \cap S = \{e\}, g = h$. Thus f is a homeomorphism, and it surely satisfies condition i above. That it also satisfies ii is the result of an exercise (see Exercise B). Hence G and its quotient, modulo a discrete normal subgroup, are locally isomorphic. In our example of the circle, the subgroup is $S = \{2\pi k: k \in \mathbf{Z}\} \subset \mathbf{R}$. (The discrete normal subgroup S is necessarily closed; since this is an intuitive truth, we defer its proof to Exercise C.)

Another example is provided by a group G and its universal covering group $\tilde{G}$. The kernel Ω/Ω_0 of the covering map $\rho: \tilde{G} \to G$ is, of course, normal; we claim that it is discrete. It is assumed, as in Chap. IX, that G is semilocally simply connected; therefore there is a nucleus M of G with every loop in M homotopically trivial. Thus there exists a nucleus N with $NN^{-1} \subset M$. Let a and b be elements of E, the path group of G, which map $\mathbf{I}$ into N, and suppose $\rho(a\Omega_0) = \rho(b\Omega_0)$. Then $p(a) = p(b)$, and the path ab^{-1} is a loop at e which lies in M. It is thus homotopic to the constant loop; that is, $ab^{-1} \in \Omega_0$, or $a\Omega_0 = b\Omega_0$. But $q(\mathbf{I},N)$ is a nucleus of $\tilde{G}$; we have shown that ρ is 1-1 on $q(\mathbf{I},N)$, so the kernel of ρ is discrete. Hence G and $\tilde{G}$ are locally isomorphic.

An alternative way of describing this situation lies in the definition that an epimorphism which restricts to a local isomorphism is called a **covering morphism.** Such a morphism is obviously open at the identity, and therefore open. *If $f: G \to G'$ is a covering morphism, then its kernel K is a discrete normal subgroup of G, and G/K is isomorphic to G'. Conversely, the quotient map of a group, modulo a discrete normal subgroup, is a covering morphism.* ■

The relation between groups of being locally isomorphic is an equivalence relation, since inverses and composites of local isomorphisms are again local isomorphisms. This makes it clear that a given local isomorphism from G to G' may not be extendible to any covering morphism at all. An example is the local inverse f for *exp*, which maps the set of angles of "absolute value" less than π onto the set of reals $(-\pi,\pi)$; since $\mathbf{R}$

is not compact it is not a quotient of S^1, so there can be no covering morphism from S^1 to $\mathbf{R}$.

We shall continue the assumption *throughout this chapter* that the group G under discussion satisfies the requirements i, ii, and iii which were set at the beginning of the previous chapter. The universal covering group $\tilde{G}$ of G is connected, since it is the continuous image of the contractible space E. Since $\tilde{G}$ and G are locally isomorphic, $\tilde{G}$ is locally path connected because G is. The simple connectivity of $\tilde{G}$ is established in the next section; thus $\tilde{G}$ will always satisfy our requirements when G does.

Exercises A, B, C, and F

THE SIMPLE CONNECTIVITY OF $\tilde{G}$

We shall now prove that *the universal covering group $\tilde{G}$ of a topological group G is always simply connected.* This is certainly the case with S^1; its universal covering group is contractible. In general, let $a\colon \mathbf{I} \to \tilde{G}$ be a loop at e in $\tilde{G}$; we use the loop $\rho \circ a$ in G to define a path $\hat{a}$ at e in E, $\hat{a}(u)(t) = \rho \circ a(tu)$. For each u, then, $\hat{a}(u)$ is the path in G which goes along the path $\rho \circ a$ from e to $\rho \circ a(u)$:

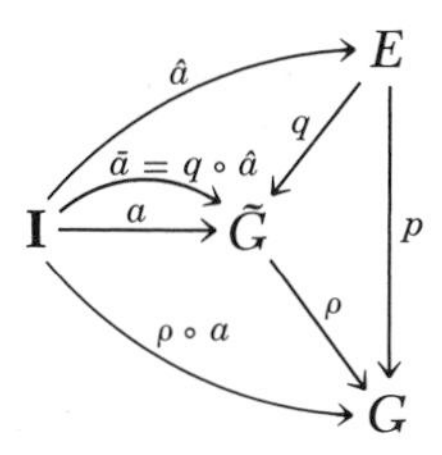

We assert that the image $q \circ \hat{a} = \bar{a}$ of $\hat{a}$ in $\tilde{G}$ is exactly the loop a with which we began. Clearly, $a(0) = \bar{a}(0) = e$, so the subset $L = \{u \in \mathbf{I}\colon a(u) = \bar{a}(u)\}$ of $\mathbf{I}$ is not empty. But the function $a\bar{a}^{-1}\colon \mathbf{I} \to \tilde{G}$ is continuous and $\{e\}$ is closed in $\tilde{G}$; the inverse image of $\{e\}$ under this function, which is L, is therefore closed. Furthermore,

$$\begin{aligned}\rho \circ \bar{a}(u) &= \rho \circ q \circ \hat{a}(u)\\ &= p \circ \hat{a}(u)\\ &= \hat{a}(u)(1)\\ &= \rho \circ a(u)\end{aligned}$$

so

$$\rho \circ \bar{a} = \rho \circ a.$$

Since ρ is a covering morphism, if $a(u_0) = \bar{a}(u_0)$, then there is a neighborhood N of $a(u_0)$ on which the restriction of ρ is a homeomorphism; it is easy to see that a agrees with $\bar{a}$ on the neighborhood $a^{-1}(N) \cap \bar{a}^{-1}(N)$ of u_0, and thus L is open as well. The interval $\mathbf{I}$ is connected, so $L = \mathbf{I}$ and $a = \bar{a}$.

Now, a is a loop, so $a(1) = \bar{a}(1) = \Omega_0 \in \tilde{G}$. And $\bar{a}(1) = \hat{a}(1)\Omega_0$; therefore $\hat{a}(1) \in \Omega_0$. A summary of the argument thus far is that if a is an arbitrary loop in $\tilde{G}$, then $\rho \circ a$ is a trivial loop in G. But we are done: if H is a homotopy of $\rho \circ a$ with e, then

$$\hat{H}: \mathbf{I}^2 \to E: \hat{H}(u,t)(s) = H(u,st)$$

is a homotopy of $\hat{a}$ with e (why?) and hence $q \circ \hat{H}$ is a homotopy of $q \circ \hat{a} = \bar{a} = a$ with e. This is true for each loop a in $\tilde{G}$; thus $\tilde{G}$ is simply connected. ■

THE UNIQUENESS OF $\tilde{G}$

We continue our study of universal covering groups with a fundamental lemma. It generalizes the fact that the path spaces of S^1 and of $\mathbf{R}$ are isomorphic via $exp^{\mathbf{I}}$. Its proof, while rather long, should be easy to follow when the example (S^1 and $\mathbf{R}$) is well understood; while reading it, keep before you a mental picture of what the proof is saying in terms of that example.

LEMMA *Suppose that G and G' are locally isomorphic via the homeomorphism f of a nucleus N of G with a nucleus N' of G'. Then there exists a unique isomorphism θ of the path space E of G with the path space E' of G' which agrees with $f^{\mathbf{I}}$ on some nucleus of E. (Hence $f^{\mathbf{I}}$ is a local isomorphism with the global extension θ.)*

Proof Let $a \in E$; we shall construct a path a' in G' and define $\theta(a)$ to be the element $a' \in E'$. The path a' is to begin at e in G' and to be an image under f of a in the sense that there exists an integer n, depending on a, such that if $|t_0 - t_1| < 1/n$, then $a(t_0)a(t_1)^{-1} \in N$ and

$$f[a(t_0)a(t_1)^{-1}] = a'(t_0)a'(t_1)^{-1}.$$

Choose a symmetric nucleus M of G with $M^3 \subset N$ (see Exercise V.F). By the compactness of $\mathbf{I}$, it is possible to find a finite set of open intervals $T_1, T_2, \ldots, T_k$ which cover $\mathbf{I}$, with $a(T_i)a(T_i)^{-1} \subset M$ for $i = 1, 2, \ldots, k$.

If δ is the length of the shortest of the nonempty intersections $T_i \cap T_j$, and n is an integer with $1/n < \delta/2$, then $|t_0 - t_1| < 1/n$ implies $a(t_0)a(t_1)^{-1} \in M$ for each pair (t_0,t_1) in $\mathbf{I}$ (why?). It is clear that we can begin a definition of a' on the interval $[0,1/n]$ which satisfies the above requirement; $a'(t) = f[a(t)]$ must work if $t \in [0,1/n]$. Inductively, presume a' is defined on $[0,m/n]$ so as to satisfy the requirement, with $1 \leq m \leq n - 1$.

For each $t \in [m/n, (m+1)/n]$ use the following trick to define a': $a'(t) = f[a(t)a(m/n)^{-1}]a'(m/n)$. This extension of the definition of a' clearly agrees with the definition of the previous step at $m/n \in \mathbf{I}$, and it is continuous on $[0, (m+1)/n]$. If $t_0 \in [m/n, (m+1)/n]$ and $|t_0 - t_1| < 1/n$, then

$$a'(t_i) = f\left[a(t_i)a\left(\frac{m}{n}\right)^{-1}\right]a'\left(\frac{m}{n}\right)$$

for $i = 0$ and 1. This is evident for $i = 0$, and also for $i = 1$ in case $t_1 \geq m/n$. But if $t_1 < m/n$, this is just the inductive assumption. Accordingly,

$$\begin{aligned}
a'(t_0)a'(t_1)^{-1} &= f\left[a(t_0)a\left(\frac{m}{n}\right)^{-1}\right]a'\left(\frac{m}{n}\right)a'\left(\frac{m}{n}\right)^{-1}f\left[a(t_1)a\left(\frac{m}{n}\right)^{-1}\right]^{-1} \\
&= f\left[a(t_0)a\left(\frac{m}{n}\right)^{-1}a\left(\frac{m}{n}\right)a(t_1)^{-1}\right] \\
&= f[a(t_0)a(t_1)^{-1}].
\end{aligned}$$

Inductively, such a path a' does exist. But if there are two paths a' and a'' which satisfy our requirement, then $a'(0) = a''(0) = e$, so that the set $L = t \in \mathbf{I}\colon a'(t) = a''(t)$ is nonvoid; L is closed by the usual argument. And if $t_1 \in L$ and $|t_0 - t_1| < 1/n$, then

$$\begin{aligned}
a'(t_0) &= f[a(t_0)a(t_1)^{-1}]a'(t_1) \\
&= f[a(t_0)a(t_1)^{-1}]a''(t_1) \\
&= a''(t_0),
\end{aligned}$$

so $t_0 \in L$ and L is open. As before, L must be all of $\mathbf{I}$, $a' = a''$, and our path a' is unique. Hence $\theta\colon E \to E'$ is a well-defined function. Furthermore, if a is a path lying in M and $a \in (\mathbf{I},M)$, then the composite path, $a'(t) = f[a(t)]$, will do for $\theta(a)$. Thus, on the nucleus $(\mathbf{I},M)$, θ agrees with $f^{\mathbf{I}}$, which is clearly a homeomorphism of $N^{\mathbf{I}}$ with $(N')^{\mathbf{I}}$. Therefore θ is continuous and open at e; it is a morphism if it preserves products.

Now, is $a'b' = (ab)'$ for all a and b in E? We first show that $(ab)' = a'b'$ whenever the image of a lies inside M, so that $a'(t) = f[a(t)]$.

As usual, the two maps agree on a closed set L containing 0. Suppose that $t_1 \in L$ and that $|t_0 - t_1| < 1/m$, where m is the greater of the two integers n_0 and n_1 used, respectively, to construct b' and $(ab)'$. Then $b(t_0)b(t_1)^{-1} \in M$ and

$$\begin{aligned}(ab)'(t_0) &= f[(ab)(t_0)(ab)(t_1)^{-1}](ab)'(t_1)\\ &= f[a(t_0)b(t_0)b(t_1)^{-1}a(t_1)^{-1}](a'b')(t_1)\\ &= f[a(t_0)]f[b(t_0)b(t_1)^{-1}]f[a(t_1)^{-1}](a'b')(t_1)\\ &= a'(t_0)b'(t_0)b'(t_1)^{-1}a'(t_1)^{-1}(a'b')(t_1)\\ &= (a'b')(t_0)(a'b')(t_1)^{-1}(a'b')(t_1)\\ &= (a'b')(t_0).\end{aligned}$$

Thus $t_0 \in L$ and L is open, $L = \mathbf{I}$ and $(ab)' = a'b'$. Here we have used the facts that $M = M^{-1}$ and $M^3 \subset N$; they imply that if $m_1, m_2, m_3, \in M$, then $f(m_1m_2m_3{}^{-1}) = f(m_1)f(m_2)f(m_3)^{-1}$. Since E is contractible, it is connected; the nucleus $(\mathbf{I},M)$ generates E, and every element a of E may be expressed as a product $a = a_1a_2 \cdots a_j$, with the image of each a_i lying in M. A simple induction now shows that if each a_i lies in $(\mathbf{I},M)$, then $(a_1a_2 \cdots a_jb)' = a_1'a_2' \cdots a_j'b'$. Taking b to be e, we have $a' = a_1'a_2' \cdots a_j'$, and then for any path $b = b_1b_2 \cdots b_k$, $(ab)' = a_1' \cdots a_j'b_1' \cdots b_k' = a'b'$.

We have proved that our function θ is a morphism. If $\bar{\theta}: E \to E'$ also agrees with $f^{\mathbf{I}}$ on some nucleus, then $\bar{\theta} = \theta$, since every nucleus generates E. Therefore θ is unique. Furthermore, by the symmetry of the local isomorphism relation, there is a morphism $\theta': E' \to E$ which agrees with $(f^{-1})^{\mathbf{I}}$ on some nucleus of E'. The unique morphism $\theta \circ \theta': E \to E$ must also agree with $(f^{-1})^{\mathbf{I}} \circ f^{\mathbf{I}} = (f^{-1} \circ f)^{\mathbf{I}}$ on some nucleus. But so does 1_E, hence $\theta \circ \theta' = 1_E$; similarly, $\theta' \circ \theta = 1_{E'}$. This shows θ to be an isomorphism. ■

There may or may not exist a morphism $\varphi: G \to G'$ such that $\theta = \varphi^{\mathbf{I}}$. For instance, there can be no nonconstant morphism of S^1 into $\mathbf{R}$; if $\varphi: S^1 \to \mathbf{R}$ and $\varphi(s) = r \neq 0$, then $\varphi^{-1}(\mathbf{R} - \{0\})$ is a neighborhood of s. But every neighborhood of s contains a rational multiple $(2k/l)\pi$ of 2π, and $l\varphi[(2k/l)\pi] = \varphi(2k\pi) = 0$, a contradiction. (Why? Does this not prove that no definition exists for the logarithmic function on all of S^1?)

In general, φ exists iff the kernel of $p' \circ \theta$ contains the kernel of p, and then $p' \circ \theta = \varphi \circ p$. The kernel of $p' \circ \theta$ is $\theta^{-1}(\Omega')$, the set of paths a

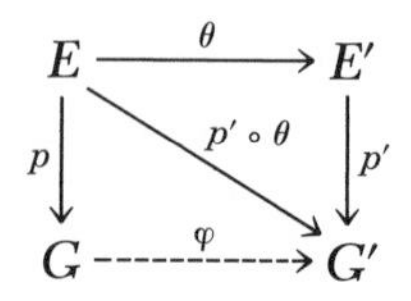

in G for which a' is a loop in G'. The kernel of p is the set of loops in G; thus φ exists iff $\theta(\Omega) \subset \Omega'$, that is, iff a' is a loop whenever a is a loop. In the example of the circle, the loop a, $a(t) = exp\ (2\pi t)$, lifts to the path a', $a'(t) = 2\pi t$, which is not a loop in $\mathbf{R}$. We now show that $a \in \Omega_0$ iff $a' \in \Omega_0'$; hence the restriction of θ to Ω_0 is an isomorphism with Ω_0', and θ defines an isomorphism of the universal covering groups.

THEOREM *If G is locally isomorphic to G' and θ: $E \to E'$ is the isomorphism of our lemma, then $\theta(\Omega_0) = \Omega_0'$. Consequently, θ induces an isomorphism ψ of the universal covering groups, ψ: $\tilde{G} \cong \widetilde{G'}$.*

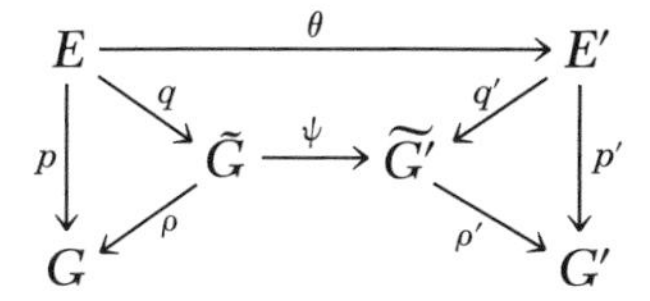

Proof Let $a \in \Omega_0$; by the symmetry of the assumption, we need only show that $a' \in \Omega_0'$, so $\theta(\Omega_0) \subset \Omega_0'$. Let h: $\mathbf{I} \to E$ be a homotopy of e with a; the continuity of h and the compactness of $\mathbf{I}$ may, in a familiar fashion, be used to find a finite set of points $0 = u_1 < u_2 < \cdots < u_k = 1$ of $\mathbf{I}$ such that if $u_i \leq u \leq u_{i+1}$, then $h(u_i)^{-1}h(u) \in (\mathbf{I},M)$. Each value $h(u)$ is a loop and thus ends at $e \in G$. Inductively, let $h(u_i)'$ end at e in G', and let $u_i \leq u \leq u_{i+1}$; then $b = h(u_i)^{-1}h(u)$ is a loop lying entirely inside M, so $b'(t) = f[b(t)]$ for all t in $\mathbf{I}$. In particular, $b(1) = e$, so

$$\begin{aligned} e = b'(1) &= [h(u_i)^{-1}h(u)]'(1) \\ &= [h'(u_i)^{-1}(1)][h'(u)(1)] \\ &= e[h'(u)(1)] \\ &= h'(u)(1) \end{aligned}$$

and $h'(u)$ is a loop, and so is $h(u_{i+1})$. Thus each value $h(u)'$ of $\theta \circ h$ is a loop, $\theta \circ h$ is a path in E' from e to a', and $a' \in \Omega_0'$.

The isomorphism of universal covering groups is now clear: the kernel of $q' \circ \theta$ is $\theta'(\Omega_0') = \Omega_0$, the kernel of q. Hence there is an isomorphism

$$\psi\colon \frac{E}{\Omega_0} = \tilde{G} \cong \widetilde{G'} = q' \circ \theta(E);$$

ψ carries $a\Omega_0$ to $a'\Omega_0'$. ∎

If, in particular, G is simply connected, then ρ: $\tilde{G} \cong G$, and the local isomorphism f agrees on some nucleus of G with the covering morphism $\varphi = \rho' \circ \psi \circ \rho^{-1}$ of G onto G'. Of course, two locally isomorphic groups which are both simply connected must be globally isomorphic. We now collect our results, stating them in seemingly greater generality and adding a few remarks which are immediate to them.

THEOREM *Let $\mathcal{G}$ be an equivalence class of mutually locally isomorphic groups (remember that the groups of this chapter are all path connected, locally path connected, and semilocally simply connected). There is, up to isomorphism, exactly one simply connected member $\tilde{G}$ of $\mathcal{G}$, and for each $G \in \mathcal{G}$ there is a covering morphism $\rho_G: \tilde{G} \to G$. The kernel of ρ_G is isomorphic to $\pi_1(G,e)$; we denote it $\pi_1(G)$. It is a discrete normal subgroup of $\tilde{G}$, and $G \cong \tilde{G}/\pi_1(G)$.*

The following statements are equivalent:

i *There exists a morphism $\varphi: G \to G'$ with $\rho_{G'} = \varphi \circ \rho_G$.*
ii $\pi_1(G) \subset \pi_1(G')$.
iii *Every loop of G' is corresponded, via the local isomorphism determined by $\rho_{G'}$ and ρ_G, to a loop of G.*

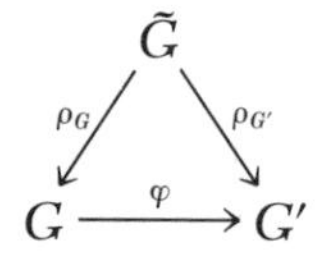

If these statements are true, then the kernel of φ is isomorphic to $\pi_1(G')/\pi_1(G)$. ■

Exercises G and N

THE CLASS OF R

We could now prove that **R** was (isomorphic to) the universal covering group of S^1 by simply observing that *exp* is a local isomorphism and **R** is simply connected. Consider the class $\mathcal{G}$ of all groups locally isomorphic to **R**. For each member G of $\mathcal{G}$ there is a corresponding discrete subgroup $\pi_1(G)$ of **R**. Clearly, if $\pi_1(G) = 0$, then G is globally isomorphic to **R**. Now assume that $\pi_1(G)$ has a nonzero member and therefore a positive member. If d is the greatest lower bound of the set of positive members of $\pi_1(G)$, then $d \in \pi_1(G)$, since this is a discrete set. We claim that $\pi_1(G) = \{kd: k \in \mathbf{Z}\}$, the subgroup generated by d. For if $r \in \pi_1(G)$, then let k_0 be such that $|k_0d - r|$ is minimal; then both $k_0d - r$ and $r - k_0d$ are in $\pi_1(G)$, so $|k_0d - r| \in \pi_1(G)$; $0 \leq |k_0d - r| < d$ implies $k_0d = r$.

Hence each nontrivial discrete subgroup $\pi_1(G)$ of **R** is isomorphic to **Z**, and each quotient of **R**, modulo such a subgroup, is isomorphic to S^1. This means that $\mathcal{G}$ has, up to isomorphism, just two members, **R** and S^1.

Let $G = \mathbf{R}/\{2k\pi: k \in \mathbf{Z}\}$ and $G' = \mathbf{R}/\{k\pi: k \in \mathbf{Z}\}$; then $\pi_1(G') \supset \pi_1(G)$, and there is a morphism $\varphi: G \to G'$ with $\varphi \circ \pi_G = \pi_{G'}$. The value of φ at $\pi_G(r) = r + \{2k\pi: k \in \mathbf{Z}\}$ is $r + \{k\pi: k \in \mathbf{Z}\} = \pi_{G'}(r)$, and the kernel of φ is a two-element group. In fact, when both G and G' are thought

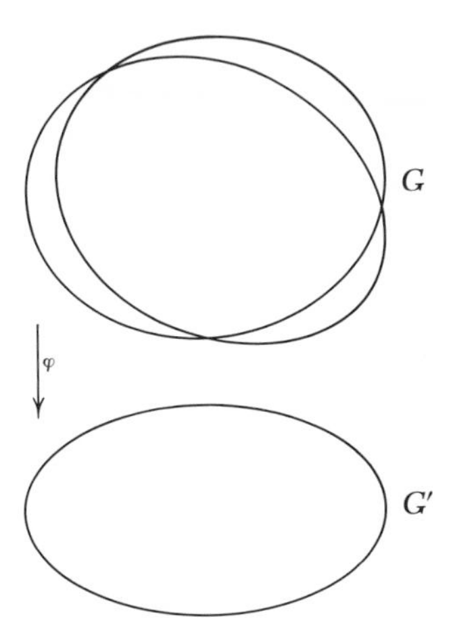

of as *being* S^1, in the obvious way, φ becomes the squaring function on the circle, $\varphi(s) = s^2$, with kernel $\{\pm 1\}$. The nonexistence of a global inverse for φ is very clear in this context; it is a two-to-one covering map which wraps the circle twice around itself.

Exercises K and L

REFERENCES AND FURTHER TOPICS

The reference closest to this chapter is

L. Pontrjagin, *Topological Groups,* chap. VIII (Princeton, N.J.: Princeton University Press, 1958).

Another presentation, which contains examples of groups of matrices, is given in

C. Chevalley, *Theory of Lie Groups,* chap. II (Princeton, N.J.: Princeton University Press, 1946).

All the results and the entire theoretical structure of this chapter have been based on the idea of local isomorphism. There is a more general notion: a "local homomorphism" f from a group G to G' is a map f, defined on a nucleus of G and having values in G', which preserves products in the same sense as does a local isomorphism; that is, if a, b, and ab are in the domain of f, then $f(a)f(b) = f(ab)$. The techniques of our proofs apply to this situation in the appropriate way to yield the statement that if G is a connected, locally path-connected, and *simply connected* group (and G' is arbitrary), then there exists a unique morphism $\varphi: G \to G'$ which extends f. If f is open (in the obvious sense), then φ is open. For a discussion of such matters, see Pontrjagin, cited above.

While we have defined the fundamental group for an arbitrary space, our emphasis has been on group spaces, where the construction of the path group and the universal covering group provided us with machinery for the computation of the fundamental group. However, a space of paths may be defined for any based space (X,x_0). The analogous construction of a "universal covering space" for spaces satisfying requirements i, ii, and iii may be carried out but is not as simply described without a group structure on X. We may define two paths p_1 and p_2 at x_0 to be equivalent if $p_1 \cdot \bar{p}_2$ is a trivial loop; if X is a group space this is exactly equivalence modulo Ω_0. The quotient space of the path space modulo this equivalence relation is a simply connected space, and there is a "covering map" of this space onto X which is very like a covering morphism. Specifically, a map $p: Y \to X$ is a "covering map," and Y is a "covering space" of X, if p is onto and for each $x \in X$ there is a connected open neighborhood N of x such that every component of $p^{-1}(N)$ is an open subset of E on which the restriction of p is a homeomorphism with N. Results for covering spaces are only a little weaker than those for covering groups; for instance it is still true that every simply connected covering space of a given space is homeomorphic to the universal covering space of the given space. The two references cited above have introductions to this theory; a more complete discussion of covering spaces is given by

W. S. Massey, *Algebraic Topology,* chap. 5 (New York: Harcourt, Brace & World, 1967).

Let $\varphi: H \to G$ be a covering morphism of topological groups; the kernel K of φ is discrete, so there exists a nucleus M with $M \cap Ker(\varphi) = \{e\}$, and we can find an open nucleus N with $NN^{-1} \subset M$. It is easy to show that $\varphi(N) = N'$ is an open nucleus of G with the property that there exists a homeomorphism $f: N' \times K \to \varphi^{-1}(N')$ with $\varphi \circ f(n',k) = n'$ for all pairs

$(n',k) \in N' \times K$. As usual, there is an appropriate translation of this situation to any point $g \in G$; it is usually described by saying that φ is a "local product projection" and that H has a "local product structure" over G. The same statements may be made for a covering map from one space to another. There is a fixed discrete space D such that every point of the range has a neighborhood N whose inverse image under the covering map is homeomorphic to $N \times D$, as above. More generally, if it is not required that D be discrete, then the map in question is said to have the "bundle property" or to be a "bundle space." An example you know is the projection of the Möbius strip onto its median circle; locally it is like the projection of the direct product of an interval with a neighborhood on the circle, but globally it is not. This gives a good intuitive feel for bundle spaces (and a special sort of bundle space called a "fiber bundle"); they are *twisted* direct products. This situation is not rare; for instance, every quotient map of a group manifold modulo a closed subgroup is a bundle space. The study of these matters is a natural next step for you as you finish this text; a well-written introduction is

N. E. Steenrod, *The Topology of Fiber Bundles* (Princeton, N.J.: Princeton University Press, 1951).

A somewhat more general presentation is given in the text of S. -T. Hu cited at the bottom of page 206. Another treatment on a high level is

D. Husemoller, *Fiber Bundles* (New York: McGraw-Hill, 1966).

EXERCISES

A Prove that if G and G' are locally isomorphic and G is a group manifold, then so is G'. In particular, the universal covering group of a group manifold must also be a group manifold.

B Prove that condition ii in the definition of a local isomorphism may be deleted. That is, show that if G and G' satisfy the definition with condition ii deleted, then G and G' are locally isomorphic.

C Show that a discrete normal subgroup of G must be a closed subset of G.

D Let s be a fixed element of the topological group G, and argue that the function $K: G \rightarrow G$: $K(g) = gsg^{-1}$ is continuous. Use this argument to show that a discrete normal subgroup S of a connected (or path-connected) group G must lie in the center of G; that is, each element of S must commute with every element of G. (*Hint:* The components of S are singletons.)

Now use this result to make a new proof that, at least if G satisfies the conditions set forth at the beginning of this chapter, the fundamental group of G is abelian. (Note the more general result for H-spaces in Exercise IX.F.)

E Prove that each contractible group is locally path connected. (*Hint:* You will need to use the continuity of the contracting map and the topology of its domain.) In particular, every path group must be locally path connected.

F Prove the fact that no nontrivial subgroup of $\mathbf{R}$ is compact, and use it to show that there exists no nontrivial morphism of S^1 into $\mathbf{R}$.

G Let G be a connected, locally path-connected, and semilocally simply connected group, and let S be a subgroup of $\pi_1(G)$. Show that there exists a group H and a morphism $f: H \to G$ whose induced morphism $f_*: \pi_1(H) \to \pi_1(G)$ is an isomorphism of $\pi_1(H)$ with S.

H Imagine three cylinders arranged one inside the other, as in the diagram, where they have been bent somewhat. If the left edges A, B, and C and the right edges A', B', and C' are all thought of as closed (in 3-space), then all six edges are homeomorphic to one another. In the obvious way, let the points of A be identified with their corresponding points in A', the points of B with those of C', and the points

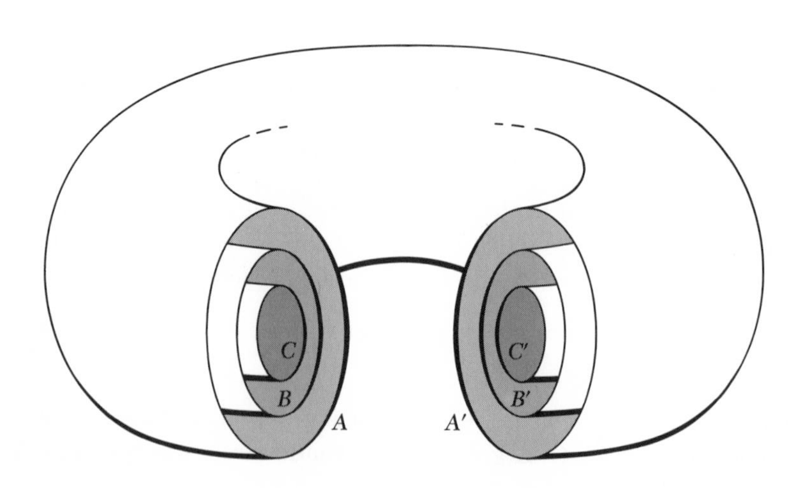

of C with those of B'. The resulting quotient space is clearly two disjoint tori; the quotient map may be visualized as a sewing together of A with A' to get one torus, and a sewing of B to C' and C to B' to get the second. (Of course, this latter sewing operation could not easily be done by a seamstress working in real 3-space.)

Intuitively, projection of the "inner," second torus onto the outer torus is a covering morphism when appropriate group structures are defined for these tori. Make this statement precise; that is, define T_1 and T_2 to be two disjoint copies of $S^1 \times S^1$ and give morphisms $\mathbf{R}^2 \xrightarrow{\varphi} T_1 \xrightarrow{\theta} T_2$, where θ arithmetically realizes the geometric picture we have offered. Discuss the relationship of the subgroups *Ker* (φ) and *Ker* $(\theta \circ \varphi)$ of $\mathbf{R}^2$, both to each other and to the geometric picture.

J If $f: G \to H$ is a covering morphism (of topological groups) with kernel K, show in detail that each element $h \in H$ possesses a connected open neighborhood N such that $f^{-1}(N)$ is homeomorphic to $N \times K$, and furthermore, that each component of $f^{-1}(N)$ is a homeomorph of N. The picture is that *locally* the map f is the projection of a direct product which collapses onto N a laminated deck of copies of N. Draw this picture for the covering morphism *exp*: $\mathbf{R} \to S^1$; arrange $\mathbf{R}$ as a helix being projected onto a circle.

K Let S be a discrete subgroup of the abelian group $\mathbf{R}^2$ such that the span of S in the real vector space $\mathbf{R}^2$ is not all of $\mathbf{R}^2$. Show that $\mathbf{R}^2/S$ is isomorphic to either $\mathbf{R}^2$ or $S^1 \times \mathbf{R}$.

L Let S be a discrete subgroup of $\mathbf{R}^2$ such that the span of S is all of the real vector space $\mathbf{R}^2$, and choose two linearly independent elements s_1 and s_2 of S. For $i = 1$ or 2, let R_i be the one-dimensional subspace containing s_i; $R_i \cap S$ is a discrete subgroup of R_i, and R_i is isomorphic to $\mathbf{R}$. Therefore $R_i \cap S$ is isomorphic to $\mathbf{Z}$. Let s_i' be a generator of $R_i \cap S$.

Prove that every element of S is of the form $n_1 s_1' + n_2 s_2'$ for some integers n_1 and n_2; then show that $\mathbf{R}^2/S$ is topologically isomorphic to the torus $S^1 \times S^1$. This result, together with that of Exercise K, shows that the local isomorphism class of $\mathbf{R}^2$ has exactly three members (up to isomorphism): the plane $\mathbf{R}^2$, the cylinder $S^1 \times \mathbf{R}$, and the torus $S^1 \times S^1$.

M Generalize Exercises K and L to find the class of all groups which are locally isomorphic to real n-space, $n \geq 0$. (*Hint:* Your proof will be inductive.)

N Prove that the universal covering group of the product $G \times H$ of two groups is isomorphic to $\tilde{G} \times \tilde{H}$; thus make a simple proof that $\pi_1(G \times H) = \pi_1(G) \times \pi_1(H)$.

PROBLEMS

AA Deck Transforms If $\varphi: G \to G'$ is a covering morphism and h is a *homeomorphism* of G such that $\varphi \circ h = \varphi$, then h is a "deck transform" for φ. Show that the deck transforms form a group (under composition) which is isomorphic to the kernel of φ.

BB Projective 3-space Consider the quotient P^n of the n-sphere S^n which you define by identifying the end points of each diameter, $P^n = \{\{x, -x\}: x \in S^n\}$. It can be thought of as the set of lines through the origin in $\mathbf{R}^{n+1}$ and is named the "projective n-space." Now, the space S^n is a topological group if $n = 0$, 1, or 3, and we have examined this quotient if $n = 1$; P^1 is a homeomorph of S^1 (and $n = 0$ is trivial).

The 3-sphere has a group structure as (a homeomorph of) the subgroup $\mathbf{Q}_1$ in the group of quaternions (see Prob. V.AA) of all those elements of norm 1. And $\{1, -1\}$ is a closed normal subgroup of $\mathbf{Q}_1$ [where 1 is the identity (1,0,0,0) of $\mathbf{Q}_1 \subset \mathbf{R}^4$], so $\mathbf{Q}_1/\{1, -1\} = P^3$ is a topological group, and the quotient map is a covering morphism. A covering morphism with a two-element kernel is often called a "double covering." You may think of this quotient map as wrapping $\mathbf{Q}_1$ twice around P^3.

Recall (from chap. VIII) that S^3 is simply connected, and use this fact to show that the class of all groups locally isomorphic to S^3 consists of just S^3 and P^3 (up to isomorphism). Furthermore, $\pi_1(P^3) = \mathbf{Z}_2$; this is our first glimpse of a space with a finite fundamental group! The nontrivial element of $\pi_1(P^3)$ is the homotopy class of the image in P^3 of any path in S^3 which goes from 1 to -1. For instance, let a' be the path in S^3 whose values are $a'(t) = [\cos(\pi t), \sin(\pi t), 0, 0]$, where $a'(t)$ is a unit quaternion presented by its four real coordinates. If $q: S^3 \to P^3$ is the covering morphism, then $a = q \circ a'$ is a nontrivial loop in P^3; can you see why $a \cdot a$ is trivial?

CC The Special Orthogonal Group SO_3 The norm morphism (see Prob. V.AA) from the group $\mathbf{Q}$ of quaternions to the reals preserves products; thus if $q_0 \in \mathbf{Q}_1$, so that $N(q_0) = 1$, then for each $q \in \mathbf{Q}$ the norm of the conjugate $q_0 q q_0^{-1}$ is $N(q_0 q q_0^{-1}) = N(q_0)N(q)N(q_0)^{-1} = N(q)$. We denote the conjugation function which q_0 determines by $K(q_0)\colon \mathbf{Q} \to \mathbf{Q}\colon K(q_0)(q) = q_0 q q_0^{-1}$. Regarded as a transformation on $\mathbf{R}^4$, $K(q_0)$ is orthogonal. A quaternion q is called a "pure imaginary" if its first coordinate in $\mathbf{R}^4$ is zero, or equivalently, if $q = (c_1,c_2) \in \mathbf{C}^2$ and the real part of c_1 is zero. The transform $K(q_0)$ carries each pure imaginary to another such, and accordingly, $K(q_0)$ may be regarded as a member of $O_3 \subset O_4$. It is easy to check that K is continuous from the connected set $\mathbf{Q}_1$ to O_3; hence K is a function from $\mathbf{Q}_1$ to SO_3 (the group of rotations of real 3-space).

The kernel of K is $\{1,-1\}$. If $q_0 = [\cos(\theta/2), \sin(\theta/2), 0, 0]$ in $\mathbf{R}^4$, then $K(q_0)$ is a rotation about the (1,0,0)-axis in $\mathbf{R}^3$. Similarly, there are elements of $\mathbf{Q}_1$ which K carries to the rotations about the other two axes, and these rotations generate SO_3; hence K is onto SO_3.

Finally, K is open. Let N be an open symmetric nucleus small enough that $-1 \notin N^3$. The family $\mathfrak{U} = \{N \cup (-N) \cup K^{-1}(U)\colon U$ is open in SO_3 and its complement U' is a nucleus$\}$ is an open cover of $\mathbf{Q}_1$; furthermore, the union of a finite subcover of $\mathfrak{U}$ is itself a member of $\mathfrak{U}$. Hence there is a nucleus U' of SO_3 for which $K^{-1}(U') \subset N$, and $K(N)$ must therefore be a nucleus. This implies that K is open, and is thus a covering morphism which is two-to-one (a "double covering"). The kernel of K is isomorphic to $\mathbf{Z}_2$, SO_3 is homeomorphic to P^3, and $\pi_1(SO_3) = \mathbf{Z}_2$.

DD Spinor Groups, and More about Orthogonal Groups The special orthogonal group SO_4 is a transitive ttg on S^3. If $p\colon SO_4 \to S^3$ is the map which evaluates members of SO_4 at the "pole" $N = (0,0,0,1) \in S^3$, so that $p(M) = M(N)$, then p has a global cross section $\sigma\colon S^3 \to SO_4$, $p \cdot \sigma = 1_{S^3}$. Obtain the value of σ at an element s of S^3 by considering s to be a unit quaternion; $\sigma(s)$ acts on $\mathbf{R}^4$ as the left-multiplication function L_s acts on $\mathbf{Q}$. The existence of σ implies that SO_4 is homeomorphic to $SO_3 \times S^3$; hence $\pi_1(SO_4) = \mathbf{Z}_2$.

The universal covering groups of the special orthogonal groups are called "spinor groups"; the n-th spinor group is

$$Spin\,(n) = \widetilde{SO}_n.$$

Thus we have that

$$\begin{aligned} &Spin\ (1) \cong \{1\}, \\ &Spin\ (2) \cong \mathbf{R}, \\ &Spin\ (3) \cong \mathbf{Q}_1, \text{ a 3-sphere,} \end{aligned}$$

and *Spin* (4) is homeomorphic to the direct product $S^3 \times S^3$.

Recall (from Prob. VII.EE) that O_n has two components, each of which is a coset of SO_n; hence $\pi_0(O_n) \cong \mathbf{Z}_2$ and $\pi_1(O_n) = \pi_1(SO_n)$.

[It may easily be shown by the techniques of fiber-bundle theory, discussed in the references for this chapter, that the inclusion of SO_3 in SO_n induces an isomorphism of fundamental groups for all n; thus $\pi_1(SO_n) = \mathbf{Z}_2$. Each spinor group is thus a double covering of the corresponding orthogonal group.]

EE Some Unitary Groups Show that the special unitary group SU_2 (see Prob. VII.DD for a definition) is simply connected, and is, in fact, a 3-sphere.

The unitary group U_1 is isomorphic to S^1.

We have seen that the coset space U_2/U_1 is homeomorphic to S^3. Because there exists a cross section to the quotient map $q: U_2 \to U_2/U_1$, the space U_2 is homeomorphic to $S^1 \times S^3$, and

$$\pi_1(U_2) = \mathbf{Z}.$$

[Again, fiber-bundle techniques quickly show that the inclusion of U_2 in U_n induces an isomorphism of fundamental groups; hence, for all n, $\pi_1(U_n) = \mathbf{Z}$.]

FF Relative Homotopy Groups and an Exact Sequence Let S be a subgroup of the path-connected group G; the **group of loops in** (G,S) is $\Omega(G,S) = \{a\colon \mathbf{I} \to G\colon a(0) = e \text{ and } a(1) \in S\}$; it is a subgroup of the path space of G. The group of path components of $\Omega(G,S)$ is

$$\pi_1(G,S) = \frac{\Omega(G,S)}{\Omega_0(G,S)},$$

the **fundamental group of** (G,S).

The inclusion map $i\colon S \to G$ induces morphisms $i_*\colon \pi_0(S) \to \pi_0(G) = \{1\}$ and $i_*\colon \pi_1(S) \to \pi_1(G)$. Furthermore, the inclusion j of e into S, or of (G,e) into (G,S), induces a morphism $j_*\colon \pi_1(G) \to \pi_1(G,S)$; here $j_*[a\Omega_0(G)] = a\Omega_0(G,S)$. There is also a morphism $\partial\colon \pi_1(G,S) \to \pi_0(S)$ which carries the class $[a]$ to $\partial[a]$, the path component of $a(1)$ in S. Each morphism in the sequence

$$\pi_1(S) \xrightarrow{i_*} \pi_1(G) \xrightarrow{j_*} \pi_1(G,S) \xrightarrow{\partial} \pi_0(S) \xrightarrow{i_*} \pi_0(G) \rightarrow \{1\}$$

has as its image exactly the kernel of the subsequent morphism (such a sequence is called "exact").

The proof of this requires some ingenuity; it will be helpful to imagine that (G,S) is $(\mathbf{R}^3,\mathbf{R}^2)$; though homotopically trivial, this is a good context in which to visualize paths. An easy consequence of the exactness of the sequence is that if S is contractible, then j_* is an isomorphism.

If S is discrete and normal, then $q\colon G \rightarrow G/S$ induces a morphism $q^{\mathrm{I}}\colon \Omega(G,S) \rightarrow \Omega(G/S)$, which in turn induces an isomorphism q_* of $\pi_1(G,S)$ onto the fundamental group $\pi_1(G/S)$ of the quotient group. Thus the sequence

$$\pi_1(S) \xrightarrow{i_*} \pi_1(G) \xrightarrow{k_*} \pi_1\left(\frac{G}{S}\right) \xrightarrow{\Delta} \pi_0(S) \xrightarrow{i_*} \pi_0(G) \rightarrow \{1\}$$

is exact if $k_* = q_* \circ j_*$ and $\Delta = \partial \circ q_*^{-1}$. In that case, we have that k_* is monic and

$$S \cong \frac{\pi_1(G/S)}{k_*\pi_1(G)},$$

a result not made explicit in the text. This formula reduces to $S \cong \pi_1(G/S)$ when G is simply connected.

GG Let $\tilde{G}$ be the universal covering group of G; we have seen that there is an isomorphism of the path space of $\tilde{G}$ with that of G and that this isomorphism carries the subgroup of (trivial) loops in $\tilde{G}$ onto the subgroup of trivial loops in G. This implies that $\pi_n(\tilde{G},e) \cong \pi_n(G,e)$ for all $n \geq 2$. [The group $\pi_n(G,e)$ is defined in Prob. VIII.EE.]

Greek Alphabet

Alpha	Α	α
Beta	Β	β
Gamma	Γ	γ
Delta	Δ	δ
Epsilon	Ε	ε
Zeta	Ζ	ζ
Eta	Η	η
Theta	Θ	θ
Iota	Ι	ι
Kappa	Κ	κ
Lambda	Λ	λ
Mu	Μ	μ
Nu	Ν	ν
Xi	Ξ	ξ
Omicron	Ο	o
Pi	Π	π
Rho	Ρ	ρ
Sigma	Σ	σ
Tau	Τ	τ
Upsilon	Υ	υ
Phi	Φ	φ
Chi	Χ	χ
Psi	Ψ	ψ
Omega	Ω	ω

Symbol Index

Author Index

Subject Index